Analytical Testing for the Pharmaceutical GMP Laboratory

Analytical Testing for the Pharmaceutical GMP Laboratory

Edited by

Kim Huynh-Ba
Pharmalytik LLC
Newark, DE, USA

WILEY

This edition first published 2022

Registered Office(s)
John Wiley & Sons, Inc., 111 River Street, Hoboken, NJ 07030, USA

Editorial Office
111 River Street, Hoboken, NJ 07030, USA

For details of our global editorial offices, customer services, and more information about Wiley products visit us at www.wiley.com.

Wiley also publishes its books in a variety of electronic formats and by print-on-demand. Some content that appears in standard print versions of this book may not be available in other formats.

Library of Congress Cataloging-in-Publication Data

Names: Huynh-Ba, Kim, author.
Title: Analytical testing for the pharmaceutical GMP laboratory / Kim Huynh-Ba, M.Sc., PMP, FAAPS, Chief Executive Officer and Managing Director, Pharmalytik LLC Newark, DE, USA.
Description: First edition. | Hoboken, NJ : Wiley, 2022. | Includes index.
Identifiers: LCCN 2021031990 (print) | LCCN 2021031991 (ebook) | ISBN 9781119120919 (hardback) | ISBN 9781119680468 (adobe pdf) | ISBN 9781119680437 (epub)
Subjects: LCSH: Pharmaceutical industry–Quality control. | Analytical chemistry. | Drugs–Testing. | Drug development.
Classification: LCC RS189 .H89 2022 (print) | LCC RS189 (ebook) | DDC 615/.19–dc23
LC record available at https://lccn.loc.gov/2021031990
LC ebook record available at https://lccn.loc.gov/2021031991

Cover Design: Wiley
Cover Image: © DocsR/iStock/Getty Images

MIX
Paper from responsible sources
FSC® C013604

This book is dedicated to the memory of Dr. James Shea, who passed suddenly after completing the technical review of several chapters of this book.

Contents

"Education is not the learning of facts, but the training of the mind to think."

Albert Einstein

Preface

Pharmaceutical analysis is an important and integral part of determining drug quality, including identity, purity, and stability of the active drug substance and drug product. In addition, related programs are necessary to ensure the physical and chemical performance of the medicines. With such critical functions, it requires analysts to acquire a solid understanding of analytical chemistry and a thorough appreciation of pharmaceutical requirements to address these challenges.

The pharmaceutical industry is one of the major employers of science graduates, especially analytical chemistry majors. However, many students graduate with a limited background in pharmaceutical analysis or related programs and are not prepared for employment in this industry. Those who do will find the transition from academia to this type of industry difficult due to a lack of formal training in most academic institutions. Rather, the training/mentoring program is often performed by the pharmaceutical industry, formally or informally, to develop these individuals into productive employees as soon as possible. Therefore, this type of training program is conducted as a part of the company's regulatory and quality programs.

In addition, many scientists are interested in changing their career to join the healthcare industry and they found that there is a lack of introductory materials in applying GMP regulations and connecting it with industry best practices for the pharmaceutical laboratory.

This book is intended to be an introductory book for pharmaceutical scientists, who are directly or indirectly involved with the laboratories supporting drug development and manufacturing processes. It covers the main activities for new and experienced scientists, including analytical chemists, pharmaceutical scientists, pharmacists, quality control professionals, and quality assurance specialists. This book will be useful for students in undergraduate or graduate Schools of Pharmacy in the United States. It could also benefit the pharmaceutical analysis course or online regulatory or quality programs.

There are other books that cover Good Manufacturing Practices; however, there is not a good mix of regulations, guidelines, and practical procedures in these books. My goal is to deliver a concise and comprehensive book, yet affordable, for this area that can be used in an industrial or academic environment. It focuses on the smaller molecular weight drug substances and products. The biological/biotechnological field and related bioanalytical analyses are beyond the scope of this introductory book and thus are not covered; however, you might find some similarities in GMP practices.

The quality of pharmaceutical products must meet the required regulatory specifications, related guidelines, good manufacturing, and laboratory practices prior to commercialization. Therefore, this book starts with the structure and roles of the Food and Drug Administration (FDA) and the International Council for Harmonisation of Technical Requirements for Pharmaceuticals for Human Use (ICH) in Chapter 1, *Drug Regulations and the Pharmaceutical Laboratories*, with core guidelines and critical regulations. Chapter 2, *Good Manufacturing Practices (GMPs) and the Quality Systems*, introduces the Good Manufacturing Practices (GMPs), specifically the Code of Federal Regulations (CFR) Title 20 and 21, and furthers the discussion leading to laboratory quality systems that support the pharmaceutical industry.

Chapter 3, *Common Methods in Pharmaceutical Analysis*, introduces the common analytical methods used in pharmaceutical laboratories. The emphasis of this chapter is more on specific and stability-indicating, and instrument techniques rather than classical wet chemistry methods. This chapter also contains several examples of analytical techniques used in the laboratory for the release and stability testing of drug substances and drug products. Chapter 4, *Control Strategies for Pharmaceutical Development*, discusses control strategies established from quality systems with case studies and calculation examples.

Analytical data are important in the evaluation of the drug substance and drug product; thus, the qualification of analytical procedures is critical. Chapter 5, *Development and Validation of Analytical Procedures*, covers the validation of the analytical procedures, and Chapter 6, *Transfer of Analytical Methods*, discusses the different types of transfer strategies to ensure the procedure can be successfully transferred to another laboratory. Chapter 7, *Dissolution Testing in the Pharmaceutical Laboratory*, introduces dissolution testing as a performance test for pharmaceutical products, such as the extended-release dosage forms, aerosols, or inhalers.

Chapter 8, *Analytical Data and the Documentation System*, discusses the product specifications and challenges of investigations. Good documentation and data reporting practices are the focus of this chapter. Specifications must be set for the release time and the shelf-life of the drug in the market. The heavy workload in any pharmaceutical quality control laboratory is to support the stability program.

Therefore, Chapter 9, *Stability Program Supporting Pharmaceutical Products*, covers the most critical regulations and practices necessary to set up an effective stability program. The final chapter is Chapter 10, *Laboratory Information Management System (LIMS) and Electronic Data*, covering the analytical data, including electronic data. This chapter emphasizes the integrity of the results and how they are used in the evaluation of pharmaceutical products.

The order of the chapters is carefully laid out to provide the reader with a logical flow of materials to explore Good Manufacturing Practices (GMPs) that apply to critical quality systems supporting the pharmaceutical laboratories. However, each chapter is written in a way that it can stand on its own for ease of reading or downloading. Therefore, each chapter contains its own table of content, a list of acronyms, a reference list, a list of tables and/or figures. The included regulations and guidances are up to date at the time of publishing. However, readers should be aware that these guidelines may change due to the dynamic of the drug development and commercialization process.

I hope you enjoy reading these chapters as much as we enjoyed writing them. It is my sincere desire that this book will help those who work in the pharmaceutical laboratories seeking to learn more about GMP regulatory compliance today and for years to come.

Kim Huynh-Ba

About the Editor

Kim Huynh-Ba, M.Sc., PMP, FAAPS, has almost 30 years of experience in analytical development, project management, strategic drug development, and stability sciences. She is currently the Chief Executive Officer and Managing Director of Pharmalytik LLC, where she provides consulting and training services to pharmaceutical companies, including companies operating under FDA's Consent Decree, on harmonization and optimization of analytical practices since 2003. Before Pharmalytik, Kim was the Director of Pharmacopeial Education Department at the United States Pharmacopeia (USP), where she was responsible for their onsite and online education programs worldwide. Kim has held several technical and quality positions with increased responsibilities at Astra Zeneca (formerly ICI Americas), DuPont Merck, DuPont Pharmaceuticals, Bristol Myers Squibb, and Wyeth Vaccines.

As an education advocate, Kim teaches cGMP compliance and quality topics for several global organizations such as the American Chemical Society (ACS), American Association of Pharmaceutical Scientists (AAPS), Pittsburgh Conference, and many other international training groups. She is an Adjunct Professor at Temple University School of Pharmacy, Widener University, and Illinois Institute of Technology (IIT), teaching Pharmaceutical Analysis, Good Manufacturing Practices (GMP), ICH Quality Guidelines, and Quality Audit and Inspection for their graduate programs of regulatory compliance.

Kim is a member of the Executive Committee of the Governing Board of Eastern Analytical Symposium (EAS) and was their 2013 President. She is a member of the United States Pharmacopeia (USP) Council of Experts (2015–2025), chairing the Small Molecules 4 Expert Committee, responsible for the psychiatric, psychoactive, neuromuscular, radio-pharmaceutical and imaging products. She was the Chair of the USP *Good Documentation Practices* Expert Panel and a member of USP's *Impurities of Drug Products* Expert Panel. From 2010 to 2013, Kim also participated in the Consumer Healthcare Product Association's (CHPA) Stability and Impurities Breakout Groups as a technical advisor. Kim is a member of the AAPS Publication

Committee, Chair of the Stability Focus Group, and serves on the Steering Committees of Chemistry, Manufacturing and Controls (CMC) and Pharmaceutical Impurities Focus Groups. Kim is a recipient of the 2008 Service Award of Analysis and Pharmaceutical Quality Section and the 2008 Recognition Award of Regulatory Section of AAPS. She received the 2001 DuPont Pharmaceutical Company Asian American Leadership Award. In 2020, Kim was named a Fellow of the AAPS organization. In 2021, she received the EAS Distinguished Services Award.

Kim has authored over 30 technical publications, 17 book chapters, and delivered over 400 presentations, both domestic and internationally, in the areas of compliance and quality. She is the editor of the *Handbook of Stability Testing in Pharmaceutical Development: Regulations, Methodologies and Best Practices* (2008), which is known as a notable reference in Stability sciences, and *Pharmaceutical Stability Testing to Support Global Markets* (2010).

Kim completed her graduate degree in Analytical Chemistry from Villanova University with the late Dr. Robert Lee Grob. She earned the Project Management certification from the University of Delaware. Kim has a Bachelor of Science in Chemistry and a Bachelor of Arts in Mathematics from the Millersville University of Pennsylvania.

For corresponding, contact: kim.huynhba@pharmalytik.com or www.pharmalytik.com.

Biographies of Contributing Authors

Susan Cleary, BSc., MBA, is the Director of Product Development at Novatek International. Prior to this, Susan was the Program Manager and Lead Software Engineer for three major applications: Finished Products Analyzer, Raw Materials Management, and Environmental Monitoring software programs. Susan has more than 20 years of experience in designing, developing, implementing, validating, and managing large-scale LIMS and Quality Management software projects, while working with both Pharmaceutical and Biotech companies to capture their requirements and provide fully validated turnkey systems. Susan specializes in Laboratory Information Management, Environmental Monitoring, and Cleaning Validation processes and works with clients to streamline their procedures and manage their data more effectively.

Parsa Famili, MSc., is the President and CEO of Novatek. He has held senior management positions in quality departments of several North American pharmaceutical companies before joining Novatek. He has participated and completed several FDA, EMEA, TGA, and Health Canada audits. Parsa was also an instructor of Chemistry and Biochemistry at Vanier College in Montreal and has shared his more than 25 years of practical experience in many published chapters and articles in various journals and books. He is an international speaker at ISPE, IVT, PDA, IPA, CBI, among others.

Vivian A. Gray, B.Sc., has spent the last 42 years involved in all aspects of dissolution testing and evaluating new dissolution technology. She enjoyed an extensive career with the United States Pharmacopeia and DuPont-Merck Pharmaceutical Company. She has over 50 publications, including 6 book chapters. Vivian has co-authored a book on dissolution testing called *Handbook of Dissolution Testing*, Third Edition, published in 2004. She served on the PhRMA Dissolution Committee from 1997 to 2001. Vivian received the American Society of Hospital Pharmacists Research and Education Foundation 1982 Research Award. In 2002, she formed V.A. Gray, Consulting, LLC to support dissolution testing and became the Managing Director of *Dissolution Technologies in June 2003.*

Walter Holberg, B. Sc., is the Director of Chemistry, Manufacturing, and Controls for Rockwell Medical, where he oversees and manages new product development, API and drug product manufacturing, and analytical development. Walter has worked in or managed GMP laboratories for 30 years and has extensive knowledge of current analytical methodologies, US, and international regulatory requirements, and GLPs and GMPs as they relate to analytical development and manufacturing. He currently lives near Philadelphia and spends his free time restoring a classic air-cooled Porsche 911.

Kim Huynh-Ba, MSc., PMP, FAAPS, is the Managing Director of Pharmalytik, LLC, where she provides consulting and training services to pharmaceutical companies since 2003. She has almost 30 years in analytical development, project management, strategic drug development and stability sciences. Kim has held several technical and quality positions at Astra Zeneca (formerly ICI Americas), DuPont Merck, DuPont Pharmaceuticals, Bristol Myers Squibb, and Wyeth Vaccines. She is an Adjunct Professor at Temple University-School of Pharmacy, and Illinois Institute of Technology (IIT) teaching a variety of topics on GMPs and compliance audit. She serves on the Board of Eastern Analytical Symposium and the editorial board of Journal of Validation Technology. She has published over 30 publications and 17 book chapters. She was named Fellow of American Associate of Pharmaceutical Scientists in 2020.

Judy Lin, MSc., RAC., Associate Director of Regulatory Affairs – CMC, Novartis Pharmaceutical Corp. Prior to joining Novartis, she worked in the analytical development departments of Bristol-Myers Squibb Company and Merck & Co. She has more than 20 years of experience in pharmaceutical development with numerous successful worldwide approvals of new chemical/ biologics drugs. She is an expert in CMC development strategy for projects ranging from early development to post-approval submissions. Judy serves on the Board of Eastern Analytical Symposium and was their 2020 President.

Linda L. Ng, Ph.D., is a Senior Director, Strategic Regulatory Affairs and Quality Management with Fresenius-Kabi. Previously she was with FDA, CDER for 25 years, serving 20 years in the CMC New Drugs Review Divisions and last 5 years in the Compliance Division. Linda was the FDA Analytical Method Expert with numerous nationally and internationally invited presentations through the years. In addition, she served for two terms on a USP Expert Committee and AOAC International Task Force and Advisory Committee.

Editorial Notes

The technical information presented, and practical recommendations, are drawn from our experience and knowledge. Review perspectives may vary depending on individual technical backgrounds, personal experiences, and discussion preferences. In addition, information may be drawn from industry practices or web links that appear to be valid at the time of publication. Great care has been taken in the individual development of each chapter to ensure accuracy as much as possible. However, we assume no responsibility nor can we endorse the material referenced in this publication.

Acknowledgments

As an educator, I have found it difficult to find a textbook that is not only to be used in the classroom, but one that an analyst can also use in his/her day-to-day work. Therefore, the writing and publication of this book was undertaken. After 30 years in the pharmaceutical industry, I would have thought that GMP regulations would have been outdated by now. However, I am constantly amazed with the new scientific programs developed based on fundamental GMPs to effectively manage the regulatory compliance and build the quality systems. All of this is possible because of the collaboration and transparency of the regulatory authorities and the industry practitioners.

This work cannot be accomplished without the encouragement and support from many individuals. Each chapter was reviewed by at least three individuals of various disciplinaries to provide clarity and technical applicability to the readers. I want to express my sincere thanks and appreciation to the following individuals: D.J. Doan, Yangming Lin, Oscar Liu, Karen Lucas, Greg Martin, Mariann Neverovitch, Richard Nguyen, Leonel Santos, Ani Sarkahian, James Shea (late), Thomas Rosanske, and Martin Williamson. I also want to thank Mr. Richard Nguyen and Ms. Karishma Kumar for their assistance in formatting and proofing my chapters.

Lastly and most importantly, I want to thank my family for their support: my husband, Thai Huynh-Ba, and my sons, John Huynh-Ba and James Huynh-Ba. I also want to thank my parents, Mr. Cao Nguon Hong and Mrs. Bach Ngoc Kim Hoa, who have always stood by me to make sure I do not give up on what I started. To them, I am indebted with my gratitude.

Winter 2021,
Kim Huynh-Ba

1

Drug Regulations and the Pharmaceutical Laboratories

Kim Huynh-Ba

Pharmalytik LLC, Newark, DE, USA

TABLE OF CONTENTS

Analytical Testing for the Pharmaceutical GMP Laboratory, First Edition. Edited by Kim Huynh-Ba.

1.1 Introduction

The pharmaceutical industry is a complex business in the global economy. As well as having to meet profit margins, issues such as safety considerations, effectiveness assurance, regulatory compliance, patent protection, labeling, and the supply chain, can significantly affect the development and commercialization of pharmaceutical products. Regulatory agencies define testing requirements, and these requirements vary from country to country where the products are marketed. This chapter explains different regulatory bodies and how they affect the testing of pharmaceutical products.

1.2 Food and Drug Administration: Roles and Its Regulations

The regulatory body in the United States (US), the Food and Drug Administration (FDA) (www.fda.gov), is within the Department of Health and Human Services (HHS). The primary responsibility of the FDA is to protect the public health, and pharmaceutical companies must maintain the quality of the drug products through their shelf life based on current regulations. The FDA consists of the Office of the Commissioner and four Directorates overseeing the core functions of the agency: Medical Products and Tobacco, Foods, Global Regulatory Operations and Policy, and Operations. Each of these offices has numerous centers and offices with the goal of protecting public health. Figure 1.1 shows a few centers that interact closely with the pharmaceutical industry.

The Center for Drug Evaluation and Research (CDER) ensures that safe and effective drugs are available to improve the health of people in the United States. This center regulates over the counter (OTC) and prescription drugs, including biological therapeutics, brand drugs, and generic drugs. The Center for Biologics Evaluation and Research (CBER) regulates biological products for human use. The Center for Devices and Radiological Health (CDRH) oversees the medical devices and radiation-emitting products. The Center for Veterinary Medicine (CVM) regulates the manufacture and distribution of drugs, devices, and food additives that will be given to animals, including animals from which human foods are derived, as well as food additives and drugs that are given to companion animals. The Center for Food Safety and Applied Nutrition (CFSAN) provides support for safety and labeling issues relating to food and cosmetics. For additional information, visit the FDA webpage at www.fda.gov/ [1, 2].

Although residing under the same umbrella, these FDA centers may have different policies specific to the products they review, and pharmaceutical laboratories interact with the different centers based on their products. Since FDA issues regulations that are legally enforced, the agency works closely with the industry to develop various guidances and initiatives that support the development of new medicines for US consumers. For the past several years, the FDA has gone through many reforms to be more effective in providing support to the industry (Figure 1.2).

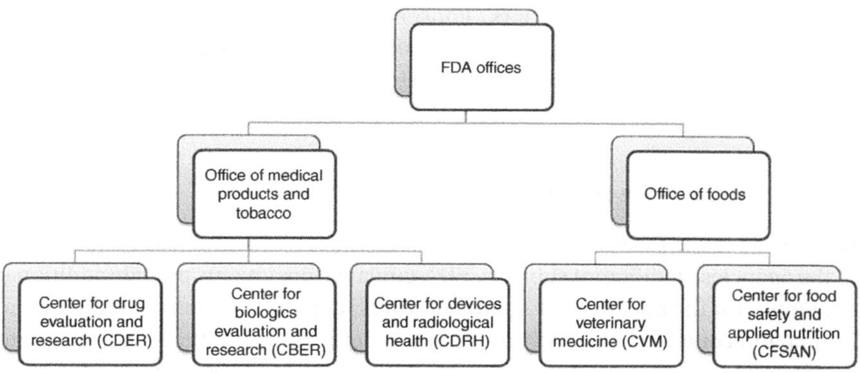

Figure 1.1 FDA offices that interact closely with the pharmaceutical industry.

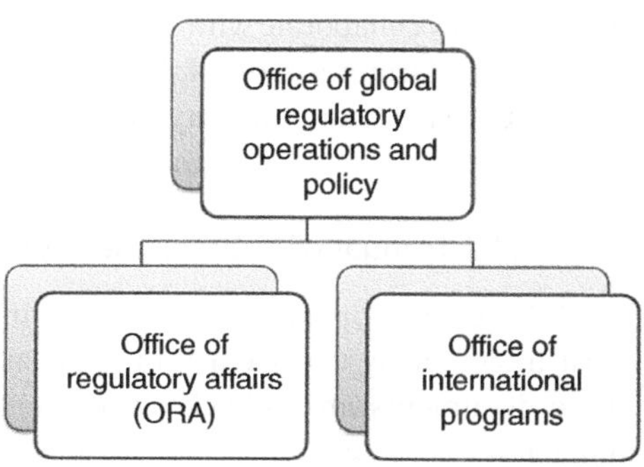

Figure 1.2 FDA offices overseeing the development of policies and guidelines.

1.2.1 Code of Federation Regulations

The code of federal regulations (CFR) is the set of final regulations that are published in the federal register, *which* companies must comply with while distributing their products in the United States [1]. The CFR encompasses 50 broad areas subject to federal regulations. Section 21 of the CFR contains regulations pertaining to food and drug products. The federal law governing pharmaceutical labs can be found in 21 CFR Part 210 – Current Good Manufacturing Practices (CGMP) in Manufacturing Process, Packing, or Holding of Drug; General, and 21 CFR Part 211 – CGMP for Finished Pharmaceuticals [2].

Given these regulations are written into law and are legally enforceable, the FDA audits manufacturers and laboratories and conducts inspections of total quality systems. When a manufacturer fails to meet CGMP requirements, its products are considered adulterated and FDA can take legal action without proving that the product is contaminated. Any violations in CGMPs can result in observations, warning letters, consent decree, or criminal prosecution. These regulations also include the requirement that manufacturers wanting to market a prescription product in the United States submit an appropriate application to the FDA.

1.2.2 FDA Guidance Documents

In addition to the CFR, the FDA also issues guidances to introduce current views of the agencies on different subjects. FDA guidance documents provide additional guidelines for the operation of manufacturing facilities, laboratories, etc., and the format and content of the application that is not included in the CFR, helping to maintain consistency among agency reviewers or application owners. The guidance documents are not laws and thus are not enforceable; however, many find that these documents provide more specific details and better clarity than the requirements listed in the regulations. Scientific justifications are necessary if the application does not follow this guidance. The FDA typically issues these guidelines after they collaborate with other regulatory agencies, such as the International Council for Harmonisation of Technical Requirements for Pharmaceuticals for Human Use (ICH) and industry [3–5].

1.2.3 FDA Manual of Policies and Procedures

The FDA creates operating manuals listing internal policies and procedures to assist in achieving consistency in application review and operational activities. These documents are publicly available to provide transparency and a better understanding of regulatory policies and agency responsibilities.

When a pharmaceutical manufacturer develops a new drug, it must understand the regulations of quality systems supporting their processes. The manufacturer must conduct its work correctly the first time, as the cost of noncompliance can be enormous. A manufacturer can request meetings with the appropriate center to discuss various scientific and regulatory aspects of their application prior to submission to assure that their development strategy is acceptable. Since 2016, the FDA has required all regulatory submissions to be filed electronically for ease of review and communication [4, 6, 7].

1.3 International Council for Harmonisation of Technical Requirements for Pharmaceuticals for Human Use (ICH) and Its Role

In the late 1980s, to harmonize regional regulatory requirements, many industry expert groups collaborated with regional agencies to discuss technical requirements for drug submissions. This section goes into further detail on the collaboration within the global pharmaceutical industry [6].

1.3.1 ICH Background

The fundamental purpose of a regulatory agency is to protect public health; therefore, it can deny market access to pharmaceutical products if the product's development does not meet the agency's requirements. These requirements usually vary from country to country and often with similar but slight, and sometimes significant, variations in registration requirements. This key factor drives up the cost of development, expands resources, and delays market access to high-quality medicines. Recognizing this problem, in the early 1990s, regulators and industry representatives from the European Union (EU), Japan, and the United States formed a group called the *International Conference on Harmonisation of Technical Requirements for Registration of Pharmaceuticals for Human Use* (ICH) (www.ich.org) [5, 6, 8]. On 23 October 2015, this group was reformed as a nonprofit legal entity under Swiss Law and named *International Council for Harmonisation of Technical Requirements for Pharmaceuticals for Human Use* (ICH). The objective of the ICH is to provide guidelines for harmonizing their testing requirements, application, and approval processes of a pharmaceutical product intended for human use. These three regions represent 85% of the worldwide pharmaceutical consumption. The ICH members, collectively representing 18 nations, have now harmonized over 50 guidelines to ensure the quality, safety, and efficacy of their pharmaceutical products. ICH guidelines have been widely used and adopted in several countries voluntarily, and several non-ICH members have also adopted these guidelines [8].

The purpose of the ICH is to streamline the process of registration requirements and avoid unnecessary duplication of work. Additionally, ICH provides a forum for dialogue between health authorities and industry on the disparities concerning requirements of the United States, Europe, and Japan. This group has developed several guidelines that are acceptable for regulatory review by these regions.

1.3.2 ICH Structure

Each ICH member has two active parties, one representing the regulatory agencies and the other representing industry manufacturers. The two parties work closely together to ensure the smooth development of guidelines that address industry concerns while maintaining standards for people's safety. The ICH members consist of the European Union (EU) and the European Federation of Pharmaceutical Industries' Associations (EFPIA), the Ministry of Health, Labor and Welfare (MHLW), the Japan Pharmaceutical Manufacturers Association (JPMA), the US Food and Drug Administration (FDA), and Pharmaceutical Research and Manufacturers of America (PhRMA). ICH has a nonvoting member, the International Federation of Pharmaceutical Manufacturers & Associations (IFPMA), and three observers whose roles are to act as liaisons between ICH and non-ICH countries and regions, the World Health Organization (WHO), the European Free Trade Area (EFTA), and Canada [6, 8].

1.3.3 ICH Organization

The ICH organization has six main sections as illustrated in Figure 1.3. They are described as follows.

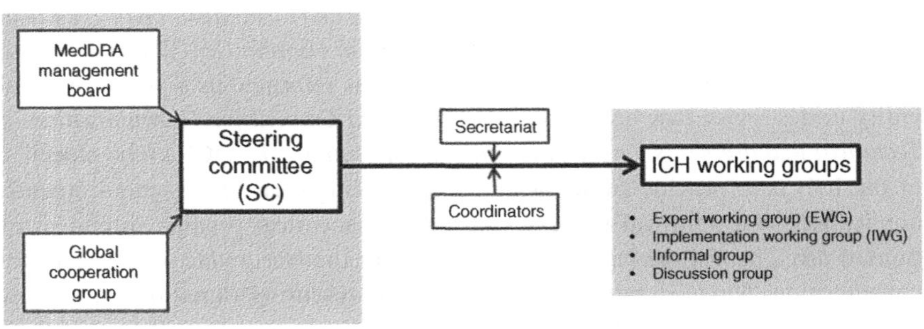

Figure 1.3 The ICH organization.

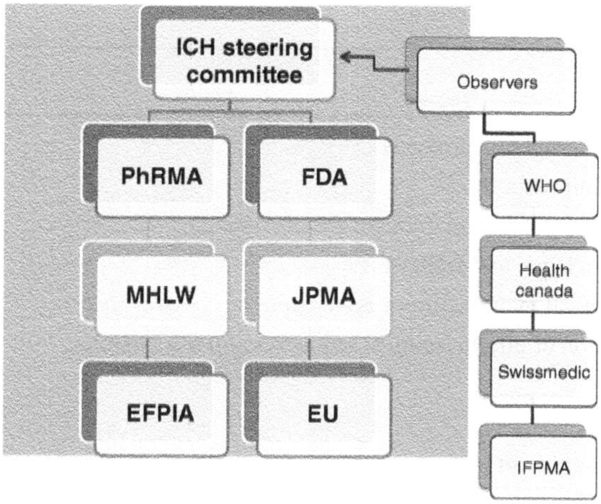

Figure 1.4 The ICH steering committee. *Source:* Based on the International Council on Harmonisation of Technical Requirements for Registration of Pharmaceuticals for Human Use.

1.3.3.1 Steering Committee

The steering committee (SC) has two representatives from each of the six parties and one representative from each of the four observing parties. Their roles are to determine the policies and procedures, select topics for harmonization, and monitor the progress of these initiatives. Figure 1.4 illustrates the composition of the ICH steering committee and Table 1.1 provides a list of the organizations [8].

1.3.3.2 Global Cooperation Group

Representatives from eight Drug Regulatory Authorities/Department of Health (Australia, Brazil, China, Taipei, India, Republic of Korea, Russia, and Singapore) and several regional harmonization groups (Table 1.2) are invited to the ICH meetings to listen to ICH technical topics discussed and invited to nominate technical experts for the expert working groups or the implementation working groups.

1.3.3.3 MedDRA Management Board

This board is responsible for standardizing the dictionary of medical terminology and overseeing the activities of the MedDRA.

1.3.3.4 Working Groups

There are two different working groups within the ICH structure: Expert Working Groups (EWGs) and Implementation Working Groups (IWGs). The EWG is responsible for developing harmonized guidelines that meet the objectives according to the

Table 1.1 Organizations involved in the ICH process.

Abbreviation	Organization
Swissmedic	Swiss agency for therapeutic products
FDA	Food and drug administration
PhRMA	Pharmaceutical research and manufacturers of America
MHLW	Ministry of health, labor, and welfare
JPMA	Japan pharmaceutical manufacturers association
EFPIA	European federation of pharmaceutical industries and associations
EU	European Union
WHO	World Health Organization
IFPMA	International federation of pharmaceutical manufacturers & associations

Source: Based on the International Council on Harmonisation of Technical Requirements for Registration of Pharmaceuticals for Human Use.

Table 1.2 The regional harmonization groups.

Abbreviation	Group
APEC	Asia-Pacific Economic Cooperation
ASEAN	Association of Southeast Asian Nations
EAC	East African Community
GCC	Gulf Central Committee
PANDRH	Pan-American Network for Drug Regulatory Harmonization
SADC	Southern African Development Community

Concept Paper and Business Plan. The Steering Committee assigns each selected topic to an EWG. The EWG represents six parties, the observers, along with any interested Industry members, such as the International Generic Pharmaceutical Alliance (IGPA), World Self-Medication Industry (WSMI), and other Pharmacopoeias. These groups are industry specialists on the topics discussed, and they participate in the timeframe in which their topic is being discussed and reviewed. Membership of an EWG can vary depending on the need. The IWGs are tasked to recommend the process to facilitate the implementation of existing guidelines.

Other than the above two Working Groups, other groups such as Informal Working Groups or Discussion Groups are formed to either develop harmonization activity or discuss scientific considerations.

1.3.3.5 Secretariat

The primary duty of the Secretariat involves the preparation of documentation for all ICH meetings. The International Federation of Pharmaceutical Manufacturers and Associations (IFPMA) provides the Secretariat in Geneva, Switzerland.

1.3.3.6 Coordinators

The coordinators are the main contact between the six-member parties and the Secretariat, ensuring that ICH documents are distributed properly.

The SC and EWGs develop the work and process of ICH, while the Secretariat and Coordinators have administrative duties.

ICH Conferences take place every two to three years with a series of meetings taking place twice a year. The location of these meetings rotates among the three-member regions. Table 1.3 lists the ICH members. More detailed information on the ICH structure can be found at www.ich.org.

1.3.4 ICH Topics

In the early 1990s, the ICH initiated 11 topics for harmonization. Approximately 20 years later, ICH developed close to 100 guidelines for the three regions, divided into four main categories: quality (Q), safety (S), efficacy (E), and

Table 1.3 ICH membership.

Membership type	Member organizations
Voting members	• Pharmaceutical Research and Manufacturers of America (PhRMA) • Food and Drug Administration (FDA) • European Federation of Pharmaceutical Industries and Associations (EFPIA) • European Union (EU) • Ministry of Health, Labor, and Welfare (MHLW) • Japan Pharmaceutical Manufacturers Association (JPMA)
Nonvoting members	• International Federation of Pharmaceutical Manufacturers and Associations (IFPMA)
Official observers	• Canada, • World Health Organization (WHO) • The European Free Trade Association (EFTA)
Interested parties	• Pharmacopeia(s) • International Generic Pharmaceutical Industry (IGPA) • World Self-medication Industry (WSMI)

Table 1.4 Categories of the ICH guidelines.

Category	ICH guideline
Quality	Chemical and pharmaceutical quality assurance
Safety	*In vitro* and *in vivo* preclinical studies
Efficacy	Clinical studies involving human subject(s)
Multi-disciplinary	Cross-cutting topics that do not fit in the above

multidisciplinary (M), as listed in Table 1.4. The following is a list of guidelines that ICH started and currently has available. New guidelines continue to be developed [8].

1.3.4.1 Quality Guidelines

The quality guidelines cover various aspects, such as stability, temperature, trial duration, light sensitivity, residual solvent, impurities, etc. Table 1.5 lists the currently available quality guidelines.

Table 1.5 ICH quality guidelines.

Guidelines	Quality topics
Q1A	Stability
Q2	Analytical validation
Q3A–Q3D	Impurities
Q4–Q4B	Pharmacopeia
Q5A–Q5B	Quality of biotechnological products
Q6A–Q6B	Specifications
Q7	Good Manufacturing Practices
Q8	Pharmaceutical development
Q9	Quality risk management
Q10	Pharmaceutical quality system
Q11	Development and manufacture of drug substance
Q12	Lifecycle management
Q13	Continuous manufacturing of drug substances and drug products
Q14	Analytical procedure development

1.3.4.2 Safety Guidelines

The safety guidelines deal with scientific issues, which include carcinogenicity, genotoxicity, pharmacokinetics, and toxicokinetics (including reproductive toxicity testing), in a detailed manner. The guidelines in this section cover toxicity issues during the testing phase and preclinical safety evaluations. Table 1.6 lists the ICH safety guidelines.

1.3.4.3 Efficacy Guidelines

The efficacy guidelines address technical and administrative issues. Technical issues include the effectiveness of long-term treatment for non–life-threatening conditions and dose–response information. Administrative issues include clinical safety data management and standards for successful expedited reporting, maintenance of the ICH guidelines, structure, and content of clinical safety reports, and sets of ethnic factors for the acceptability of foreign clinical data. Table 1.7 lists the ICH efficacy guidelines.

1.3.4.4 Multidisciplinary Guidelines

The multidisciplinary guidelines contain topics that do not fit into one of the above three sections. Many of these sections (Table 1.8) are tools that the ICH used to help create a medical dictionary with appropriate terminology and to create a common marketing application for new pharmaceutical products.

Table 1.6 ICH safety guidelines.

Guidelines	Safety topics
S1A–S1C	Carcinogenic studies
S2	Genotoxicity studies
S3A–S3B	Toxicokinetics and pharmacokinetics
S4	Toxicity testing
S5	Reproductive toxicology
S6	Biotechnological products
S7A–S7B	Pharmacology studies
S8	Immunotoxicology study
S9	Nonclinical evaluation for anticancer pharmaceuticals
S10	Photo-safety evaluation
S11	Nonclinical pediatric safety
S12	Nonclinical biodistribution studies for gene therapy products

Table 1.7 ICH efficacy guidelines.

Guidelines	Efficacy topics
E1	Clinical safety for drugs used in long-term treatment
E2A–E2F	Pharmacovigilance
E3	Clinical study reports
E4	Dose–response studies
E5	Ethnic factors
E6	Good clinical practices
E7	Clinical trials in geriatric population
E8	General considerations for clinical trials
E9	Statistical principles for clinical trials
E10	Choice of control group in clinical trials
E11	Clinical trials in pediatric population
E12	Clinical evaluation by therapeutic category
E14	Clinical evaluation
E15	Definitions in pharmacogenetics/pharmacogenomics
E16	Qualification of genomic biomarkers
E17	Multiregional clinical trials
E18	Genomic sampling
E19	Safety data collection
E20	Adaptive clinical trials

1.4 Pharmaceutical Analysis

Although the quality of the manufacturing process defines a drug product, pharmaceutical analysis is crucial to verify drug product quality. CFR defines the quality of a drug product as identity, purity, and stability of the drug, whereas analytical procedures are developed to monitor physical, chemical, biological, and microbiological properties at release and throughout its shelf life. In essence, Current Good Manufacturing Practices (CGMP) are critical aspects of analytical functions that apply to pharmaceutical laboratories.

Figure 1.5 simplifies the development of a drug formulation. All formulations contain at least one active pharmaceutical ingredient (API), also known as a drug

Table 1.8 ICH multidisciplinary guidelines.

Guidelines	Multidisciplinary topics
M1	MedDRA terminology
M2	Electronic standards
M3	Nonclinical safety studies
M4	Common technical document
M5	Data elements and standards for drug dictionaries
M6	Gene therapy
M7	Genotoxic impurities
M8	Electronic common technical document (eCTD)
M9	Biopharmaceutics classification system-based biowaivers
M10	Bioanalytical method validation
M11	Clinical electronic structured harmonized protocol (CeSHarP)
M12	Drug interaction studies
M13	Bioequivalence for immediate-release solid oral dosage forms

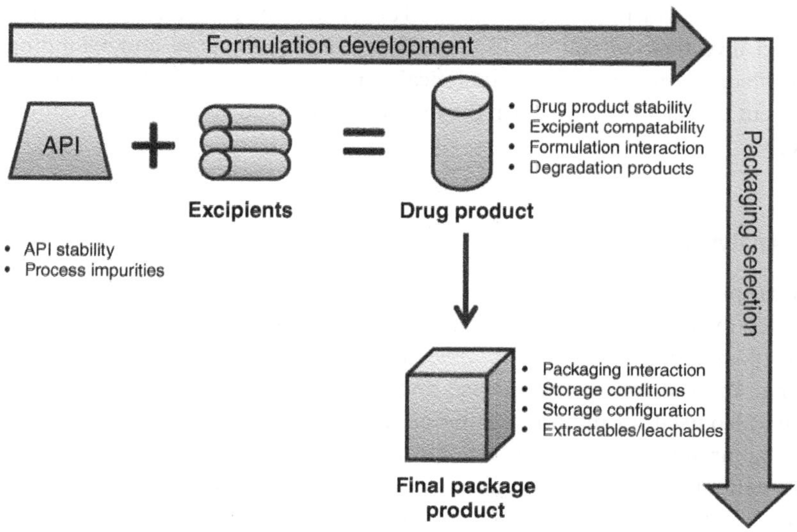

Figure 1.5 Development of a typical drug product.

substance. Formulation studies are performed by combining this API with excipients (usually non-active inert materials) to define the best version of the finished product. A drug product can be a solid dosage form (such as a tablet or capsule) or a liquid dosage form (such as syrup or injectable), or it can also be in the form of a transdermal patch or inhaler. The packaging system is studied for commercialization and the exact nature of which depends on the country or geographic region of interest.

Many factors affect the quality of drug products during their shelf life. The main factor is the quality of the API at the time of manufacture and during storage. Also, the interaction between the API and excipients is critical. The selection of the dosage form and the container closure packaging system are important to maintain the quality of the drug product. Environmental storage conditions and handling of the finished products are important to maintain such that the quality of the product remains within its established criteria [3, 7, 9].

If the drug quality changes, it may put patient safety at risk. Drug products may become subpotent with time, and new or unknown impurities that have developed may be toxic.

While developing the drug product, the analytical department must establish appropriate analytical procedures to monitor the quality of the products, and operational programs are also needed to check for the performance of the drug. When the manufacturing process is not well developed and controlled, it may lead to product investigation and recalls. Therefore, a comprehensive understanding and knowledge of regulatory and analytical requirements are imperative for any pharmaceutical scientist to perform the manufacturing and analytical procedures effectively.

1.4.1 Analytical Testing

Analytical procedures are developed in pharmaceutical laboratories to verify acceptance testing of APIs, drug products, in-process samples, raw materials, devices, packaging components, investigational samples, and equipment cleansing. In addition to the release and stability testing, analytical procedures are also used for method validation, method transfer, instrument qualification, laboratory qualification, laboratory investigation, technology transfer, process validation, and investigation. The key information section below lists the common terms used in the pharmaceutical industry for analytical analysis.

Common terms used in pharmaceutical analysis

- Act: A written ordinance from the governing body. For example, federal food, drug, and cosmetic act.
- Batch: A specific manufactured quantity that is intended to have uniform character and quality within specified limits. It is produced from a single manufacturing order during the same cycle of manufacture. For a continuous manufacturing process, it is a specific identified amount of material produced in a unit of time or a quantity in a manner that assures it has a uniform character and quality within specified limits.
- Lot: A batch or a specifically identified portion of the batch.
- A lot or batch number: A unique, distinctive combination of letters, numbers, or symbols from which the complete history of the manufacturing, packing of the batch/lot can be determined.
- Theoretical yield: A calculated quantity that would be produced based upon the number of components used in the absence of any loss or error.
- Percent of theoretical yield: Actual yield divided by the theoretical yield.
- Representative sample: A number of units that are drawn randomly based on established criteria and intended to represent the material being sampled.
- Component: Any ingredient intended for use in the manufacture of the drug product.
- Drug product: Finishe d dosage form that contains an API in general, but not necessarily, in association with inactive ingredients.
- Placebo: Finished dosage form that contains no API in general.

1.4.2 Interaction of the Analytical Development Department and Other Functional Areas

Developing a drug product is a concerted team effort within a pharmaceutical manufacturer. Figure 1.6 shows the six different areas within a typical pharmaceutical company, and as can be seen, there is close collaboration among these departments. The primary focus of the formulation development department is to develop the formulation of the drug product used for clinical studies. The chemical process department synthesizes the API and is responsible for scaling up the process of manufacturing the API and drug products. The packaging development department selects the appropriate container and closure for the drug product. The stability department sets up stability studies to support the storage condition and shelf life. The raw materials department is responsible for acquiring and releasing raw materials used. The quality assurance group is responsible

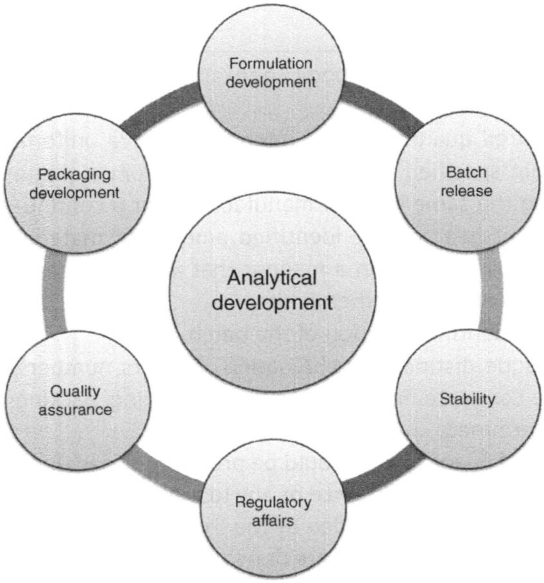

Figure 1.6 Collaboration of analytical development department with other functions.

for the compliance status of the site, while the regulatory affairs group is responsible for all regulatory submissions. Shipping and handling areas are important, too, especially for materials that require specific handling and storage conditions.

To provide adequate controls and support to the development chain, the analytical department develops testing procedures to release and monitor raw materials and drug products at all stages. Moreover, the analytical development department can also interact with the toxicological, drug metabolism and pharmacokinetics, process development, and clinical supplies manufacturing departments.

Analysts and their managers use the analytical results to make business decisions such as setting specifications, determining the dosage forms, monitoring processes, releasing raw materials or finished products, or determining product expiration and storage conditions. Therefore, the analysts must have a broad understanding of the pharmaceutical operations such as the drug development process, handling of samples, and applications of analytical methods, so that they can apply their experience to develop appropriate methods, troubleshoot the problems if needed, and investigate analytical issues as it arises.

As well, analysts must keep up with changes in regulatory and compliance requirements. Regulations continue to change as the pharmaceutical industry goes global in its operations. Analysts must be aware of new regulations and changes in industry practices to stay current and in compliance [3, 10, 11].

Although most medicines, especially solid dosage forms, such as tablets, are stable for many years, most drug products typically carry an expiration date of three to five years, unless warranted by special circumstances to minimize the risk of compromising the drug quality during storage. Analytical procedures used for stability samples must be validated for stability-indicating. Most manufacturers set specifications for their products at release time and shelf life. Typically, release specifications are more stringent than stability specifications to ensure that the products continue to remain within acceptable specifications during their shelf life. More information on validation is available in Chapter 5.

1.4.3 Drug Development Process

Drug development is the process of developing a drug product that is stable, safe, efficacious, and acceptable to the patient. Regulatory agencies, such as the FDA, are concerned about the ultimate quality of the drug that is available for the patients. Therefore, many regulations and guidelines are developed for the development, manufacturing, and commercialization of drug products. Pharmaceutical manufacturers must adhere to different regulations specific to a geographical region to gain market access to that region. These regulations usually require that those wanting to market products submit an appropriate application to the regulatory authorities in the respective country or region. Once the product is approved, the applicant must manufacture the product in such a way that it meets its acceptance criteria and is fit for its intended use. A drug product may need to comply with numerous sets of regulations to be marketed in other countries and to be available to their consumers. An understanding of the drug development process helps to support the pharmaceutical analysis effort. Figure 1.7 illustrates regulatory requirements considered during product development [5–7].

The drug development process consists of three stages: discovery/toxicology, clinical development, and commercialization. Figure 1.8 demonstrates the different phases of the drug development process.

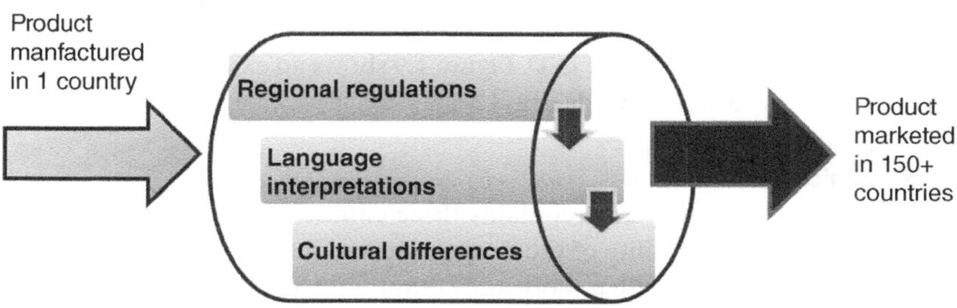

Figure 1.7 Regulatory requirements considered during product development. *Sources:* Adapted from Nally [5]; Lee [7]; Cortez [6].

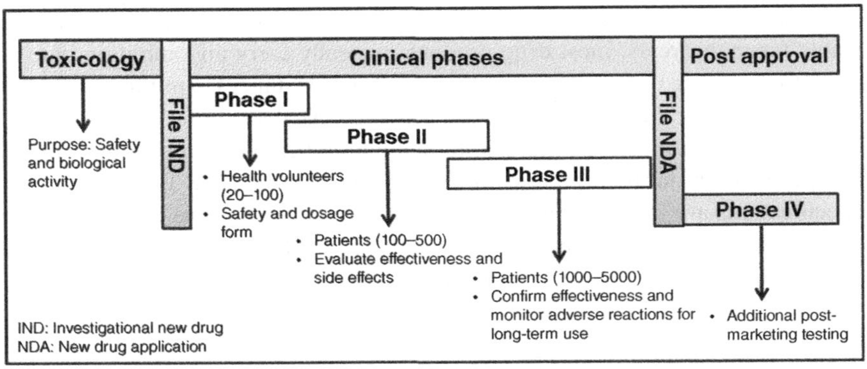

Figure 1.8 The drug development process.

1.4.3.1 Toxicological Phase

As a compound enters the drug development process, animal studies are conducted to learn about its safety and biological activity. This early development phase is called the toxicological or discovery phase. When the submission is for the US market, it is also called the pre-IND (investigational new drug) phase.

Characterization of the API and drug product must be well understood and characterized to support the IND submission. During this phase, the analytical chemist works closely with the process chemist to develop analytical methods to characterize the compound as well as to determine the purity of the API.

1.4.3.2 Investigational New Drug

An investigational new drug (IND) application begins the first step in the drug development process as the manufacturer starts formal clinical phases to study this compound in humans. The submission is also the first interaction the manufacturer has with the FDA on a new drug. The analytical scientist will be responsible for the analytical information included in the chemistry, manufacturing, and control (CMC) submission of the IND. Figure 1.9 shows an example of IND application (FDA Form 1571) (www.fda.org) [2–5].

1.4.3.3 Clinical Phase

The clinical development period contains three clinical phases. Phase I investigates the safety and tolerability of the drug product on healthy volunteers. Phase I studies are usually small with a short study duration. The subjects in this clinical phase are healthy volunteers and the population could range from 20 to 100 subjects.

Next Page | Export Data | Import Data | Reset Form

DEPARTMENT OF HEALTH AND HUMAN SERVICES Food and Drug Administration **INVESTIGATIONAL NEW DRUG APPLICATION (IND)** *(Title 21, Code of Federal Regulations (CFR) Part 312)*	Form Approved: OMB No. 0910-0014 Expiration Date: **March 31**, 2022 *See PRA Statement on page 3.* NOTE: No drug/biologic may be shipped or clinical investigation begun until an IND for that investigation is in effect (21 CFR 312.40)

1. Name of Sponsor

2. Date of Submission *(mm/dd/yyyy)*

3. Sponsor Address

Address 1 *(Street address, P.O. box, company name c/o)*

Address 2 *(Apartment, suite, unit, building, floor, etc.)*

City | State/Province/Region

Country | ZIP or Postal Code

4. Telephone Number *(Include country code if applicable and area code)*

6A. IND Number *(If previously assigned)*

6B. Select One: ☐ Commercial ☐ Research

5. Name of Drug *(Include all available names: Trade, Generic, Chemical, or Code)*

Continuation Page for #5

7A. (Proposed) Indication for Use

Is this indication for a rare disease (prevalence <200,000 in U.S.)? ☐ Yes ☐ No

Does this product have an FDA Orphan Designation for this indication? ☐ Yes ☐ No

If yes, provide the Orphan Designation number for this indication:

Continuation Page for #7

7B. SNOMED CT Indication Disease Term *(Use continuation page for each additional indication and respective coded disease term)*

8. Phase of Clinical Investigation to be conducted ☐ Phase 1 ☐ Phase 2 ☐ Phase 3 ☐ Other *(Specify)*: ____________

9. List numbers of all Investigational New Drug Applications (21 CFR Part 312), New Drug Applications (21 CFR Part 314) , Drug Master Files (21 CFR Part 314.420) , and Biologics License Applications (21 CFR Part 601) referred to in this application.

10. IND submission should be consecutively numbered. The initial IND should be numbered "Serial number: 0000." The next submission (e.g., amendment, report, or correspondence) should be numbered "Serial Number: 0001." Subsequent submissions should be numbered consecutively in the order in which they are submitted..

Serial Number _ _ _ _

11. This submission contains the following *(Select all that apply)*

☐ Initial Investigational New Drug Application (IND)
☐ Response to Clinical Hold
☐ Response To FDA Request For Information
☐ Request For Reactivation Or Reinstatement
☐ Annual Report
☐ General Correspondence
☐ Development Safety Update Report (DSUR)
☐ Other *(Specify)*: ____________

Protocol Amendment
☐ New Protocol
☐ PMR/PMC Protocol
☐ Change in Protocol
☐ New Investigator
☐ Human Factors Protocol

Information Amendment
☐ Chemistry/Microbiology
☐ Pharmacology/Toxicology
☐ Clinical/Safety
☐ Statistics
☐ Clinical Pharmacology

Request for
☐ Meeting
☐ Proprietary Name Review
☐ Special Protocol Assessment
☐ Formal Dispute Resolution

IND Safety Report
☐ Initial Written Report
☐ Follow-up to a Written Report

12. For Originals, is the product a combination product (21 CFR 3.2(e))? ☐ Yes ☐ No | Combination Product Type *(See instructions)* | Request for Designation (RFD) Number

13. Select the following only if applicable. *(Justification statement must be submitted with application for any items selected below. Refer to the cited CFR section for further information.)*

☐ Emergency Research Exception From Informed Consent Requirements, 21 CFR 312.23 (f)
☐ Charge Request, 21 CFR 312.8

Expanded Access Use, 21 CFR 312.300
☐ Individual Patient, Non-Emergency 21 CFR 312.310
☐ Intermediate Size Patient Population, 21 CFR 312.315
☐ Individual Patient, Emergency 21 CFR 312.310(d)
☐ Treatment IND or Protocol, 21 CFR 312.320

For FDA Use Only

CBER/DCC Receipt Stamp	DDR Receipt Stamp	Division Assignment
		IND Number Assigned

Figure 1.9 FDA Form 1571 – investigational new drug (IND) application. *Sources:* Adapted from Weinberg [3]; Nally [5]; Huynh-Ba [4].

Previous Page | Next Page

14. Contents of Application – This application contains the following items *(Select all that apply)*

- ☐ 1. Form FDA 1571 *(21 CFR 312.23(a)(1))*
- ☐ 2. Table of Contents *(21 CFR 312.23(a)(2))*
- ☐ 3. Introductory statement *(21 CFR 312.23(a)(3))*
- ☐ 4. General Investigational plan *(21 CFR 312.23(a)(3))*
- ☐ 5. Investigator's brochure *(21 CFR 312.23(a)(5))*
- ☐ 6. Protocol *(21 CFR 312.23(a)(6))*
  - ☐ a. Study protocol *(21 CFR 312.23(a)(6))*
  - ☐ b. Investigator data *(21 CFR 312.23(a)(6)(iii)(b))* or completed Form FDA 1572
  - ☐ c. Facilities data *(21 CFR 312.23(a)(6)(iii)(b))* or completed Form FDA 1572
- 6. Protocol *(Continued)*
  - ☐ d. Institutional Review Board data (21 CFR 312.23(a)(6)(iii)(b)) or completed Form FDA 1572
- ☐ 7. Chemistry, manufacturing, and control data *(21 CFR 312.23(a)(7))*
  - ☐ Environmental assessment or claim for exclusion *(21 CFR 312.23(a)(7)(iv)(e))*
- ☐ 8. Pharmacology and toxicology data (21 CFR 312.23(a)(8))
- ☐ 9. Previous human experience *(21 CFR 312.23(a)(9))*
- ☐ 10. Additional information *(21 CFR 312.23(a)(10))*
- ☐ 11. Biosimilar User Fee Cover Sheet *(Form FDA 3792)*
- ☐ 12. Clinical Trials Certification of Compliance *(Form FDA 3674)*

15. Is any part of the clinical study to be conducted by a contract research organization? ☐ Yes ☐ No

If Yes, will any sponsor obligations be transferred to the contract research organization? ☐ Yes ☐ No

If Yes, provide a statement containing the name and address of the contract research organization, identification of the clinical study, and a listing of the obligations transferred *(use continuation page)*.

Continuation Page for #15

16. Name and Title of the person responsible for monitoring the conduct and progress of the clinical investigations

17. Name and Title of the person responsible for review and evaluation of information relevant to the safety of the drug

I agree not to begin clinical investigations until 30 days after FDA's receipt of the IND unless I receive earlier notification by FDA that the studies may begin. I also agree not to begin or continue clinical investigations covered by the IND if those studies are placed on clinical hold or financial hold. I agree that an Institutional Review Board (IRB) that complies with the requirements set forth in 21 CFR Part 56 will be responsible for initial and continuing review and approval of each of the studies in the proposed clinical investigation. I agree to conduct the investigation in accordance with all other applicable regulatory requirements.

18. Name of Sponsor or Sponsor's Authorized Representative

19. Telephone Number *(Include country code if applicable and area code)*	20. Facsimile (FAX) Number *(Include country code if applicable and area code)*

21. Address

Address 1 *(Street address, P.O. box, company name c/o)*	
Address 2 *(Apartment, suite, unit, building, floor, etc.)*	
City	State/Province/Region
Country	ZIP or Postal Code

22. Email Address

23. Date of Sponsor's Signature *(mm/dd/yyyy)*

24. Name of Countersigner

25. Address of Countersigner

Address 1 *(Street address, P.O. box, company name c/o)*	
Address 2 *(Apartment, suite, unit, building, floor, etc.)*	
City	State/Province/Region
Country	ZIP or Postal Code

26. Email Address

WARNING : A willfully false statement is a criminal offense (U.S.C. Title 18, Sec. 1001).

27. Signature of Sponsor or Sponsor's Authorized Representative	28. Signature of Countersigner
Sign	Sign

Figure 1.9 (Continued)

The information below applies only to requirements of the Paperwork Reduction Act of 1995.

The burden time for this collection of information is estimated to average 100 hours per response, including the time to review instructions, search existing data sources, gather and maintain the data needed and complete and review the collection of information. Send comments regarding this burden estimate or any other aspect of this information collection, including suggestions for reducing this burden to the address to the right:

"An agency may not conduct or sponsor, and a person is not required to respond to, a collection of information unless it displays a currently valid OMB number."

Department of Health and Human Services
Food and Drug Administration
Office of Operations
Paperwork Reduction Act (PRA) Staff
PRAStaff@fda.hhs.gov

Please do NOT send your completed form to this PRA Staff email address.

Figure 1.9 (Continued)

Phase II evaluates the efficacy and extended safety assessment of patients. Studies of this phase typically contain approximately 100–500 patients. This phase also gathers the side effects and adverse events.

In the United States, a meeting at the end of Phase II with the FDA is recommended to discuss strategies of the application before entering Phase III development. This is an important go or no-go marker in the development process since Phase III studies would require a much greater investment of resources than earlier phases.

Phase III is an expansion of Phase II with a larger patient population to link the safety, efficacy, and effectiveness of the drug. Studies typically include 1000–5000 patient volunteers. The main purpose of Phase III is to confirm the effectiveness of the drug and monitor adverse reactions from long-term usage with a larger population.

After approval, sometimes there is still a need to conduct additional clinical studies. Phase IV studies may focus on different purposes, such as supplemental categories and different populations. The size of these studies varies depending on the objectives [6, 8].

1.4.3.4 Registration Phase

Once Phase III is completed successfully, a New Drug Application (NDA) or Marketing Authorization Application (MAA) is filed with the regulatory agency. It normally takes in the range of six months to a year for the initial review process, after which the FDA will issue a complete response letter detailing the deficiencies in the application. The Key Information section below lists the steps in the FDA Review Process. The successful approval rate for a new drug is very small, and in general, only one out of five applications will get approved.

Once approved, additional postmarketing testing is likely to be needed to study the long-term effects, adverse events from continued use, or additional packaging types or formulations. This testing can be required by the regulatory agencies or initiated by the company. Manufacturers may want to expand the packaging configuration or add dosage strengths [3, 7, 12].

Analytical testing is performed on all batches, and stability studies are conducted to monitor the quality of clinical materials during storage. The safety and efficacy of drug products are established during the development phase via clinical studies; therefore, if the testing profile changes beyond the acceptance criteria, then the established safety and efficacy are no longer applicable, and thus, the safety and efficacy of the drug product may need to be re-established. This information can be very critical when changes are made during the lifecycle of the products. These changes may affect product stability; thus, additional studies may be necessary, and further data are needed to support these changes [4, 5, 13, 14].

The drug review process

1) Preclinical (animal) testing
2) An Investigational New Drug (IND) Application outlines what the sponsor of a new drug proposes for human testing in clinical trials
3) Phase 1 studies involve less than 80 people with focus on safety
4) Phase 2 studies involved up to 300 people with focus on efficacy
5) The FDA and drug sponsor may have an end-of-Phase 2 face-to-face meetings
6) Phase 3 studies include several hundred to thousands of people to continue to study the efficacy and monitor side effects
7) Submission of an NDA
8) FDA has 60 days to decide if it can be filed (accepted without major revisions)
9) If the FDA files the NDA, an FDA review team is assigned to evaluate the sponsor's research on safety and effectiveness.
10) FDA reviews information that goes on the drug's labeling
11) FDA inspects the manufacturing facilities
12) FDA reviewers will approve or issue a complete response letter detailing the deficiencies in the application.

1.4.3.5 New Drug Application

The goal of the NDA is to obtain the license to market the drug product in the United States. The information contained in an NDA covers 10–12 years of research and development; therefore, it does take the agency a significant amount of time to review these submissions. Typically, one can expect six months to a year for this initial review. Safety and efficacy data are closely scrutinized since the FDA focuses on public health. The cost of taking a new chemical entity (NCE) through the drug development processes can cost more than $1.5–$2.0 billion. Therefore, optimizing the drug development process by understanding the key factors that affect the development of a drug product is very important for product commercialization.

Analytical data included in these submissions, and the pharmaceutical laboratories, will be inspected before the launch of the new products.

1.4.3.6 Post-Approval Phase

Post-approval studies expand the drug product application to different packages, strengths, different suppliers, and so forth. Analytical procedures must be verified for their intended use for any of these changes.

1.5 Summary

Pharmaceutical laboratories play a critical role in the development and manufacturing of drug products as analytical data provide the control strategies to develop, optimize, and monitor the finished products. Regulatory compliance is increasingly visible in the healthcare environment; more emphasis on GMPs regulations becomes evident as regulatory agencies are more stringent with new applications. Different countries have developed their own GMPs regulations; however, these regulations support a similar concept in the global environment. Chapter 2 will discuss GMPs in detail which is necessary to support an effective quality management system [1–3, 10, 12].

List of Abbreviations

CBER	Center for Biologics Evaluation and Research
CDER	Center for Drug Evaluation and Research
CDRH	Center for Devices and Radiological Health
CFR	Code of Federation Regulations
CFSAN	Center for Food Safety and Applied Nutrition
CGMP	Current Good Manufacturing Practices
CVM	Center for Veterinary Medicine
EFPIA	European Federation of Pharmaceutical Industries and Associations
EFTA	European Free Trade Association
EU	European Union
EWGs	Expert Working Groups
FDA	Food and Drug Administration
GLP	Good Laboratory Practices
ICH	International Council for Harmonisation of Technical Requirements for Pharmaceuticals for Human Use
IFPMA	International Federation of Pharmaceutical Manufacturers & Associations
IWGs	Implementation Working Groups
IPA	International Generic Pharmaceutical Alliance
IND	Investigational New Drug Application
JPMA	Japan Pharmaceutical Manufacturers Association
MAA	Marketing Authorization Application
MHLW	Ministry of Health, Labor and Welfare
NCE	New Chemical Entity
NDA	New Drug Application

PhRMA Pharmaceutical Research and Manufacturers of America
SC Steering Committee
WHO World Health Organization
WSMI World Self-medication Industry

References

1 The Office of Federal Register (OFR) (2021). *Federal Register | National Archives*. Rockville, MD: OFR www.ofr.gov (accessed 24 November 2021).
2 The U.S. Food and Drug Administration (2021). U.S. Food and Drug Administration (fda.gov). Rockville, MD: Food and Drug Administration www.fda.gov. (accessed 24 November 2021).
3 Weinberg, S. (2009). *Guidebook for Drug Regulatory Submissions*. Hoboken, NJ: Wiley.
4 Huynh-Ba, K. (ed.) (2009). *Handbook of Stability Testing for Pharmaceutical Products: Regulations, Methodologies, and Best Practices*. New York: Springer Science & Business Media.
5 Nally, J.D. (2007). *Good Manufacturing Practices for Pharmaceuticals*, 6e. New York: Informa Healthcare.
6 Cortez, A. (2002). Introducing the international conference on harmonization guidelines into the World Trade Organization: a strategy to remove technical trade barriers for pharmaceutical products. MACS Project.
7 Lee, C. (1993). *Development and Evaluation of Drugs: From Laboratory Through Licensure to Market*. Boca Raton, FL: CRC Press, Inc.
8 The international council on harmonisation of technical requirements for registration of pharmaceuticals for human use. www.ich.org.
9 Hansen, S.H., Pedersen-Bjergaard, S., and Rasmussen, K.E. (2012). *Introduction to Pharmaceutical Chemical Analysis*. Hoboken, NJ: Wiley.
10 Tobin, J.J. and Walsh, G. (2008). *Medical Product Regulatory Affairs: Pharmaceuticals, Diagnostics, Medical Devices*. Hoboken, NJ: Wiley.
11 Kealey, D. and Haines, P.J. (2002). *Instant Notes Analytical Chemistry*. Oxford, UK: Bios Scientific Publishers Ltd.
12 Mathieu, M. (2000). *New Drug Development: A Regulatory Overview*. Waltham, MA: PAREXEL International Corporation.
13 Huynh-Ba, K. (ed.) (2010). *Pharmaceutical Stability Testing to Support Global Markets*. New York: Springer Science & Business Media.
14 Nilsen, C.L. (1998). *The QC Laboratory Chemist: Plain and Simple*. Buffalo Grove, IL: Interpharm Press, Inc.

2

Good Manufacturing Practices (GMPs) and the Quality Systems

Kim Huynh-Ba

Pharmalytik LLC, Newark, DE, USA

TABLE OF CONTENTS

Analytical Testing for the Pharmaceutical GMP Laboratory, First Edition. Edited by Kim Huynh-Ba.

2.1 Introduction to Good Manufacturing Practices

Medicines and pharmaceutical products must be manufactured in a controlled and consistent manner to meet specific quality standards for their intended use. All facets of the manufacturing and packaging processes must be concisely established, monitored, and controlled to assure that medicines consistently meet the product specifications. All facilities and equipment must be well designed, installed correctly, and in good working order. Personnel are trained to execute their job responsibilities, understand the processes, and be able to recognize and troubleshoot issues that could potentially affect the product quality. Raw materials and packaging components must meet the approved acceptance criteria. Testing procedures are established to monitor the quality of products from the receipt of raw materials through the completion of finished products. Quality systems are developed in such a way as to provide absolute controls of all production batches; whether the batches meet the acceptance criteria and are available for distribution, or the batches do not meet the acceptance criteria and not to be distributed. The manufacturing and packaging processes are documented at the time of completion according to the procedures used. All records and documentation for all necessary steps are collected and securely retained, and the process is in place to recall or investigate any complaints. The pharmaceutical companies and regulatory agencies control complex processes of the drug industry through *Current Good Manufacturing Practices (CGMPs).*

Good Manufacturing Practices (GMPs) are the minimum required sets of regulations, which are enforced by the FDA or other local regulatory agencies, and provide systems that assure the proper design, monitoring, and control of facilities and manufacturing processes to ensure that quality products are consistently produced for their intended use. In the United States (US), the Food and Drug Administration (FDA) (www.fda.org) has the authority to access any facility and records of all pharmaceutical manufacturers. As stated in Section 501(B) of the 1938 Food, Drug, and Cosmetic Act, manufactured drugs must be in conformance with CGMPs. Pharmaceutical and medical device companies must follow these regulations to ensure that products are produced and controlled according to the quality standards pertinent to their intended use. The specified requirements do not distinguish between prescription drugs and over the counter (OTC) drugs; therefore, CGMP regulations apply to all drug products. It is still under debate whether pharmacy compounding operations fall within the scope of CGMPs. Dietary Supplements are considered as food products and therefore are not in the scope of CGMPs. Homeopathic drugs are considered as drugs within the Code of Federal Regulations (CFR) Title 21 Sec. 211.137(a) and are subjected to CGMPs with some reservations, for example, the CGMPs controls and registrations are different for drug and the food products. Many countries have developed their own CGMP requirements or adopted the FDA CGMPs to regulate their pharmaceutical industry.

The core regulations that apply to the manufacturing, processing, packaging, and holding of drugs are found under Title 21 of the CFR, in which Part 211 is primarily applicable to pharmaceutical analysis. Title 21 CFR Part 210 details the status and applicability of CGMPs regulations and contains a useful section on definitions. The purpose of these regulations is to assure the quality of medicines that are distributed to the population within different global markets. In the United States, complying with CGMPs is mandatory for the pharmaceutical and medical device industries, because a drug product is considered adulterated if it does not follow CGMP guidelines, regardless of meeting acceptance criteria [1, 8].

GMPs establish a wide range of requirements for different functions such as laboratory controls, production, facilities, and personnel. In the manufacturing process, they are neither menu-driven nor provide a list of procedures; but rather remain flexible to allow the practices to adapt new technologies and advancements over time. These regulations also allow individual pharmaceutical manufacturers to determine the most suitable procedures to produce safe and effective drug products. GMPs must be current, effective, and continue to evolve through the product lifecycle. GMPs remain flexible but are considered minimum requirements by the regulatory agencies. Therefore, pharmaceutical manufacturers must not select certain practices from the CGMPs to adopt or only apply them to certain products in their best interest. It is also common for a drug product to be developed in one country and marketed worldwide. For this reason, pharmaceutical manufacturers must be aware of the FDA CGMPs, as well as international GMPs, that would apply to their product, and develop quality systems that comply. Since 2005, many countries have developed their own GMP systems. In 1968, the World Health Organization (WHO) also issued a GMP guideline. Most of these guidelines are similar in principle because their purpose is to assure the quality of commercial products. It is helpful to understand the similarities and to recognize the differences among CGMPs worldwide.

The principles of GMPs are to ensure the safety and effectiveness of commercially available medicines. These principles and practices help predict the manufacture of drug products and track the process consistency to assure uniformity from batch to batch. The GMP guidelines are general principles to establish quality systems with high integrity within the industry. The owner of each pharmaceutical company is responsible for the development of its infrastructure to support its products by incorporating the GMP requirements into its Standard Operating Procedures (SOPs).

2.2 Objectives of GMPs

From a regulatory perspective, the drug product quality must be established based on data obtained from safety and efficacy studies during its development. The quality of a pharmaceutical product covers its identity, strength, quality, and purity. Analytical, microbiological, and physical data are used to demonstrate the drug product quality against predefined specifications. If the drug product quality changes beyond the established specifications, the established safety and efficacy data are no longer adequately applicable. The focus of this section will be the CGMPs that impact the pharmaceutical laboratory [8].

2.2.1 Definitions

First, we need to review and understand the definition of "drug" in the United States Pharmacopeia (USP) to which the GMPs in Title 21 CFR apply, as shown in the Key Information section below.

Current Good Manufacturing Practices – definitions

The term "drug" means *(A) articles recognized in the official United States Pharmacopeivwwa, official Homeopathic Pharmacopeia of the United States, or official National Formulary, or any supplement to any of them; and (B) articles intended for use in the diagnosis, cure, mitigation, treatment, or prevention of disease in man or other animals; and (C) articles (other than food) intended to affect the structure or any function of the body of man or other animals; and (D) articles intended for use as a component of any article specified in clause (A), (B), or (C).*

For a drug to be approved, Section CFR 211.137 Expiration Dating indicates that testing is necessary to assure that the drug product meets its specifications at release and that it meets applicable standards of identity, strength, quality, and purity at the time of use. Expiration dates shall appear on the labeling as per the requirements of 21 CFR 201.17. To have this assurance, GMP programs SOPs must be developed.

Stability studies are conducted according to 21 CFR Sec. 211.166 Stability Testing, to support the expiration dating and the storage conditions of the drug product. Additional information on the stability program is available in Chapter 9.

If the drug product needs to be reconstituted at the time of dispensing, then stability studies are conducted on the product both before and after reconstitution. The reconstituted form of the drug product must have a satisfactory stability profile through its intended expiry period. Typically, reconstitution is performed after a certain period (e.g. 24 or 48 hours) specified by the manufacturer to allow for the use of the reconstituted form. Specifications must be established for the product both before and after reconstitution [8].

The expiration-dating period of drug products is established from the manufacturing date of each batch. Most companies consider the *manufacturing date* as the date when the active ingredient is added to the excipient formula during manufacturing. Stability data are used to support the *expiration period* listed on the product label, and the quality of the drug product continues to meet the approved specifications. Additional information regarding expiration dating is listed in the Key Information section below.

21 CFR Sec. 211.137 expiration dating

a) To assure that a drug product meets applicable standards of identity, strength, quality, and purity at the time of use, it shall bear an expiration date determined by appropriate stability testing described in 211.166.
b) Expiration dates shall be related to any storage conditions stated on the labeling, as determined by stability studies described in 211.166.
c) If the drug product is to be reconstituted at the time of dispensing, its labeling shall bear expiration information for both the reconstituted and un-reconstituted drug products.
d) Expiration dates shall appear on labeling in accordance with the requirements of 201.17 of this chapter.
e) Homeopathic drug products shall be exempt from the requirements of this section.
f) Allergenic extracts that are labeled “No U.S. Standard of Potency” are exempt from the requirements of this section.
g) New drug products for investigational use are exempt from the requirements of this section, provided that they meet appropriate standards or specifications as demonstrated by stability studies during their use in clinical investigations. Where new drug products for investigational use are to be reconstituted at the time of dispensing, their labeling shall bear expiration information for the reconstituted drug product.
h) Pending consideration of a proposed exemption, published in the Federal Register of 29 September 1978, the requirements in this section shall not be enforced for human OTC drug products if their labeling does not bear dosage limitations and they are stable for at least three years as supported by appropriate stability data.

2.2.2 Organization of 21 CFR Regulations

In the broad range of CFR regulations, Titles 20 and 21 of the CFR (20 CFR and 21 CFR) apply to food and medicines consumed by humans and animals (see Key Information below). These titles provide essential information as described below [1, 2].

21 CFR Part 211 – subparts

A) Scope
B) Organization and personnel
C) Buildings and facilities
D) Equipment
E) Control of components and containers/closures
F) Production and process controls
G) Packaging and Labeling Controls
H) Holding and Distribution
I) Laboratory Controls
J) Records and Reports
K) Returned and Salvaged Drug Products

Subpart I shows the regulatory requirements for the pharmaceutical laboratory, in addition to Subpart B (Organization and Personnel) and Subpart D (Equipment). Other Subparts are directly related to the manufacturing area rather than the laboratory area. The laboratory facility must be designed and maintained to permit the testing conducted according to the CGMPs. The quality control area for chemical, physical, biological, and/or microbiological analysis must be separated from the production area. The location for the manufacturing of drug products must meet the requirements outlined in 21 CFR Part 211 to avoid the risk of contamination. Certain drug products may not be allowed to be manufactured in the same plant to avoid contamination [2].

In the Key Information section below, the regulations in this part contain the minimum Current Good Manufacturing Practice for preparation of drug products for administration to humans or animals. Therefore, inspectors can ask for more than the listed requirements. It allows the inspector to evaluate the pharmaceutical manufacturer's practices against current industry guidelines or any new initiatives from the agency, and to ensure that the SOPs are aligned with current regulations and the agency's current thinking. This scope clearly explains the realm of the regulations and how these regulations are applied to finished pharmaceutical drug products.

21 CFR Sec. 211.1 scope

a) The regulations in this part contain the minimum Current Good Manufacturing Practice for preparation of drug products (excluding positron emission tomography drugs) for administration to humans or animals.
b) The Current Good Manufacturing Practice regulations in this chapter as they pertain to drug products; in parts 600 through 680 of this chapter, as they pertain to drugs that are also biological products for human use; and in part 1271 of this chapter, as they are applicable to drugs that are also human cells, tissues, and cellular and tissue-based products (HCT/Ps) and that are drugs (subject to review under an application submitted under Section 505 of the act or under a biological product license application under Section 351 of the Public Health Service Act); supplement and do not supersede the regulations in this part unless the regulations explicitly provide otherwise. In the event of a conflict between applicable regulations in this part and in other parts of this chapter, or in parts 600 through 680 of this chapter, or in part 1271 of this chapter, the regulation specifically applicable to the drug product in question shall supersede the more general one.
c) Pending consideration of a proposed exemption, published in the Federal Register of 29 September 1978, the requirements in this part shall not be enforced for OTC drug products if the products and all their ingredients are ordinarily marketed and consumed as human foods, and which products may also fall within the legal definition of drugs by virtue of their intended use. Therefore, until further notice, regulations under part 110 of this chapter, and where applicable, parts 113–129 of this chapter, shall be applied in determining whether these OTC drug products that are also foods are manufactured, processed, packed, or held under Current Good Manufacturing Practice.

Paragraph (c) excludes OTC drug products from these regulations; however, many programs developed for OTC products are very similar. The challenge of the OTC industry is that the profit margin is quite small; therefore, the regulatory requirements are not as vigorous for products with raw materials that are *generally recognized as safe and effective* (GRASE), as they are for typical prescription drug products.

2.3 Personnel Qualification and Responsibilities – Subpart B

Personnel qualification is a critical component for any quality system, and it must be structured to support the organization. This section provides needed guidance on the expectations of a quality unit and dictates the vigorous training program developed for the analyst.

2.3.1 Responsibilities of the Quality Control Unit

Title 21 CFR Sec. 211.22 lists the responsibilities of a quality control (QC) unit. The regulations indicate that the QC unit is responsible for maintaining the quality of the drug product by approving or rejecting raw materials, components, or the drug product itself. This unit is also responsible for controlling the quality system of the company; thus, it approves all SOPs or quality initiatives of the company. There is no guidance on how the QC unit should be structured. In many pharmaceutical manufacturers, it includes a quality control testing laboratory, a quality assurance unit, or a compliance unit. The quality of a product is determined during the product development stage and tested before release. Confirmation testing is typically done on a representative number of samples. The QC unit must collaborate with many other functional areas such as production, packaging, and manufacturing to provide controls and monitor critical quality attributes over the entire process, beginning from raw materials to finished commercial products. These quality functions will recommend how efficiency and consistency can be improved throughout the organization.

Title 21 CFR Sec. 211.22(b), as shown in the Key Information section below, indicates that the laboratory must provide an adequate working area for equipment and analysts to allow operations. It must have sufficient and suitable storage space for test samples, reference standards, reagents, and retained samples. Records should be sufficient to identify the raw material, the in-process, intermediate, and finished product samples if testing is necessary to avoid any further contamination or mix-ups. The environment of the laboratory should have a suitable air handling system, a regular supply of water for appropriate cleaning and testing purposes, and any other requirements for specialty laboratories.

21 CFR Sec. 211.22 subpart B – responsibilities of quality control unit

a) There shall be a quality control unit which shall have the responsibility and authority to approve or reject all components, drug product containers, closures, in-process materials, packaging material, labeling, and drug products, and the authority to review production records to assure that no errors have occurred or, if errors have occurred, that they have been fully investigated.

The quality control unit shall be responsible for approving or rejecting drug products manufactured, processed, packed, or held under contract by another company.

b) Adequate laboratory facilities for the testing and approval (or rejection) of components, drug product containers, closures, packaging materials, in-process materials, and drug products shall be available to the quality control unit.
c) The quality control unit shall have the responsibility for approving or rejecting all procedures or specifications impacting the identity, strength, quality, and purity of the drug product.
d) The responsibilities and procedures applicable to the quality control unit shall be in writing; such written procedures shall be followed.

2.3.2 Personnel Qualifications and Responsibilities

Personnel qualifications are critical and a focal point of many GMP inspections. Because pharmaceutical manufacturers are required to have a robust and compliant training program for their personnel, the pharmaceutical industry spends millions of dollars annually on training. Each person must have up-to-date training records that include in-house training, outside training, education, and experience to show his/her qualifications to perform the activities. These records must be kept up to date as they will prove that the personnel are qualified to perform the tasks assigned. It is the responsibility of management to provide sufficient trained personnel for each specific task, as well as supervision. Training must be refreshed periodically, typically every two years, depending on the GMP task required, and metrics must be developed to ensure that the provided training is effective (see Key Information).

21 CFR Sec. 211.25 subpart B – personnel qualifications

a) Each person engaged in the manufacture, processing, packing, or holding of a drug product shall have education, training, and experience, or any combination thereof, to enable that person to perform the assigned functions. Training shall be in the particular operations that the employee performs and in Current Good Manufacturing Practice (including the Current Good Manufacturing Practice regulations in this chapter and written procedures required by these regulations) as they relate to the employee's functions. Training in Current Good Manufacturing Practice shall be conducted by qualified individuals on a continuing basis and with sufficient frequency to assure that employees remain familiar with CGMP requirements applicable to them.

b) Each person responsible for supervising the manufacture, processing, packing, or holding of a drug product shall have the education, training, and experience, or any combination thereof, to perform assigned functions in such a manner as to provide assurance that the drug product has the safety, identity, strength, quality, and purity that it purports or is represented to possess.
c) There shall be an adequate number of qualified personnel to perform and supervise the manufacture, processing, packing, or holding of each drug product.

The GMP regulation requires that the pharmaceutical laboratory must be segregated from the production area to avoid mix-ups and contamination. The testing must be conducted under the direct supervision of trained and competent analysts who are qualified to perform the procedures. The laboratory must also have an adequate number of analysts to complete the workload and carry out the tasks assigned to them. Therefore, a training program is required that includes technology-oriented, task-oriented, and procedure-oriented topics. Training can be done as a group or individually to ensure that the analyst learns the skill set required to perform his/her job functions. Training programs usually provide a certificate of completion, or the result from a quiz taken at the end of the training, as proof of proficiency. The documentation should also include the trainer's name and the date of completion of the training. Safety and sanitation instructions must be included as part of the training program. If any person is sick, he/she must not handle any GMP task to avoid contamination that might eventually affect the quality of the product.

It is important to have sufficient qualified trainers to support the training program. Trainers must understand the subject matter, the specific course content, and the regulations involved, and possess the knowledge necessary to train personnel. The best technical person may not be the best trainer. Training materials are equally important and need to be designed to maximize retention of the subject material. Quality metrics are used to measure the effectiveness of the training program. Evaluation of the training material is necessary to provide feedback to the trainer and to adjust the course content and training program, as needed. The QC unit should review the training materials to assure consistency with CGMPs, regulations, and corporate standards.

GMP training is a continuous process to keep analysts abreast of industry practices; thus, training is incorporated into their individual professional

Name:	Date of Hire/Transfer:
Employee Number:	Supervisor:
Department/Section:	
Training Type (Check One): ☐ Technical ☐ CGMP ☐ Safety ☐ Other	

Description/Title	SOP # Training Materials	Training completed (Initials /Date) Trainee	Trainer	Supervisor
New Employee Orientation (Tour of company and Introduction to Staff)	SOP QA-1000			
Curriculum Vitae	SOP QA-1001			
New Employee Training	SOP QA-1002			
New Employee Security and Safety Training	SOP QA-1003			
New Employee GMP Training	SOP QA-1004			
Metrology Program Training	SOP QA-1005			
Out-of-Specification Investigation	SOP QA-1006			

QA Signature:________________________ Date: ____________

Figure 2.1 Example of a personnel training record.

development. Training can be completed through in-house GMP training programs, webinars, or attending technical conferences. Three main areas that training can focus on are:

- New employee orientation program, including fundamental information such as the industry, regulations, company and its infrastructure, benefits, and safety.
- Specific training for different functional areas, including GMPs or specific instrumentation skills for analytical laboratory personnel.
- Other skills that are relevant to jobs, such as time management, project management, communication.

Figure 2.1 illustrates an example of a personnel-training record to fulfill GMP training. Other specific training or skills should also be tracked and evaluated annually as part of the personnel development plan. An example of external training or an outside workshop is shown in Figure 2.2.

Title 21 CFR Sec. 211.28 lists personnel responsibilities for all employees to prevent contamination of drug products. Only trained employees should perform the required functions. Contractors or temporary staff should acquire sufficient training to perform their job functions. The Key Information section below lists the personnel responsibilities found in Subpart B.

EMPLOYEE TRAINING RECORD				
Name:		Date of Hire/Transfer:		
Employee Number:		Supervisor:		
Department/Section:				
Training Type (Check One):	☐ Technical	☐ CGMP	☐ Safety	☐ Other

Session Title:
Date:
Location:
Trainer: (include biography, if available)
Summary of Training Session: (Attach syllabus, notes, outlines, etc. as appropriate)

Attendee:____________________ Date: __________

Supervisor:____________________ Date: __________

Figure 2.2 Example of an employee training record for continued GMP training.

21 CFR Sec. 211.28 subpart B – personnel responsibilities

a) Personnel engaged in the manufacture, processing, packing, or holding of a drug product shall wear clean clothing appropriate for the duties they perform. Protective apparel, such as head, face, hand, and arm coverings, shall be worn as necessary to protect drug products from contamination.
b) Personnel shall practice good sanitation and healthy habits.
c) Only personnel authorized by supervisory personnel shall enter those areas of the buildings and facilities designated as limited-access areas.
d) Any person shown at any time (either by medical examination or supervisory observation) to have an apparent illness or open lesions that may adversely affect the safety or quality of drug products shall be excluded from direct contact with components, drug product containers, closures, in-process materials, and drug products until the condition is corrected or determined by competent medical personnel not to jeopardize the safety or quality of drug products. All personnel shall be instructed to report to supervisory personnel any health conditions that may have an adverse effect on drug products.

2.4 Equipment - Subpart D

In a GMP laboratory, evaluation of a study depends on the reliability and accuracy of the analytical results; therefore, the equipment plays a critical role. The following two Key Information sections describe the requirements in 21 CFR Sec. 211.63 Subpart D.

In the Key Information sections below, it is indicated that all pieces of equipment used in the manufacturing, processing, packing, or holding of a drug product, must be included in the GMP system. The pharmaceutical process must be defined and validated; thus, appropriate pieces of equipment are designed for manufacturing, packing, holding, or testing. User requirements are established before an equipment evaluation can be done, and the piece of equipment must be qualified and maintained regularly. Equipment logbooks, documenting daily use, are also recommended.

21 CFR Sec. 211.63 subpart D - equipment design, size, and location

Equipment used in the manufacture, processing, packing, or holding of a drug product shall be of appropriate design, adequate size, and suitably located to facilitate operations for its intended use and for its cleaning and maintenance.

21 CFR Sec. 211.65 subpart D - equipment construction

a) Equipment shall be constructed so that surfaces that contact components, in-process materials, or drug products shall not be reactive, additive, or absorptive so as to alter the safety, identity, strength, quality, or purity of the drug product beyond the official or other established requirements.
b) Any substances required for operation, such as lubricants or coolants, shall not come into contact with components, drug product containers, closures, in-process materials, or drug products so as to alter the safety, identity, strength, quality, or purity of the drug product beyond the official or other established requirements.

Most drug manufacturers have metrology functions to qualify and maintain their equipment. However, some manufacturers opt to outsource the calibration and maintenance of equipment to either the instrument manufacturers or certified service providers if the equipment requires expertise or resources that are not available.

2.4.1 Metrology Functions

Established SOPs should be in place for the overall metrology program as well as for specific equipment. Responsibilities must be clearly defined whether internal analysts or contractors will maintain equipment in the metrology program. Manufacturers' manuals should be reviewed and can be used to establish procedures for the new equipment. The overall design, size, and location of different parts of the equipment, need to be adequate and suitable for the operation. SOPs should include the maintenance, calibration, and cleaning of equipment to allow the equipment to function properly when used and to avoid contamination that may adversely affect the quality of the products.

Critical measurement pieces of equipment (e.g. balances, pH meters) must be available in the laboratory, and qualified for the range, accuracy, and precision appropriate to the specifications of the products to be tested. The equipment must be calibrated periodically and checked daily according to their SOPs. Analysts must be trained on these procedures before using the equipment. Analysts must not use equipment that is out of calibration. Equipment that exceeds its calibration due date should be tagged and taken out of operation for service. Figures 2.3 and 2.4 provide an example of a calibration form used for pH meter and analytical balance, respectively.

In recent years, cleaning and maintenance programs have become a critical GMP area in the pharmaceutical industry. Product components and other materials should not come into contact with other substances that may result in contamination or could affect the safety, identity, strength, quality, and purity of the drug product. There should be an SOP for the cleaning program, where effective cleaning procedures are developed and validated. Acceptance criteria are

INSTRUMENT LOGBOOK

Instrument: pH meter
Make: ______________
Serial or Code no: ______________
pH Buffer used: ______________
SOP for Calibration: ______________

Equipment No.: ______________
Model no: ______________
Calibration due: ______________
pH Buffer Lot #/Expiration date: ______________
Calibration SOP Effective date: ______________

Date	Calibrated by	Result/Observation	Pass/Fail	Remarks/Correction Action

Notes:

Figure 2.3 Example of a form used for a pH meter in the analytical lab.

INSTRUMENT LOGBOOK

Instrument:	Analytical Balance	Equipment No:	________
Make:	________	Model no:	________
Serial or Code No:	________	Calibration due:	________
SOP for Calibration:	________	SOP Effective date:	________

Date	Calibrated by	Standard weight Used	Observed weight	Pass/Fail	Remarks/Correction Action

Figure 2.4 Example of a form used for an analytical balance.

determined based on the products before and after each production run. Swabbing techniques are important for verification of cleaning. The potential for low-level detection and robustness of these procedures are also critical.

Subpart D – Sec. 211.67 equipment cleaning and maintenance

a) Equipment and utensils shall be cleaned, maintained, and, as appropriate for the nature of the drug, sanitized and/or sterilized at appropriate intervals to prevent malfunctions or contamination that would alter the safety, identity, strength, quality, or purity of the drug product beyond the official or other established requirements.

b) Written procedures shall be established and followed for cleaning and maintenance of equipment, including utensils, used in the manufacture, processing, packing, or holding of a drug product. These procedures shall include, but not necessarily limited to, the following:
   1) Assignment of responsibility for cleaning and maintaining equipment;
   2) Maintenance and cleaning schedules, including, where appropriate, sanitizing schedules.
   3) A description in sufficient detail of the methods, equipment, and materials used in cleaning and maintenance operations, and the methods of disassembling and reassembling equipment as necessary to assure proper cleaning and maintenance.
   4) Removal or obliteration of previous batch identification.
   5) Protection of clean equipment from contamination prior to use.
   6) Inspection of equipment for cleanliness immediately before use.

c) Records shall be kept of maintenance, cleaning, sanitizing, and inspection as specified in 211.180 and 211.182.

Title 21 CFR Sec 211.68 refers to the requirements for automatic, mechanical, and electronic equipment. Written procedures and records for calibration, inspection, or verification of the areas should also be developed, and validated processes to control inadvertent or intentional changing of electronic records are critical to the laboratory operations. The Key Information section below lists the requirements for the different types of equipment requirements. Additionally, information about electronic data can be found in Chapter 10.

21 CFR Sec. 211.68 subpart D – automatic, mechanical, and electronic equipment

a) Automatic, mechanical, or electronic equipment or other types of equipment, including computers, or related systems that will perform a function satisfactorily, may be used in the manufacture, processing, packing, and holding of a drug product. If such equipment is so used, it shall be routinely calibrated, inspected, or checked according to a written program designed to assure proper performance. Written records of those calibration checks and inspections shall be maintained.

b) Appropriate controls shall be exercised over computer or related systems to assure that changes in master production and control records or other records are instituted only by authorized personnel. The input and the output from the computer, related system of formulas, or other records or data shall be checked for accuracy. The degree and frequency of input/output verification shall be based on the complexity and reliability of the computer or related system. A backup file of data entered in the computer or related system shall be maintained except where certain data, such as calculations performed in connection with laboratory analysis, are eliminated by computerization or other automated processes. In such instances, a written record of the program shall be maintained along with appropriate validation data. Hard copy or alternative systems, such as duplicates, tapes, or microfilm, designed to assure that backup data are exact and complete and that it is secure from alteration, inadvertent erasures, or loss shall be maintained.

c) Such automated equipment used for performance of operations addressed by 211.101(c) or (d), 211.103, 211.182, or 211.188(b)(11) can satisfy the requirements included in those sections relating to the performance of an operation by one person and checking by another person whether such equipment is used in conformity with this section, and one person checks that the equipment properly performed the operation.

One of the key factors is the use of filters at any step of the process to ensure it does not affect the quality of the product. In pharmaceutical analysis, filters are also used in the sample preparation part of the analytical procedures; thus, it is necessary to include the filters in the validation of analytical procedures, such as dissolution or assay method. Testing should be done for any changes to the filters to verify there has been no impact on the analysis (see Key Information).

21 CFR Sec. 211.72 subpart D – filters

Filters for liquid filtration used in the manufacture, processing, or packing of injectable drug products intended for human use shall not release fibers into such products. Fiber-releasing filters may be used when it is not possible to manufacture such products without the use of these filters. If use of a fiber-releasing filter is necessary, an additional no fiber-releasing filter having a maximum nominal pore size rating of 0.2 μm (0.45 μm if the manufacturing conditions so dictate) shall subsequently be used to reduce the content of particles in the injectable drug product. The use of an asbestos-containing filter is prohibited.

The above requirements are typically applied to instruments used in the production area. For the laboratory, additional requirements are listed in 21 CFR Sec. 211.160(b)(4) in Subpart I of Laboratory Controls. This section states that a metrology program should be available for all the equipment, and each piece of instrumentation operated under GMPs must contain proper procedures and specifications. Within most pharmaceutical manufacturers, there is a team of analysts comprising the metrology group. This group could be part of the manufacturer's facility department, an engineering group, or a quality group. In any case, all procedures and activities need to be reviewed by the QA department. The Key Information section lists the functions of the metrology program.

Functions of a metrology program

- Develop calibration procedures and specifications for all equipment
- Track, schedule, and calibrate the instrument
- Coordinate and determine Vendor Contracts
- Review and approve qualification and calibration data
- Work with contractors and internal personnel using internal or external procedures

2.4.2 Qualification Phases

Analytical instrument qualification is a collection of documents to show that an instrument is properly installed, is functioning suitably for its intended purpose, and is producing reliable results. It contains four phases: design qualification (DQ), installation qualification (IQ), operational qualification (OQ), and performance qualification (PQ). These qualifications must be completed using approved protocols with predetermined acceptance criteria.

2.4.2.1 Design Qualification (DQ)

This activity is performed to verify that the proposed design is suitable for the intended purpose. Specifications, manuals, procedures, equipment connections, sketches, and drawings are usually retained by the instrument manufacturer. The manufacturer should have adequate support for installation, service, and training. The purchaser should confirm that the design is consistent with the needs of the test, process, or product for which it will be used as compared to a set of user requirements for this purpose.

2.4.2.2 Installation Qualification (IQ)

The IQ documents that the equipment was received, and all components have been installed for the intended use in a manner consistent with manufacturer's recommendations. The task will be done at the purchasing laboratory and performed by the equipment manufacturer or representative.

2.4.2.3 Operational Qualification (OQ)

This qualification provides assurance that equipment operates properly according to the manufacturer's specifications for the intended use. This requirement is completed in a selected environment at the purchasing laboratory. Selected instrument parameters will be used for certain functional tests, and acceptance criteria will be set based on the manufacturer's recommendation.

2.4.2.4 Performance Qualification (PQ)

The PQ helps the user verify that the equipment performs within the process parameters and is consistent with the specifications for the defined intended use. It also demonstrates that the equipment consistently performs according to the predetermined specifications on a routine basis. This activity is performed in case the equipment is out-of-service. Procedures for operation, calibration, maintenance, and change control are also required.

A *preventive maintenance program* is developed to ensure that all pieces of equipment perform properly. The calibration and maintenance procedure are done based on the manufacturer's recommended time or laboratory usage. Individuals must be trained to perform this function properly. The qualification frequency runs typically every six months to two years, based on the manufacturer's recommendation, the reliability of the instrumentation, and the usage of equipment. The calibration and maintenance procedures must be robust. Activities for each component of the equipment must be documented in designated unique logbooks. There should be identification and status labels on each component of equipment. For an out-of-specification (OOS) investigation, it is necessary to provide evidence that equipment is within its calibration period and is well maintained.

All equipment must have a unique identification number and must have a logbook. Information such as daily usage, qualification, calibration, or maintenance, will be recorded in this logbook and subjected to quality audits (see Key Information).

Section 21CFR 211.160 (b) 4 – general procedures

4) The calibration of instruments, apparatus, gauges, and recording devices at suitable intervals in accordance with an established written program containing specific directions, schedules, limits for accuracy and precision, and provisions for remedial action in the event accuracy and/or precision limits are not met. Instruments, apparatus, gauges, and recording devices not meeting established specifications shall not be used.

The Key Information section below lists the qualification strategies to be performed on analytical equipment. For new instrumentation, all four studies (DQ, IQ, OQ, and PQ) should be completed. Qualification protocols are designed with proper acceptance criteria based on the intended use of the equipment. For the equipment that has extended its use or requires service, OQ and PQs are done once the service is complete. Additional information can be found in USP General Chapter <1058> *Analytical Instrument Qualification.*

If an equipment component must be relocated, even in the same facility, installation and PQs must be repeated. PQ should be done before and after the piece of equipment is moved. An approved protocol should be written detailing the relocation and address any impact on the equipment with the new location.

Qualification strategy for equipment

- For new instrumentation: DQ, IQ, OQ, and PQ
- For extended use and repairs: OQ and PQ
- For relocated equipment: IQ and PQ (if damaged/changed during the move then OQ may also be needed)
- PQ must be verified before and after being moved

2.5 Laboratory Controls

All drug products are approved with specifications to assure identity, strength, quality, and purity. These specifications are established based on the clinical studies conducted with the products. Therefore, a change control system is necessary to ensure that the quality of drug products is not being compromised.

2.5.1 General Requirements

The analytical department establishes specifications and test procedures during the drug development process for new drug products. The established specifications are based on clinical data. Therefore, if there are changes in the product or process or analytical methods, then these should be reviewed and updated to make sure that the safety and efficacy of the product are not affected. Each department develops its plan to collect their samples for the necessary data and establishes control mechanisms for their respective functional areas. Official compendia such as the USP, as well as other global Pharmacopeia, include specifications and analytical procedures for approved drug products. Analytical procedures are validated for their intended purposes according to regulatory requirements. All validation parameters are evaluated based on testing purposes [3, 4].

Standard Operating Procedures (SOPs) are written by the organization that is responsible for the activities. Personnel are trained to follow the SOPs as written. 21 CFR Sec. 211.160(a) requires the QC unit to review and approve all documents and changes generated by the laboratories for their technical content, as well as against current regulations and industry practices. This section also requires the analyst to record information directly into the final format of reporting (e.g. notebook, worksheet, and executed protocol) at the time of performance to eliminate possible errors in transcription and to ensure data integrity. The second analyst verifies the data for accuracy, completeness, and compliance of the record. The information must be signed and dated as the activities are completed [5].

A robust change control system should be available to record all changes and assess the impact of these changes to ensure they do not unintentionally affect the quality of the drug product. The QC unit must be aware of these changes and is responsible for making sure that the changes are implemented properly (see Key Information).

21 CFR Sec. 211.160(a) General requirements
a) The establishment of any specifications, standards, sampling plans, test procedures, or other laboratory control mechanisms required by this subpart, including any change in such specifications, standards, sampling plans, test procedures, or other laboratory control mechanisms, shall be drafted by the appropriate organizational unit and reviewed and approved by the quality control unit. The requirements in this subpart shall be followed and shall be documented at the time of performance. Any deviation from the written specifications, standards, sampling plans, test procedures, or other laboratory control mechanisms shall be recorded and justified.

The following Key Information section lists requirements pertaining to laboratory controls. All written controls, including sampling plans, procedures, and a complete set of specifications, are established for drug products, any product components, drug product containers, closures, in-process materials, and labeling. Statistical justification is used to determine the appropriate representative samples for the batch of the drug product. All released batches must meet the approved specifications at release and through expiry.

21 CFR 211.160(b) General requirements
b) Laboratory controls shall include the establishment of scientifically sound and appropriate specifications, standards, sampling plans, and test procedures designed to assure that components, drug product containers, closures, in-process materials, labeling, and drug products conform to appropriate standards of identity, strength, quality, and purity. Laboratory controls shall include: 1) Determination of conformity to applicable written specifications for the acceptance of each lot within each shipment of components, drug product containers, closures, and labeling used in the manufacture, processing, packing, or holding of drug products. The specifications shall include a description of the sampling and testing procedures used. Samples shall be representative and adequately identified. Such procedures shall also require appropriate retesting of any component, drug product container, or closure that is subject to deterioration. 2) Determination of conformance to written specifications and a description of sampling and testing procedures for in-process materials. Such samples shall be representative and properly identified.

3) Determination of conformance to written descriptions of sampling procedures and appropriate specifications for drug products. Such samples shall be representative and properly identified.
4) The calibration of instruments, apparatus, gauges, and recording devices at suitable intervals in accordance with an established written program containing specific directions, schedules, limits for accuracy and precision, and provisions for remedial action, in the event of accuracy and/or precision limits not being met. Instruments, apparatus, gauges, and recording devices not meeting established specifications shall not be used.

Title 21 CFR Sec. 211.160(b)(1) requires that each lot within each shipment of components, drug product containers, closures, and labeling, must meet the established specifications. Pharmaceutical manufacturers must also establish a sample plan to ensure samples are representative and adequately identified. Retest dates are set for any component, container, and closure that may be subject to deterioration. Similar requirements of specifications and sampling plans are defined for in-process materials and drug products based on 21 CFR Sec. 211.160 (b)(2) and 21 CFR Sec. 211.160(b)(3).

Title 21 CFR Sec. 211.160(b)(4) lists requirements for equipment used in GMP activities. A metrology program is required to maintain equipment such as instruments, apparatus, gauges, and other recording devices.

2.5.2 Testing and Release for Distribution

Testing procedures provide data confirming identification, strength, and other critical attributes of a drug product. These methods are developed, validated, and documented with appropriate sampling and testing plans. Acceptance and rejection levels must also be justified in accordance with the statistical quality control criteria (see Key Information).

21 CFR Sec. 211.165 testing and release for distribution

a) For each batch of drug product, there shall be appropriate laboratory determination of satisfactory conformance to final specifications for the drug product, including the identity and strength of each active ingredient, prior to release. Where sterility and/or pyrogen testing are conducted on specific batches of short-lived radiopharmaceuticals, such batches may be released prior to completion of sterility and/or pyrogen testing, provided such testing is completed as soon as possible.

b) There shall be appropriate laboratory testing, as necessary, of each batch of drug product required to be free of objectionable microorganisms.
c) Any sampling and testing plans shall be described in written procedures that shall include the method of sampling and the number of units per batch to be tested; such written procedure shall be followed.
d) Acceptance criteria for the sampling and testing conducted by the quality control unit shall be adequate to assure that batches of drug products meet each appropriate specification and appropriate statistical quality control criteria as a condition for their approval and release. The statistical quality control criteria shall include appropriate acceptance levels and/or appropriate rejection levels.

Control of the product quality depends on the available analytical data; therefore, data generated must be accurate and reliable to evaluate the sample tested. According to 21 CFR Sec. 211.165(b)(e), validation of these procedures is conducted and documented. More information on method validation is discussed in Chapter 5. Method transfer, a process by which a laboratory is qualified to perform the approved analytical procedures, is discussed further in Chapter 6.

Title 21 CFR Sec. 211.165(b)(f) requires that all analytical results meet the approved acceptance criteria for the batch to be released; otherwise, it is rejected. Reprocessing or rework is done with an approved reprocessing procedure and evaluation of product impact. If the reprocessed materials are acceptable, then they can be released.

21 CFR Sec. 211.165(e)(f) testing and release for distribution (*cont'd*)

e) The accuracy, sensitivity, specificity, and reproducibility of test methods employed by the firm shall be established and documented. Such validation and documentation may be accomplished in accordance with 211.194(a) (2).
f) Drug products failing to meet established standards or specifications and any other relevant quality control criteria shall be rejected. Reprocessing may be performed. Prior to acceptance and use, the reprocessed material must meet appropriate standards, specifications, and any other relevant criteria.

2.5.3 Stability Program

Most of the routine tests conducted in the pharmaceutical laboratory provide data to support stability testing and establish the expiry of the drug product. The stability program is documented based on 21 CFR Sec. 211.166 to establish the expiry date and storage conditions for the drug product using an adequate number of batches. Samples are tested at different storage conditions, such as intended storage, accelerated, and stress conditions, at an appropriate length of time. More information is available in Chapter 9.

2.5.4 Retention Program

A written SOP is necessary to describe the retention program of all approved products. A representative amount of drug product sample must be kept and stored under the labeled conditions for every released batch as a part of the retention program as defined in the product monographs. The appearance or visual inspection of the samples is tested at least annually, and any change in their appearance is documented and investigated. These data will be reviewed and reported as a part of the annual report of the product.

2.6 Records and Reports

GMPs for the pharmaceutical laboratory are essential to assure the safety and quality of pharmaceutical products. Established programs, test procedures, or operating procedures are necessary are part of the company's written documentation. Subpart J of 21 CFR Sec. 211 presents requirements for records and reports. The Key Information section below lists the topics discussed in this section. Due to the importance of the documentation system, Chapter 8 [5].

21 CFR Part 211 subpart J – records and reports

- 211.180 – General requirements.
- 211.182 – Equipment cleaning and use log.
- 211.184 – Component, drug product container closure, and labeling records.
- 211. 186 – Master production and control records.
- 211. 188 – Batch production and control records.
- 211. 192 – Production record review.
- 211. 194 – Laboratory records.
- 211. 196 – Distribution records.
- 211. 198 – Complaint files.

2.7 Pharmaceutical Quality

The QC department is a critical cornerstone of the pharmaceutical quality system, with the responsibility for developing analytical procedures for controlling manufacturing processes and monitoring the quality of the drug product. The QC laboratory establishes procedures for sampling, specifications, and analytical procedures for release and stability testing of raw materials, reagents, reference standards, in-process, intermediate materials, bulk finished products, and packaged drug products. Within the QC department are several specialized laboratory areas such as chemical analysis, physical analysis, dissolution, biological, and microbiological testing; and several support groups to oversee critical programs such as reference standards, metrology, and stability. The purpose of all the groups is to manage the materials and their laboratory information.

2.7.1 Quality Manual

The quality manual describes the quality policy, scope of the pharmaceutical quality systems, and identification of all processes within the pharmaceutical manufacturer. This manual includes the responsibility of management, which will be held accountable for ensuring the quality and performance of the quality systems, the overall direction of the company, and applicable regulations used.

Deviations must be reported, investigated, and documented to prevent unplanned changes. The investigation should include a root cause of analysis to identify suspected product defects and the development of an appropriate corrective and preventive actions (CAPA) plan, to avoid similar potential deviations recurring in the future. All deviations should be monitored and trended periodically during the product lifecycle to identify potential issues and unanticipated impact to the product or process.

Regular evaluation of the quality of the products is done through testing to verify the consistency of the process and consider opportunities for continuous improvement. Management is responsible for ensuring that the products are fit for the intended use and meet the appropriate quality, safety, and efficacy.

Each pharmaceutical manufacturer goes through vigorous self-audit and self-inspection programs to ensure that the quality systems remain suitable and effective. These programs also help to identify critical areas to be maintained or improved.

2.7.2 Quality Risk Management

Quality risk management (QRM) establishes the control strategy that maintains the quality of drug products. The QRM process identifies control parameters and critical attributes for drug substance and drug products, raw materials, components, facilities, equipment-operating conditions, in-process controls, finished product specifications, and associated analytical procedures [6].

The Key Information section below lists five essential steps of a QRM system. This system will provide data management and statistical tools for measurement and analysis of parameters and attributes defined in the control strategy. It also identifies sources of variations affecting process performance and product quality for potential continual improvement activities to reduce or control variation.

Five steps of quality risk management
1) Risk identification: What can go wrong 2) Risk evaluation: Determine levels for evaluation of severity, occurrence, detectability 3) Risk analysis: Gather quantitative or qualitative data 4) Risk control: Determine steps to control the risk by reduction or acceptance 5) Risk communication: Review and communicate the plan

The key to effective risk management is to understand the product and its underlying critical process attributes. Data collection is done based on scientific evaluation and its impact on the existing system. These impacts are determined based on patient safety, product quality, data integrity, and the risks involved. Once the functional risks are assessed, controls are identified and implemented for the studied process. These controls should be closely monitored and verified periodically to maintain the approved quality or to identify areas of improvement.

The QRM process provides a systematic way to assess, control, communicate, and review the risks concerning the quality of the drug product. Depending on the level of risks, the mitigation efforts and documentation will be oriented accordingly. The objective is to manage risks, allow efficient allocation of resources, and enable documentation of rational decision-making quantitatively and consistently.

2.7.3 Product Quality Review

Management assures product quality and process performance through regular review of the products to verify the consistency of the existing process, appropriateness, and reliability of current specifications, and identify product and process improvements, as necessary. The Key Information section below lists several data systems used in the product quality review; they typically are part of the Annual Product Reviews (APR) or are used more often depending on product needs. Quality Assurance must verify the Product Quality Review through the audit program.

Systems included in product quality review

- Results of critical raw materials and or packaging materials
- Results of critical in-process or finished products
- Results of audits/inspections and finding
- Conclusions of process performance and product quality monitoring
- Batches that do not meet established specifications and their investigation
- Effectiveness of process and product changes
- Changes and validation of analytical procedures
- Results of the stability monitoring program and any adverse trends
- Changes from corrective actions and preventive actions
- Review of post-marketing commitment and technical agreements
- Product quality complaints and recalls
- Follow up on previous quality reviews.

Emerging regulations, guidance, and quality issues are critical factors that can impact the pharmaceutical quality system. Innovations that come along with new technologies bring significant changes that can enhance the pharmaceutical quality system. Therefore, adapting to these changes is necessary to acclimatize to the new concepts. Changes in the business environment and product ownership can also have an impact. Based on the quality reviews, management must be sure to provide resources and training to maintain and improvise quality systems and any related process. The documentation system records changes and decisions made. Chapter 8, will further discuss key considerations of documentation systems in pharmaceutical laboratories.

2.7.4 Pharmaceutical Quality Systems

GMP regulations have evolved significantly; therefore, quality concepts are continuously changing. While original GMP regulations emphasize quality through end-product testings, the current quality system provides a framework to develop drug products as it shifts to process focus. Quality units, quality assurance and quality control, ensure quality design built into the product and various other operations associated with all systems being planned, approved, executed, and monitored. The International Conference on Harmonisation of Technical Requirements for Registration of Pharmaceuticals for Human Use (ICH) Q10 guidance presents an effective pharmaceutical quality system based on the International Organization for Standardization (ISO) quality concepts and GMP regulations that support each of the product's life-cycle stages, recognizes the differences among these stages and their goals at each stage. The life cycle of each product consists of four main phases: product development, technology transfer, commercialization, and discontinuation [7–10].

Starting in 2006, the FDA initiated a six-system inspection model, of which the Quality system overarches the other five operational systems that include: facilities and equipment, production, materials, packaging and labeling, and laboratory controls (see Key Information).

Six quality systems for inspection
1) Quality system 2) Facilities and equipment 3) Production 4) Materials 5) Packaging and labeling 6) Laboratory controls

The Key Information section below lists the areas where the Quality System Management is implemented. The FDA will focus on these systems during their inspection to determine the level of compliance attained by the company. More information on audit and inspection is available in Chapter 8.

Four key areas to implement the quality management system
1) Responsibilities of management 2) Allocation of resources 3) Manufacturing operations 4) Activity evaluations

2.7.4.1 Responsibilities of Management

In the QMS, management is held responsible for the effectiveness and robustness of the quality system. They set implementation priorities and develop action plans based on risk analysis. They allocate adequate resources to support the objectives of quality systems and establish measurable goals to be regularly monitored for the operation [11, 12].

2.7.4.2 Allocation of Resources

Management allocate adequate resources for all areas, such as facilities and equipment, available raw materials, laboratory analysis of in-process, stability, and reserve samples. The training program is in place to ensure all employees are proficient in their operational functions and CGMP regulations. If outsourced resources are used in operations, the contracted employees are qualified and accredited.

2.7.4.3 Manufacturing Operations

Manufacturing process is defined and documented from design to delivery. All sources of variations are identified and monitored. Written procedures are in place to control all critical steps of the process. QC establishes procedures for the receipt, production, and storage of all raw materials. Suppliers must be qualified and audited periodically. Operational changes must be controlled and verified to determine the impact on processes or product quality.

2.7.4.4 Activity Evaluations

It is crucial to analyze manufacturing data regularly to identify trends and maintain continuous improvement. This allows the pharmaceutical manufacturer to detect potential issues or problems early and be able to plan for CAPA. Corrective action addresses the root cause of the investigation; preventive action prevents the occurrence or reoccurrence of the issue. Internal audit and inspection procedures identify compliance gaps or areas of improvement [12].

The above factors are essential to build a robust quality system and ensure that continuous improvement is addressed and developed throughout the product lifecycle [13, 14].

List of Abbreviations

APR	Annual Product Review
CAPA	Corrective Action and Preventive Action
CBER	Center for Biologics Evaluation and Research
CDER	Center for Drug Evaluation and Research
CFR	Code of Federation Regulations
CGMP	Current Good Manufacturing Practices
DQ	Design Qualification
FDA	Food and Drug Administration
GRASE	Generally Recognized as Safe and Effective
ICH	International Council for Harmonisation of Technical Requirements for Pharmaceuticals for Human Use
IQ	Installation Qualification
ISO	International Organization for Standardization
OOS	Out-of-specification
OQ	Operational Qualification

OTC	Over the Counter
PQ	Performance Qualification
QA	Quality Assurance
QC	Quality Control
QMS	Quality Management System
QRM	Quality Risk Management
SOP	Standard Operating Procedure
USP	United States Pharmacopeia
WHO	World Health Organization

References

1. Food and Drug Administration (2019). Current good manufacturing practice for finished pharmaceuticals in manufacturing, processing, packing, or holding of drugs; general. Code of Federal Regulations. Title 21, Subchapter C, Part 210. Rockville, MD: Food and Drug Administration. https://www.accessdata.fda.gov/scripts/cdrh/cfdocs/cfcfr/CFRSearch.cfm?CFRPart=210 (accessed 13 July 2020).
2. Food and Drug Administration (1978). Current good manufacturing practice for finished pharmaceuticals. Code of Federal Regulations. Title 21, Part 211. Rockville, MD: Food and Drug Administration. https://www.ecfr.gov/cgi-bin/text-idx?SID=a3df941b23a5efab57746e7fb497bb5b&mc=true&node=pt21.4.211&rgn=div5 (accessed 13 July 2020).
3. Food and Drug Administration (2015). *Guidance for Industry: Analytical Procedures and Methods Validation for Drugs and Biologics.* Rockville, MD: Food and Drug Administration https://www.fda.gov/media/87801/download (accessed 13 July 2020).
4. USP <1225> (2011). Validation of analytical procedures. *USP 34–NF 29.* Rockville, MD: United States Pharmacopeial Convention.
5. USP <1029> (2011). Good documentation principles. *USP 34–NF 29.* Rockville, MD: United States Pharmacopeial Convention.
6. Food and Drug Administration (2006). *Guidance for Industry: Q9 Quality Risk Management.* Rockville, MD: Food and Drug Administration https://www.fda.gov/media/71543/download (accessed 13 July 2020).
7. Food and Drug Administration (2009). *Guidance for Industry: Q10 Pharmaceutical Quality System.* Rockville, MD: Food and Drug Administration https://www.fda.gov/media/71553/download (accessed 13 July 2020).
8. Nally, J.D. (2007). *Good Manufacturing Practices for Pharmaceuticals*, 6e. New York, NY: Informa Healthcare.
9. Lee, C. (1993). *Development and Evaluation of Drugs: From Laboratory through Licensure to Market.* Boca Raton, FL: CRC Press, Inc.

10 Weinberg, S. (2009). *Guidebook for Drug Regulatory Submissions*. Hoboken, NJ: Wiley.

11 Hansen, S.H., Pedersen-Bjergaard, S., and Rasmussen, K.E. (2012). *Introduction to Pharmaceutical Chemical Analysis*. Hoboken, NJ: Wiley.

12 Tobin, J.J. and Walsh, G. (2008). *Medical Product Regulatory Affairs: Pharmaceuticals, Diagnostics, Medical Devices*. Hoboken, NJ: Wiley.

13 World Health Organization (2013). Proposed updated text for good manufacturing practices for pharmaceutical products: main principles. https://www.who.int/medicines/areas/quality_safety/quality_assurance/GMP-update-QAS13-517Rev1_11072013.pdf (accessed 13 July 2020).

14 Edwards, A. (2008). Quality systems for drugs and biologics. *Pharmaceutical Technology* 32 (2): 60–74.

3

Analytical Techniques Used in the GMP Laboratory

Walter Holberg[1] *and Kim Huynh-Ba*[2]

[1]*Rockwell Medical, Quakertown, PA, USA*
[2]*Pharmalytik LLC, Newark, DE, USA*

TABLE OF CONTENTS

Analytical Testing for the Pharmaceutical GMP Laboratory, First Edition. Edited by Kim Huynh-Ba.

3.1 Introduction

Throughout the first two chapters, Current Good Manufacturing Practices (CGMPs) and the regulatory process introduce the requirements that a pharmaceutical company must maintain to consistently produce drug products that aim to support the quality of life. This chapter will cover the analytical laboratory, where the pharmaceutical products are thoroughly tested and evaluated before they are considered safe, meet quality standards, and are ready for release to the market for consumer use. The goal of this chapter is to provide you with an overview of various analytical test procedures and an understanding of some common laboratory skills used in the pharmaceutical laboratories for releasing and monitoring the quality of drug products throughout its expiry. It is not suggested that this is the complete list of all the possible tests used in the pharmaceutical laboratory. This chapter also introduces the concept of the good manufacturing practice (GMP) pharmaceutical lab to new analysts or students who are interested in analytical chemistry for their careers in the pharmaceutical industry. Through understanding and familiarization with common apparatus in the laboratory, students will be equipped with better techniques to work more efficiently, accurately, and in compliance with CGMP regulations. A few procedures are included as examples to familiarize analysts with a few Standard Operating Procedures used.

3.2 Definitions

The GMP regulations are purposely written to be general; thus, analysts may find different interpretations of terminology in the analytical laboratory. This section will provide definitions to clarify better the concepts discussed.

3.2.1 Raw Data and Analytical Data

Data are records appearing on any form of media, such as paper, laboratory notebook, laboratory worksheets, memos, notes, automatic printouts from an electronic device, or computer-generated spreadsheets. *Raw data* are results of original observations captured either by an analyst or an instrument – these activities are necessary for the reconstruction and evaluation of the report of that study [1]. *Metadata* is the contextual information required to understand electronic data.

Examples of raw data include the original recording of observation for a visual description of sample appearance made at the time of conducting the test and the original electronic signals from a high-performance liquid chromatography (HPLC) detector captured by data acquisition software. Examples of metadata could include the wavelength that the HPLC detector signal used to capture data, the time/date stamp for when the data were acquired, the user ID of the person who generated the data, audit trails, etc. Data and raw data are essential evidence for the reconstruction and evaluation of an experiment, which proves the authenticity and integrity of the analysis. Therefore, the creation, maintenance, and preservation of raw data are crucial for the registration and successful launch of the product for a company. Raw data with timestamps are the evidence that will win or lose your case in a litigation process. As modern instrumentation relies more and more heavily upon computerized data acquisition systems and database storage of acquired data, "raw data" is increasingly being defined as the electronic data files and their associated metadata.

Analytical data or reported results derived from the raw data are collected from the study. Analytical data can be numerical or confirmatory results. Analytical data are calculated manually or automatically using a computer program. If the computer is used to generate analytical data, then the calculations of the programs must be verified and documented.

3.2.2 Analyses

Analytical scientists often use terms like *qualitative analysis* and *quantitative analysis* to classify their testing types. These terms are characteristic descriptions of the physical and chemical testing of a substance to determine its identity, purity, and quality. Most, if not all, of the analytical techniques used in the GMP pharmaceutical laboratory, are either qualitative or quantitative analysis.

Qualitative analysis is testing that defines a characteristic without quantifying it. It can be as simple as visual observation of material at the ambient condition or it can be as a conformation to a nuclear magnetic resonance (NMR) spectrum of the chemical composition of a material.

"The sky is blue" is a qualitative statement. "The eye perceives the sky to be a monochromatic blue of a wavelength of 474–476 nm" is a quantitative statement. Quantitative analysis is testing that provides a numerical value of the test material. Quantitative analysis can also be as simple as determining the density (mass/volume) of material to as complex as again the NMR of, or more specifically, quantitative NMR (qNMR) of the composition of a sample. Both qualitative and quantitative analyses are needed to provide evidence of what a material is and how potent it is using a unit of measurement. The most common quantitative techniques in pharmaceutical development are chromatographic techniques such as HPLC, gas chromatography (GC), and ion chromatography (IC).

3.2.3 Analytical Documents

Analytical documents are used for compilation and evaluation of analytical data. They can be protocol, procedures, work instructions, and reports, providing useful information to the GMP-regulated industries and ensure record integrity and control. These activities are important to the evaluation and reconstruction of GMP activities, and ultimately for ensuring the quality and safety of regulated drug products.

The analytical protocol is a document that gives directions on how a study will be conducted. Protocols should contain acceptance criteria to evaluate the outcome of a study. The *analytical procedure* is a process to evaluate a chemical, physical, biological, or microbiological of material using analytical chemistry techniques. Analytical procedures used in the pharmaceutical laboratories must be validated for the intended use of the method. Validation is discussed in Chapter 5.

Work instructions document step-by-step instructions on how a study is to be performed. Work instructions are used for routine solution preparations, glassware cleaning instructions, etc.

An *analytical report* is a document that contains the evaluation of the study against the predetermined acceptance criteria and communicates the solutions or conclusions of a study. The analytical report typically contains a description of the experiment, a compilation of data, an evaluation of the data set, and a conclusion.

3.3 Basic Laboratory Procedures

Many tools are critical in pharmaceutical laboratories to obtain accurate analytical data. Qualitative analysis is subjective because it depends on the training and experience of the analyst. Quantitative analysis is typically based on data generated using analytical instrumentation.

3.3.1 Balances

The balance is one of the most important pieces of equipment in the analytical laboratory to measure mass depending on the desired accuracy and precision. There are various types of balances that are mechanically designed to operate at different weight ranges. Understanding the engineering default of the balance will allow an analytical scientist to minimize errors in quantitative analyses. The common balance types present in every pharmaceutical testing laboratory are the top-loading balance and the analytical balance.

The top-loading balance is for weighing material equal to or greater than 5 g. An analytical balance is used to weigh material from 10 mg to 100 g within 0.01 mg variance. For weighing a limited, expensive, hard-to-obtain substance, an analytical scientist can utilize a micro-balance and even an ultra-micro balance. These types of balances are equipped to deliver an accurate weight of between 2 and 10 mg, or less than 1 mg, respectively, with the readability of 0.0001 mg.

The semi-micro balance is the workhorse of most analytical labs with a range of 10 mg–100 g with a variance of 0.01 mg. Besides understanding the mechanical capability of each type of balance, the weighing technique contributes significantly to the accuracy of analyses, which consequently increases the reliability of data collection. Although the balance is a simple tool, the use of a balance should be well-thought-out, and weighing techniques should be understood. Therefore, training on the use of the balance is an important skill to master to obtain a consistent weighing with minimum error. An analytical scientist needs to understand the construction of the balance, causes of dynamic mechanical errors, and the differences in readability at each location on the pan of the balance to place the weighing object. Besides, the surrounding environment, static effects, and vibration can also play a significant role in obtaining an accurate measurement.

With so many models and manufacturers of balance available to choose from, a decision on the right balance should still be the focus of needs for your laboratory. The right choice of a balance is one that delivers the most accurate consistent results, as well as accommodates the needs of your laboratory with the least amount of downtime due to maintenance. Using a balance properly requires good housekeeping and good weighing practices. Identify a good location to set up the balance where it should be least disturbed by any sources of vibration, from the movement of personnel, near-by instruments, the movement of air when doors are opened or closed, or by ambient conditions change. These factors can adversely affect the operation of the balance. The area around the balance should be well kept, clean, and clutter-free, to allow the analysts to conduct the weighing task with minimal chance of contamination.

Gravimetric techniques are more accurate than volumetric methodologies for quantitative analytical experiments. The advancement of balance engineering and electronics has *improved* the accuracy of the balance, in comparison to earlier, mechanical balances. Other peripheral pieces of equipment like a glovebox, an enclosed box that isolates the balance from the laboratory environment, also contributes to the accuracy of the balance when used with hygroscopic or difficult to weigh samples.

3.3.1.1 Qualification of a Balance

In the pharmaceutical industry environment, qualifications must be performed for every balance. These qualifications include the minimum installation, operational, and performance qualifications (PQs), which are often referred to as IQ/OQ/PQ. These qualification protocols are required for most analytical instruments to adequately demonstrate that the instrument has been correctly installed and operated and performs as intended by the manufacturer. The balance's manufacturers or authorized vendors often do the balance qualification. The vendors are certified and approved by the balance manufacturers. Other companies rely on internal Metrology departments to perform these activities. Once the IQ/OQ/PQ is completed, a Standard Operating Procedure (SOP) is then established to supersede the manufacturer's manual procedures. A complete SOP must include details on the daily check for accuracy and repeatability. An example of the use of balance is provided in the Key Information section below.

Use of an analytical balance

1) Verify that the balance is within its calibration period.
2) Perform balance-daily check if it has not yet been done on the day of use.
3) Inspect the balance to ensure it is clean and dry. Remove any residual or powder, using a soft brush.
4) Verify the level indicator by inspecting the bubble in the circle.
5) Place a weighing paper or weighing boat directly on the weighing pan.
6) Transfer material to be weighed on the weighing paper.
7) Allow the balance to stabilize.
8) Record the weight directly onto the notebook.
9) Once the material to be weighed is placed on the balance pan, allow the balance reading to stabilize.
10) Record the weights with all the displayed digits.
11) Clean the area for any loose dust or power.
12) An anti-static device or anti-static gloves must be used when weighing samples under conditions that are conducive to static charges.

3.3.1.2 Weighing a Solid Material

For solid material, the desired amount of material can be transferred to a tared weighing paper or weighing boat using a clean stainless-steel spatula. The weight of the material is recorded, and the material is transferred to the volumetric flask. The weighing paper or weighing boat is rinsed at least three times with the solvent into the volumetric flask. When the material is a limited or small amount to be weighed, a suitable volumetric flask can be placed directly onto the balance and tared (the error of the balance at higher weights must be taken into consideration). A glass tube (e.g. the end of a glass Pasteur pipette) is used to gather the solid inside the tube, then the excess on the outside glass wall is swiped off before transferring and dispensing the material into the bottom of the tared volumetric flask. The direct weighing technique minimizes the unaccounted loss of material through transferring steps.

The material can also be weighed by difference. An amount of material that is approximately close to the desired weight is placed on a weighing paper or weighing boat using a clean spatula. The initial weight, including the weighing paper or boat, is recorded. The material is transferred to the volumetric flask. The final weight of the weighing paper or boat is then recorded. The weight of the material is the difference between the initial and final weights. The analyst must record the weights according to the required significant figures dictated in the procedure.

These techniques can also be used for weighing powder or granulation.

3.3.1.3 Weighing a Liquid Material

Weighing a liquid material can also be adapted similarly by withdrawing the liquid material using a suitable syringe and needle, then dispense the material drop-by-drop from the syringe into the tared volumetric flask. Weigh-by-difference may be used for small amounts of liquid by weighing the syringe filled with liquid and after dispensing.

Accurately weighing is the foundation for accurate and reliable results achieved after the testing. The analyst must understand the balance and its functions to perform the activity. Before using the balance, the analyst must inspect the balance for proper installation and cleanliness. The digital balance should always be plugged into a power outlet. If it is turned off, it should be allowed to equilibrate before use. This equilibration time is specified in the balance SOP, and it could be several hours. If the balance is due for recalibration, then it must not be used for GMP analysis.

3.3.1.4 Performing a Daily Check

All balances must be checked daily at the beginning of each day before routine use. The daily check must be performed with a qualified set of calibration weights that must be within 5–100% of the balance capacity for the weight check procedure [2, 3]. The proper class of calibration weights must be used (e.g. class 0 for ultra-micro balances), and the weights are handled properly using gloves or tweezers, if necessary, to avoid oil residue from the hands. The weight check must demonstrate the accuracy and repeatability according to the operation of the balance. The class of calibration weight is dependent upon the accuracy and weight range of the balance. Calibration weight classes are defined in a document published by the American Society for Testing and Materials (ASTM), *ASTM E617, Standard Specification for Laboratory Weights and Precision Mass Standards*. The data from the daily balance check must be recorded in the equipment logbook and retained next to the equipment.

3.3.2 Volumetric Glassware

Volumetric glassware is used extensively in the GMP laboratory. Glassware must be certified Class A and graduated with clear markings on each unit with the certified accuracy information. The most common volumetric glassware is transferred pipettes, volumetric flasks, graduated cylinders, and burets. All are available in various sizes, and all should have been tested and warranted to have certified accuracy and be labeled clearly on the body of the glassware. Using the correct type of glassware for the experiment is crucial to perform an analytical procedure accurately.

There are two types of analytical volumetric pipettes: TC vs. TD (To Contain vs. To Deliver). When using the *To Deliver* (*TD*) pipette, the liquid must be allowed to drain in a vertical position with the tip touching the wall of the receiving flask. The material to be measured and volumetric glassware must be allowed to reach room temperature before use to minimize measurement error. The *To-Contain* (*TC*) pipette is used for viscous liquid such as suspension or syrup. The TC pipette should be washed after draining by rinsing it with the solvent and allowing the rinse to be added to the measured material [4]. The Key Information section below provides a guideline for using a pipette for volumetric transfer.

Volumetric transfer procedure using a TD pipette

1) Inspect the pipette to ensure it is clean, dry, and has no crack or chip.
2) Wet the pipette by placing the tip of the pipette below the surface of the liquid. Use the pipette bulb to draw liquid up into the pipette to about 1 cm above the mark. Ensure the tip of the pipette remains submerged in the liquid to avoid air being drawn into the pipette.
3) Allow the liquid to drain completely into a waste container.
4) Use the pipette bulb to draw liquid into the pipette to about 1 cm above the mark.
5) Wipe the outside of the pipette using a Kimwipe or absorbent tissue.
6) Slowly drop the liquid level in the pipette unto the waste container until the meniscus is level with the mark.
7) Allow the content to drain naturally into a material receiving flask. Do not apply any additional pressure to push the liquid through.
8) Do not use the bulb to blow out the liquid at the tip of the pipette out.

Proper dilution techniques must be used to prepare standard or sample solutions in the laboratory. Class A volumetric flasks and pipettes are used in the analytical laboratory. Mixed solutions must be allowed to reach room temperature before any transfer can be done, and pipettes should be soaked in a soapy solution after use if immediate cleaning is not done. If used pipettes are left to dry after use, then it may be more difficult to remove the residue and may cause contamination to future analyses.

3.3.3 Potentiometry (Ion-Selective Electrode) and pH Test

Potentiometry is the practice of measuring the potential difference between two electrodes by comparison with a known voltage. The most understood application of potentiometry in a pharmaceutical laboratory is pH measurement. However, it is important to know that a pH electrode is effective in measuring pH because it is selective for the activity of the hydrogen ion, H^+. This selectivity is achieved via a specific composition of the glass electrode, and by altering the glass composition, the selectivity of an electrode can be altered.

An ion-selective electrode (ISE) is an electrode that can measure the activity of a specific ion. Early glass ISE's were developed for common counter ions used in pharmaceutical products, such as chloride and sodium. However, there are now multiple types of non-glass electrodes available that have significantly increased the numbers and types of analytes that can be measured. The caveat with ISE's is that they are generally more susceptible to matrix interferences, such as pH and ionic strength than the ubiquitous pH electrode, so some sample modification is usually required to obtain a reliable value. Similar to a pH electrode, they must be standardized before measurement. Despite the drawbacks, ISE analysis offers a quick and reliable solution to ion measurement in some applications.

pH is a measurement of the relative activity of the hydrogen ion in an aqueous solution at a specific temperature. pH value is measured as $-\log(A_{H+})$, where A_{H+} is the hydrogen ion activity and has a scale that ranges from 0 to 14 in water, in which a pH of 7 is considered neutral. An aqueous solution that has a pH of <7.0 is considered an acidic solution and a solution with $pH > 7.0$ is known as a basic solution. A pH value of an aqueous solution is often measured using a pH meter, which consists of a measurement (pH) and a reference electrode, and electronic programs built to measure the electrical potential difference between the pH electrode and the reference electrode. Most modern pH meters utilize a combination electrode, in which the pH electrode and reference electrode are combined into a single unit. To assess the accuracy of a pH measurement of the test solution, the pH electrode first must be standardized with suitable reference pH solutions, which are commercially available, to establish the standardization curve. This process is referenced as "electrode standardization" or "electrode calibration." The electronics in a pH meter apply the potential obtained at fixed points (using pH buffers) to create an internal calibration curve utilizing the Nernst equation [5]. As electronics can drift and the hydrogen ion activity is affected by temperature, regular standardization of the pH meter is essential for an accurate measurement. A common practice to standardize the electrode is to use two or three reference solutions by bracketing with standards that are at the pH lower and upper end of the aqueous solution that is measured. For example, if the testing solution is predicting to have a pH in the range of 5–6, then appropriate reference solutions used to standardize the electrode should be pH 4.00, and pH 7.00.

Because hydrogen ion activity can be affected by temperature, to measure accurately, the pH meter is often equipped with an automatic temperature control (ATC) unit. This feature is purported to consider the temperature effect when pH electrodes submerse into a test solution with a slightly different measured temperature. If a pH meter is not equipped with ATC capability, then standardization buffers and the test solution should be at the same temperature for an accurate measurement. When measuring pH, the test solution should be homogeneously mixed, and the pH electrode tip should be submerged into the solution at a minimum of 1 cm or at a depth necessary to submerge the porous junction. The pH reading should be recorded while the aqueous solution is being stirred, and when it comes to a stable reading on the instrument display. An example of a typical instruction for using a pH meter is provided in the Key Information section that follows.

Determining the pH of a liquid solution

1) Standardize the pH meter
   - 1.1 The pH meter must be standardized before the first use of the day with two or three buffer solutions with pH that bracket the expected pH of the sample being measured.
   - 1.2 Rinse the electrode with distilled or deionized water. Blot dry with Kimwipes.
   - 1.3 Fill a clean and dry beaker with a pH 4.0 buffer solution.
   - 1.4 Immerse the tip of the electrode into the pH 4.0 buffer flask. Ensure the electrode does not touch the bottom or the side of the flask.
   - 1.5 Wait at least 30 seconds for the reading to stabilize. Enter the certified value of the reference solution.
   - 1.6 Read and record the pH of the buffer. Adjust to 4.0 if necessary.
   - 1.7 Repeat the above procedure with the second pH 7.0 buffer solution.
   - 1.8 Repeat the above procedure with a third pH buffer, if desired. Two calibration points are the minimum necessity to create a calibration curve.
   - 1.9 Press the Calibration button on the pH meter to get the slope for the standardization
   - 1.10 The acceptance criteria for the slope are specified by the manufacturer, typically 95–105%. Record the slope and its pass/fail status in the logbook.

2) pH measurement of a sample.
   - 2.1 Rinse the electrode with distilled and deionized water. Blot dry with Kimwipes.
   - 2.2 Add the sample to the flask.
   - 2.3 Add a small coated magnetic stir bar to the flask containing the tested material.

2.4 Place the flask on the stir plate and gently stir the solution.
2.5 Insert the electrode into the sample as far as possible; however, avoid touching the bottom and sides of the sample container or the stirring bar.
2.6 Press the measure sample button. Wait 30 seconds for the reading to stabilize.
2.7 Read the pH of the sample and record the value and the temperature of the solution.
2.8 Rinse the electrode with deionized or distilled water when done.
2.9 The electrode must be stored in the buffer solution or tap water.

Table 3.1 List of the typical sources of errors in pH measurements.

Source of error	The effect upon pH measurement
Sodium-ion error – The electrode becomes selective for sodium ions at high pH.	Above pH 10, the measured pH is lower than actual. The deviation is gradual but may be as large as one (1) pH point at pH 13.
Acid Error – At acid concentrations above 2 M, hydration of the glass surface is altered.	Below pH 1, the measured pH is higher than actual.
Temperature changes alter the hydrogen ion activity.	The measured pH may be higher or lower, depending upon the direction of the temperature shift. For this reason, the USP defines that pH measurement should be performed at 25 ± 2 °C.
Blocked reference junction.	Slow response time, unstable readings, inaccurate readings, changing readings with the solution is stirred.
A coating on the pH glass.	Sluggish response, low slope.

pH testing is a critical test for any liquid product, and it is important in the preparation of solutions. Table 3.1 lists the typical sources of error in pH measurements.

3.3.4 The Density Test

The density test is another test in a pharmaceutical laboratory. The density of a material is defined as a measured mass of the material over a known volume. So, a density test can be completed with an analytical balance and a certified volumetric flask or pipette, or with a specialized piece of glassware called a pycnometer.

A density meter can be used to minimize exposure to the analyst to a dangerous chemical. It is built on the principle of calculating the mass of material over a known volume, which is the fixed built-in tube of the instrument. To ensure that the meter is functioning as intended, before measuring the density of the material, the density meter should be checked for accuracy with pure water, or with a certified accuracy check solution that is recommended by the instrument manufacturer. Suitability of the density meter should be verified before any testing of the sample.

3.3.5 The Friability Test

This test is applicable for solid dosage forms such as tablets. Tablets are constantly subjected to mechanical shocks and vibration during the manufacturing, packing, and distribution. These manufacturing and transportation activities can expose tablets to capping, chipping, and breakage. It is, therefore, important that the tablet is formulated to withstand these conditions. The procedure used to conduct a friability test is often referred to as the Roche friability or tumbler test. Friability testing is designed to monitor the resistance of tablets to physical stress by subjecting them to tumbling in a rotating drum. After tumbling, the integrity of the tablets and the weight loss are evaluated to determine their suitability for further processing, such as coating.

Friability is defined as the percentage of weight loss by tablets due to mechanical action during the test. Friability is expressed as a percentage loss of pretest tablet weight and often is referred to as the ability of the compressed tablet to avoid fracture and breaking during transport.

An example of a procedure for the friability test is outlined in the Key Information section below. This procedure can be applied to single or dual drum friabilators. It is noted that left-hand and right-hand drums are not interchangeable, and the vanes inside the drums curve in the opposite direction. Therefore, it is important to identify the drums for proper installation.

Conducting the friability test for tablets

1) Turn on the equipment.
2) Select the mode of operation, either timer or counter, depending on the method.
3) Input either the number of rotations (up to 999 999) or the time duration as applicable and press ENTER.
4) Remove the drum cover carefully on the benchtop. Ensure that the inside of the drum is clean and dry.
5) Remove any loose dust from the tablets carefully with a soft brush or stream of compressed air. Accurately weigh the tablets to be tested in grams and record the initial weight (W_i).
6) Place the tablets in one of the two chamber-drums and replace the cover.
7) Hold the cover firmly in place over the drum and slide the drum onto the shaft.
8) Place the locking nut on the end of the shaft and tighten by hand. Do not overtighten, or it may result in damage to the drum.

9) Press "START STOP" to begin the test. The time duration or number of rotations count down on the display screen, and the test stops when 0 is reached.
10) Once the test is complete, remove the tablets. To prevent mix-up, carefully identify the samples from the two tests in the dual chambers drum.
11) Use a soft brush or compressed air to remove any loose dust from the tablets. Visually inspect the tablets for any chipping or capping and record your observations.
12) Weigh the tablets in grams (W_f). Calculate the percentage of weight loss using the following formula:

    $\%\text{weight loss} = (W_i - W_f) \times 100\ W_i$

    where

    W_i is the pretest weight of the tablets in grams

    W_f is the after-test weight of the tablets in grams after rotations.

The percent weight loss is reported. If obviously cracked, cleaved, or broken tablets are present after tumbling, the sample fails the test. If results are questionable or the weight loss does not meet specification, repeat the test twice, and calculate the mean of the three tests.

3.3.6 The Hardness Test

A hardness tester determines the force necessary to break a tablet. Typically, a tablet is placed on a flat surface, against a solid backrest, or anvil. A movable ram slowly presses the tablet against the anvil with steadily increasing force until the tablet fractures or breaks. The force required to break the tablet is measured by the instrument and recorded. It is dependent upon on the tablet's orientation in the tester; therefore, to ensure the repeatability of the test, the orientation of the sample placed in the tester must be consistent [6].

Typically, 10 tablets are tested, and the average result is recorded. After each batch is tested, the operator should brush the jaws and jaw plate to clean any debris and dust. The testing area should be wiped down with an alcohol solution and air-dried. The waste receptacle should be emptied into the appropriate waste disposal container and rinsed with an alcohol solution and dried. Like many other pieces of the analytical instrument, the hardness tester must have scheduled maintenance and calibration at least once every six months. The Key Information section lists the steps to conduct a hardness test for tablets.

3.3.7 The Titration Test

Titration is a wet chemistry technique that uses a solution of known concentration to determine the concentration of an unknown solution through chemical

Conducting a hardness test for tablets

1) Select the unit setting (kp or SC) by toggling the switch to the proper unit.
2) Place the tablet against the sensing jaw, make sure the tablet is centered against the face of the jaw.
3) Place the tablet fragment safety shield over the sensing jaw and crushing jaw to reduce exposure to project fragments of the tablet as it is being crushed. Align the groove of the safety shield over the sensing jaw.
4) Press the TEST button to initiate the testing process. Testing is complete when the tablet is fragmented, and the hardness value is displayed on the LED display.
5) Clean the test area using a soft brush to sweep away the dust and fragments that may interfere with the subsequent test.
6) Repeat steps 2–4 for the remaining test tablets.
7) Calculate and report the average results from all individual results.

reactions. The known solution, called the Titrant, is added from a buret to a known quantity of the unknown solution containing the analyte of interest, allowing the reaction to take place. The titration is complete, called reaching the endpoint, when the mixture solution reaches a targeted visual indication (e.g. the solution changes color and retains the color for at least 30 seconds, or reaches a defined pH, or precipitation occurs, etc.). A color indicator is often used in titration techniques to signal the completion of the reaction. The volume of Titrant used for the titration is recorded and used to calculate the actual amount of the analyte of interest through a known conversion factor from the chemical reaction equation. The accuracy of a titration depends on the precise recorded volume used.

Titration techniques used to be a powerful tool to determine the quantity value of unknown analytes with the capability of quantifying a very low, part-per-billion level. Due to its nonspecific characteristic (most titrations are specific for a functional group, not the entire molecule), its technique-dependence (titration skills vary largely among chemists), and lower throughput process, titration techniques have become less popular in modern analytical laboratories. Nonetheless, the titration technique for water contents, acid values, peroxide values is irreplaceable. Manufacturers have built high-tech titration equipment (automatic titrators) to enable users to achieve more accurate, consistent results with less labor-intensive through automation.

3.3.8 The Karl Fischer Titration–Water Determination

Another titration method is Karl Fischer titration, which is the most common chemical analysis technique used to determine the water content present in a sample. This titration method is named after the German chemist who invented the technique in 1935, hence the name, Karl Fischer Titration. It is based on the principle of chemical oxidation of sulfur dioxide by iodine in a solution of a methanolic hydroxide described by the equation below:

$$CH_3\,OH + SO_2 + RN \rightarrow [RNH]SO_3CH_3$$

$$H_2O + I_2 + [RNH]SO_3CH_3 + 2\,RN \rightarrow [RNH]SO_4\,CH_3 + 2\,[RNH]I$$

where RN is a base

Today, water determination is done using automatic titrators that may be further classified as Volumetric Titration, and Coulometric Titration. In a volumetric titration, a solution containing iodine serves as a titrating agent and is added until excess iodine is present. The amount of iodine converted is determined from the buret's volume of the iodine-containing solution. Water and iodine are consumed in equimolar amounts. Therefore, the water content of the sample is calculated using the titration volume and titer of the titrating agent. While in the coulometric titration, a coulometric reagent is used. The iodine participating in the reaction is generated electrochemically in the titration vessel by anodic oxidation of the iodide contained in the coulometric reagents until excess iodine is detected. The consumed iodine is calculated from the quantity of electricity required to generate iodine by using the principle of Faraday's law. The water content of the sample is then calculated from the quantity of iodine used with the ratio of 1 : 1, as seen in the equation above. The volumetric titration technique is reliable for samples with a higher level of water content from 10 to 100 mg of water. Coulometric titration can deliver accurate results for the sample with water content from 10 μg up to 10 mg.

Different instrument manufacturers will have slight differences in instrument operations. However, all follow the fundamental principle described above [7]. Before using an automatic Karl Fischer titration, it is recommended to carefully follow the instrument manual and be familiar with Material Safety Data Sheet (MSDS), as Karl Fischer reagents are considered highly toxic materials.

Before use, the titration vessel should be inspected according to the Key Information section. The typical operating procedure for a volumetric titration is described in the Key Information section below. The reagent needs to be standardized before any sample can be tested [8]. The water equivalence factor (*F*) for the Karl Fischer reagent must be calculated and consistent with the typical value used for the specific method. Purified water, a commercial water standard such as sodium tartrate dehydrate, or a USP reference standard with a traceable certificate of analysis, should be used to standardize the Karl Fischer reagent. The *F* value for water is 1. The manufacturer's instructions should be followed when operating the Karl Fischer titrator.

Inspect the vessel before use

- Inspect for cleanliness, adequate, and appropriate reagent,
- Inspect fitting and tubing for leak.
- Inspect the desiccant, the titration stand, the titrant, the gas drying line, the solvent, and the waste bottles for proper operation.
- Inspect the vacuum line for no leak, no bubble in the tubing line.
- Rinse the buret or dispensing tubing with the appropriate amount of titrant into the waste bottle before starting the test.
- When the oven is to be used, ensure all units are connected appropriately, and the transfer tubing is clean.
- Check all settings and parameters for proper setup configuration before use.

It is important to note that some chemical functional groups and elements can interfere with the Karl Fischer reaction, so additional methods for the determination of water content are necessary.

Perform a KF titration using a volumetric titrator

1) Standardization of the KF reagent using purified water
   1.1 Follow the manufacturer's instructions for operating the Karl Fischer titrator capable of amperometric endpoint detection.
   1.2 Record the volume of the buret in the titrator.
   1.3 Use the Karl Fischer reagent for a non-pyridine substitute, with a titer of about 5 mg water mL^{-1}.
   1.4 Transfer about 50 mL of dry methanol (or another suitable solvent) to the titration vessel. The volume must be sufficient to cover the electrodes.
   1.5 Pre-titrate the solvent in the titration vessel to the endpoint so that any water present is removed. The volume of titrant used in this pre-titration is not used in any subsequent calculations and does not have to be recorded.
   1.6 The reference water standards are available in glass ampoule with a Certificate of Analysis to provide the expected value for its water content. Open the ampoule, rinse a syringe with Water Standard, and then draw the required weight (1–1.5 mL of water standard weighing about 1–1.5 g)
   1.7 Inject the content into the titration vessel. Stir to mix and use the KF titrator to titrate the contents to the endpoint immediately.
   1.8 Record the volume of KF reagent (titrant) used in the titration.

1.9 Calculate the water equivalency factor F (in mg mL^{-1}) by dividing the weight of water added to the titration vessel (in mg) by the volume of titrant (in mL) and then multiplying by the fraction of water in the standard (1 for purified water).

F (mg mL^{-1}) = mL of Karl Fischer reagent grams of water

Notes: The water equivalency factor may be calculated automatically by the KF titrator.

(1.7) The above should be done in triplicate, and the % RSD should be less than 1%.

2) Titration of sample
   2.1 Introduce a sufficient volume of KF reagent to cover the electrodes.
   2.2 Add a stir bar. Stir gently.
   2.3 Blank about 50 mL of Methanol to ensure that any water present in the reagent is removed. Record any value for the baseline drift.
   2.4 Minimize the exposure of the titrant and sample solution to room air to avoid moisture in the atmosphere.
   2.5 For a liquid sample, inject the sample into the Karl Fischer vessel using a syringe with a long needle. Do not allow the needle to touch the sides of the titration vessel or adapter.
   2.6 For a solid sample, accurately weigh an amount of about 20–30 mg of the sample on the weighing paper and introduce directly to the Karl Fischer vessel. Weigh the weighing paper again and record the weight of the sample used.

Note: If the finished product is tablets or chewables, they must be crushed before being transferred to the titration vessel.

   2.7 Titrate with the Karl Fischer reagent. The titrator will determine the weight of water measured in the sample introduced.
   2.8 Calculate the percent of the water content of the samples:

$$\%\ \text{water} = \frac{\text{Weight of water measured in the sample} \times 100}{\text{The weight of the sample introduced}}$$

3.3.9 Loss on Drying

The Loss on Drying (LOD) test is designed to measure the volatile components (e.g. water and volatile solvents), in a sample when the sample is dried under specified conditions. LOD may be used in those instances where Karl Fischer analysis is inappropriate. The test involves a specific drying process that a test sample is exposed to, to remove the volatile components. The amount of the removed volatile compounds is expressed as a percentage of weight loss from the pre-test weight of the test material and is calculated by weighing the test material before and after it is being subjected to the drying condition. Several ways that this test can be done effectively are listed in USP General Chapter <731> *Loss on Drying* [9]. It is important to note that the materials must be allowed to cool to room temperature to be weighed. Balance, good weighing practices, and handling of materials are factors that might affect the result of this analytical test.

3.3.10 Residue on Ignition/Sulfated Ash

While the LOD test is to measure the volatile components, the residue on ignition (ROI) test is intended to measure the nonvolatile matter in a sample. This test is usually used to determine the number of inorganic impurities present in an organic substance. Concentrated sulfuric acid is added to the test sample and then ignited in an oven at 600 °C over a significant length of time. The organic matter is burned off completely, and the remaining substances are inorganic impurities. Similar to LOD, the ROI test is calculated based on the difference in weighing the material before and after the ignition process.

3.3.11 Thermo Gravimetric Analysis

Thermogravimetric analysis (TGA) test measures the amount of weight change of a test article in relation to a function of increasing temperature, or to a function of time, isothermally. The principle of TGA is to apply heat to force reactions and physical changes in materials. In recording characteristic thermogravimetric curves, TGA provides a quantitative measurement of mass changes in materials associated with decomposition, dehydration, and oxidation, of a sample in relation to time and temperature. TGA provides specific materials and chemical compounds due to a unique character from physicochemical reactions occurring over specific temperature ranges and heating rates, usually 25–900 °C in small degree increments. TGA measures the percent weight loss of a test sample while the sample is heated at a uniform rate in a defined environment (e.g. the atmosphere of nitrogen, helium, air, other gas, or in a vacuum). The weight loss over specific temperature ranges indicates the composition of the sample, as well as indications of thermal stability. These unique characteristics can be used as a "fingerprint" of the molecular structure of the test sample.

3.3.12 Differential Scanning Calorimetry

Differential scanning calorimetry (DSC) is used to characterize the physical stability of a material. The principle of this technique is based on phase transitions and chemical reactions as a function of temperature concerning heat effects. DSC measures the amount of energy absorbed or released by a sample when it is heated or cooled, providing quantitative and qualitative data on endothermic and exothermic processes. DSC is often used to measure melting temperature, the heat of fusion, reaction energy, crystalline phase transition temperature and energy, and precipitation energy.

DSC is also used for determining crystalline transition temperatures of an active pharmaceutical ingredient (API). The crystalline structure of an API can be a critical component of a GMP pharmaceutical product. It is also useful to monitor the quality of excipients used in the manufacturing process. As an example, different forms of lactose, a low cost, high stability, low hygroscopicity, and compatibility with a wide range of active ingredients, can be monitored using DSC [10].

3.3.13 The Disintegration Test

The disintegration test determines the time it takes for the solid dosage to dissolve within a liquid medium. Typically, the samples are uncoated tablets, plain-coated tablets, enteric-coated tablets, buccal tablets, sublingual tablets, hard gelatin capsules, and soft gelatin capsules. Disintegration testing has been used for decades and was a precursor to dissolution testing. Disintegration testing only measures the time required for a dosage unit to break apart into particles; it differs from dissolution testing as disintegration testing does not measure the release of the active ingredient from the dosage form and particles.

The most common device used for disintegration is the basket–rack assembly with a 1-L beaker submerged in a water bath between 35 and 39 °C. A full description of the Basket-Rack assembly can be found in USP General Chapter <701> *Disintegration*. The basket-rack assembly has six tubes; a tablet is placed in each unit. The basket is then allowed to be raised and lowered into an immersion fluid held at 37 ± 2 °C at a constant frequency rate of 29–32 cycles per minute^{-1} with a distance of 53–57 mm. A stopwatch is used to record the time it takes for all six tablets to be broken into smaller particles, and no particle remains in the basket. USP General Chapter <701> discusses in more detail specific requirements about different solid dosage forms [11]. In many countries, this test is also the required release tests and stability tests for many approved products and performed by the Quality Control laboratory.

3.3.14 Particulate Matter

In pharmaceutical applications, particulate matter is defined as extraneous mobile undissolved particles, other than gas bubbles, unintentionally present in the solutions [12]. Particulate matter is a significant concern in injectable solutions, in which undissolved particles could pose a safety risk for the patient. USP defines particulates as "mobile undissolved particles, other than gas bubbles, unintentionally present in solutions." USP <790> *Visible Particulate Matter in Injections* addresses visible particulates while USP <788> *Particulate Matter in Injections* addresses subvisible particles.

USP <790> became official in August 2014. It is a nondestructive visual inspection of drug products looking for visible particulates. It is surprisingly effective in detecting metal particles from manufacturing equipment or nonmiscible contaminants, such as silicone oil. Vials are inspected against a white background and a black background using an intense light source. This procedure results in the identification of solutions containing particles larger than roughly 200 μm.

USP <788> describes two tests for sub-visible particles, Method I, an automated instrument method that detects particles using light obscuration in a flow path, and Method II, a microscopic particle count test that uses the human eye to count particles on a filter membrane. The method I is generally preferred due to the subjectivity of Method II; however, there are drug products that are unsuitable for Method I, so Method II remains in use.

In August 2017, USP <1790> *Visual Inspection in Injections* became official with the intent of clarifying visual inspection processes to assist the industry in achieving compliance with GMP obligations. USP <1790> provides information on the visual inspection of injectable solutions and attempts to define the USP requirement for a solution being "essentially free from particles." Inspection methods and technologies are discussed as well as qualification and validation of inspection processes.

As this testing is limited by human ability, high technology strives for a sophisticated automatic methodology to replace the dependency of manual inspection. However, the main limitation of these test methods resides in the fact that many of these tests are destructive [12].

Quantification of particulate matter in injectable solutions is performed by either light obscuration or microscopic observation methods. Light obscuration methods utilize an instrument that pumps liquid through a flow path containing a light source and detector that can count the particles crossing the detector that obscure light. This process is automated but cannot distinguish between particles and gas bubbles, so thorough rinsing and the use of particle-free degassed water is essential. Qualification of liquid particle counters evaluates flow accuracy, volume accuracy, and size accuracy, in which standard solutions containing particles of known concentration and size are analyzed and the results compared to specifications.

In microscopic methods, the sample is filtered through a membrane filter, air-dried, and then placed under a microscope at 100× magnification. An analyst counts all the particles on the membrane and estimates their size using a calibrated ocular micrometer. There is no formal qualification of microscopy methods other than the use of a calibrated optical micrometer. The microscopic method is highly dependent upon analyst training and is somewhat subjective; however, it is still used when light obscuration methods are unsuitable.

3.3.15 Osmolality

Osmolality is the concentration of a solution expressed as the number of moles of solute particles per kilogram (Osmols). Osmolarity, defined as Osmoles per liter, is also used. USP <785> *Osmolality and Osmolarity* describes osmolality and osmolarity tests. It is an important characteristic of parenteral dosage forms. An injectable solution that does not have an osmolality similar to human plasma (i.e. is not isotonic), will create an osmotic gradient when injected into the body, causing discomfort and possible cellular damage. Osmolality measurement of parenteral drug solutions is typically automated, using instrumentation that measures freezing-point depression, vapor pressure, or boiling point elevation. The amount of solution required is small, typically a few drops, and the response relatively rapid. Like most modern laboratory instruments, installation, operational and PQ, and re-qualification at regular intervals are required to generate data in a GMP laboratory. Qualification of osmometers is performed by comparing the response of standard solutions to established specifications.

3.4 Chromatography

Chromatography is the dominant test used in the pharmaceutical industry. Different types of chromatography are used quantitatively or qualitatively for release, in-process, stability, or confirmatory testing of all different dosage forms. Therefore, it would be remiss to discuss analytical techniques in the pharmaceutical industry without mentioning chromatography. With numerous references available for chromatography, this section will only discuss the fundamental principles of different types of chromatography and critical information on the parts of equipment. Today, chromatography has advanced so much and has been transformed through many different applications in comparison to its early discovery of a special type of

Table 3.2 Summary of different modes of HPLC.

Separation mode	Stationary phase	Mobile phase
Normal-phase	Polar	Nonpolar
Reversed-phase	Nonpolar	Polar
Size-exclusion	Neutral	Aqueous or organic
Anion-exchange	Positively charged	Negatively charged
Cation-exchange	Negatively charged	Positively charged
Affinity	Antibody/enzyme	Buffered aqueous solution

Source: Based on Dong [14].

adsorption technique used by Mikhail Semenovich Tswett in his investigations for separation of leaf pigments in early 1900 [13]. GC gained its popularity in the 1960s from packed columns to capillary columns, with gas being the mobile phase. Liquid chromatography became a popular technique and an extremely useful tool in the 1980s for the separation of nonvolatile analytes. Further advances with better instrumentation, which are built to sustain much higher pressure and being able to handle many viscous solvents with efficiency and ease, have made HPLC an irreplaceable technique for pharmaceutical scientists [14]. Table 3.2 shows a summary of the different modes of HPLC used in the pharmaceutical industry.

3.4.1 High-Performance Liquid Chromatography

HPLC is a separation technique utilizing a precision pump to generate high pressure for transporting a liquid sample (mixed with mobile phase) containing analytes, through a packed material column (stationary phase), where the separations of components take place based on chemical interactions, polarities, and hydrophobic interactions. The separation is recorded with a detector, which captures the electronic signals and transcribes them on a form of media. The signal display chart is called a chromatogram. A complete HPLC system is composed of five essential parts: an injector, a high-pressure precision pump, a separation column, a detector, and a computer data acquisition system. The main functions of each of these five parts are described as follows:

Pump

The function of the pump is to continuously transport liquid solvent, *mobile phase*, through the system at a precise flow to carry the component interest through a separation column.

Injector

The function of the injector is to introduce a precise amount of liquid sample into the sample solvents stream of mobile phase that carries the sample through the separation column.

Separation Column

Most chromatography columns are made of steel with micro-sized packed particles that selectively separate the components of a liquid sample as they pass through. The column packing, or *stationary phase*, is where the separation takes place. The analytes travel through the column at different speeds based on the chemical interactions between the packed particle materials and the chemical structures of the analytes found in the sample.

Detector

The function of the detector is to detect the analytes after they have passed through the separation column. The detector signals record the concentration of each analyte. The commonly used detector in Quality Control (QC) departments is an ultraviolet-visible (UV/VIS) detector. Research and Development (R&D) laboratories, and a few QC laboratories, are also equipped with the mass spectroscopic detector. It is mainly used to identify and quantify organic impurities. Whether it is LC-UV/VIS or LC-MS, the output is recorded using a data acquisition system that documents the analysis on a chromatogram.

Today, the precision pump can operate precisely to deliver flow rate ranges from 0.5 to 5 $mL\,min^{-1}$ for a typical HPLC pump, or even lower at 0.1–0.4 $mL\,min^{-1}$ for an ultra-high performance liquid chromatographic (UHPLC) pump. The UHPLC can significantly reduce the amount of solvent used or significantly speed up the analysis time, thereby providing more flexibility and control to the analyst.

Advancement in columns technologies has contributed significantly to the use of HPLC in the pharmaceutical industry. Columns packed with different particle sizes and shapes and different bonded phase chemistries have made separations possible for broad ranges of chemicals and medicines, from protein purification analysis to inorganic materials. Liquid chromatography operates in two main modes: normal-phase separation mode and reversed-phase separation mode.

3.4.1.1 Normal-Phase Separation Mode

Normal-phase chromatography is a type of HPLC chromatography, where the liquid mobile phase is nonpolar solvent(s), and the stationary phase is a more polar material. The separation is based on polarity. The less polar compound would move through the column faster, thus elutes quicker than the more polar compound. Nonpolar solvents are organic solvents like hexane or benzene. The column stationary phase used in normal-phase chromatography is a more polar compound such as silica.

3.4.1.2 Reversed-Phase Separation Mode

In reversed-phase chromatography, the mobile phase is more polar than that of the stationary phase, which contrasts with normal-phase chromatography; hence the name "Reversed." Mobile phases used in reversed-phase chromatography often contain a mixture of water, which is a more polar compound, and organic solvents like acetonitrile or methanol. The stationary phase in reversed-phase chromatography is a nonpolar column, which often is silica bonded with functional groups like cyanopropylsilyl- [CN], *n*-octylsilyl-[C8], and *n*-octadecylsilyl-[C18] moieties. Most pharmaceutical compounds are polar compounds; thus, reversed-phase chromatography has broader applicability and has wider use in the pharmaceutical industry [14].

3.4.1.3 Other HPLC Separation Modes

3.4.1.3.1 Size-Exclusion Separation The principle of size-exclusion chromatography (SEC) is that the particles packing the HPLC column contain pores that allow smaller molecules to enter but exclude molecules larger than the pore diameters. Molecules will be separated based on their size, with the largest molecules eluting first. The elution volume of the molecules in SEC is a linear function of the logarithm of the molecular mass. This separation mode is widely used for biopharmaceutical applications of large proteins and enzymes, and to a lesser extent, in traditional pharma applications, typically for characterization of inactive ingredients, such as large polyethylene glycols (PEG's). When the separation uses the aqueous solution as the mobile phase, it is called "Gel Filtration," and when the mobile phase is an organic solvent, "Gel Permeation."

3.4.1.3.2 Ion-Exchange Separation Mode Ion-exchange separations (IEX) have two modes, anion-exchange, in which the stationary phase is positively charged, and the analyte is negatively charged, and cation-exchange, in which the stationary phase is negatively charged and the analyte is positively charged. HPLC systems designed for IEX are typically metal-free systems. It avoids contamination of the mobile phase with metal cations that can leach out of stainless steel and eliminates corrosion that would occur with a stainless-steel system from highly acidic or basic mobile phases. IEX separations are widely used for the analysis of metal impurities, quantitation of critical ions in formulations, and some elemental actives, such as fluoride in toothpaste.

3.4.1.3.3 Affinity Separation Mode Affinity chromatography is based on a specific interaction between a ligand immobilized on a stationary phase and a protein of interest. One example of this would be a monoclonal antibody that is covalently bound to a stationary phase. The monoclonal antibody will bind only the antigen it is specific for, so it can selectively pull the protein or enzyme from a complex mixture. The target antigen can then be eluted by changing the ionic strength of the pH of the eluent, typically preserving its biological activity. This separation mode is most commonly used in biopharmaceutical applications.

3.4.2 Ultra-High-Pressure Liquid Chromatography

The separation principle of ultra-high-pressure liquid chromatography (UHPLC) is no different than traditional HPLC. The difference is that a UHPLC system can operate at very high pressures up to approximately 15 000 psi. Similar to an HPLC system, the typical UPLC system consists of an autosampler, a high-pressure precision pump, a separation column, a detector, and a computer data acquisition system. The system may also have an in-line solvent vacuum degasser, a sample temperature control, and a column heater compartment. Although the principles are similar, the equipment of HPLC and UHPLC systems are built slightly different from being able to handle very high pressure, low flow rate, and very small injection volume. UHPLC systems contain a column with a narrower internal diameter and much smaller particle size packing materials. Therefore, more stringent UHPLC qualification and calibration programs are recommended to ensure that the instrument operates as its intended design.

3.4.3 Detectors of Liquid Chromatography

While the variable-wavelength UV/VIS detector is the most used detector in the pharmaceutical analysis, other detector types are also available and briefly introduced below.

3.4.3.1 UV/VIS Detector

A variable-wavelength UV/VIS detector works by directing polychromatic light onto a diffraction grating that splits the light and reflects a *narrow wavelength band* through a flow cell onto a sampling diode that detects the light intensity. Qualification of a UV/VIS detector typically involves wavelength accuracy, the linearity of response, baseline drift, and baseline noise. Wavelength accuracy is normally determined by measuring the accuracy of reference materials of known absorbance maxima and minima. For example, caffeine may be used for the wavelengths between 200 and 277 nm. When conducting calibration for these components, use a certified device to ensure the accuracy of results and for compliance to GMPs. For calibration testing, wavelength accuracy should be obtained at a minimum of two wavelengths settings. The specification can be set at ±2 nm from each target selected wavelength. Baseline noise is determined by injecting a blank solution (e.g. mobile phase, water, or organic diluent) with an absorbance detector set at a designed value (e.g. 254 nm). Serial dilutions of caffeine are typically used to verify the linearity of detector response over a range specified by the manufacturer. The drift in the baseline is measured over several hours and is a measurement of the stability of the electronics and lamp output.

3.4.3.2 Photodiode Array Detector (PDA)

A photodiode array detector (PDA) works by shining polychromatic light through the flow cell onto a diffraction grating that splits the light into individual wavelengths and reflects it onto a diode array where each diode receives a narrow wavelength band. Similar to the UV/VIS detector, light absorbance is measured, but the PDA measures all wavelengths at once, allowing the user to extract a full absorbance spectrum from each sample. A disadvantage is that PDA detectors are not as sensitive as UV/VIS detectors. Qualification of PDA detectors is nearly identical to that for UV/VIS detectors, with the exception that wavelength accuracy is more complex with additional wavelength settings over a broader range.

3.4.3.3 Fluorescence Detector (FLD)

Fluorescence is one of the most sensitive of all detection techniques, although the choice of samples is limited because not all molecules fluoresce. Fluorescence detectors (FLDs) shine an excitation wavelength into a flow cell, causing an analyte to fluoresce. The emitted light is detected at a 90° angle to the emission light. Qualification of an FLD covers linearity, wavelength accuracy of both excitation and emission wavelengths, baseline noise, and baseline drift.

3.4.3.4 Refractive Index Detector (RID)

A refractive index detector (RID) detects components based on the refraction of light in the sample solution. Two parallel light beams shine through a sample cell and the reference cell, and when a sample passes through the sample cell, the angle of light refraction is measured against the reference cell. RIDs are sometimes called a "Universal" detector as a chromophore is not required for detection; however, they are much less sensitive than other detector types. Qualification of a RID is relatively simple, evaluating the only linearity of response, baseline drift, and noise.

3.4.3.5 Electrochemical Detector (ECD)

In electrochemical detector (ECD), an electric current is generated resulting from oxidation or reduction reactions on a sample in a flow cell. Qualification of ECDs includes linearity of response and accuracy of the response using solutions of known composition. Qualification of the electronics in the detector is complicated and typically performed by the manufacturer or an outside vendor. ECD is typically used in the biopharmaceutical industry. Similar to an FLD, an ECD has limited utility because not all molecules have an oxidizable or reducible functional group, called an electro chromophore, necessary for detection.

3.4.3.6 Conductivity Detector (CD)

Conductivity detectors (CDs) measure changes in electrical current between two electrodes in a flow cell. It is the most common type of detector used in ion-exchange chromatography as the analytes must have a charge to affect the conductivity. CDs can have a nonlinear response, so linearity may or may not be a qualification parameter. Baseline noise and drift are also qualification parameters. Some manufacturers will use ion suppressors, which reduce background conductivity and enable linear response over a narrow range. Conductivity accuracy is determined with a solution of known ionic concentration. Similar to ECDs, accurate detection is dependent upon complicated electronics. Therefore, the manufacturer or a specialized vendor usually performs the hardware qualification.

3.4.3.7 Evaporative Light-Scattering Detector

In an evaporative light-scattering detector (ELSD), a sample coming from an HPLC is forced through a nebulizer and mixed with inert gas, thereby separating the liquid into aerosolized droplets. This mist passes through a heated drift tube, and the mobile phase is evaporated off, leaving microscopic particles of the dried analyte. These particles then pass through a detector cell where the scattering of a light beam is measured. ELSD is unlike spectroscopic detectors in that it quantifies the mass of the analyte, not the number of moles. The only qualification parameter tested with ELSDs is linearity.

3.4.3.8 Charged Aerosol Detector

When using a charged aerosol detector (CAD), the HPLC eluent is nebulized with nitrogen, and the nebulized vapor is passed through a drift tube, drying it to create a stream of particles. A secondary nitrogen stream is passed through an electrical corona, causing the nitrogen molecules to ionize, and the now positively charged nitrogen stream is mixed with the analyte particles where the charge migrates to the analyte. A signal is produced that is proportional to the weight of the analyte passing through a sensitive electrometer. Qualification of the instrument is similar to an ELSD, whereas most annual qualifications evaluate the only linearity.

3.4.3.9 Mass Spectrometry Detectors

In general, mass spectrometry detectors (MSDs) ionizes a vaporized sample, accelerates it with an electromagnetic field, and deflects it, based upon mass, into a detector that estimates the mass of the ion by the angle of deflection. There are many different types of MSDs, so it is important to know what type of MSD is in use and how it works. Qualification of an MSD typically consists of; vacuum verification, a very high vacuum is required for proper operation, scan verification, accurate quantification of masses over a dynamic range, linearity of response, injection precision, injection carry over, and signal-to-noise.

3.4.4 System Suitability Tests for Chromatographic Methods

System suitability tests and criteria have been used for the chromatographic method, such as HPLC/UHPLC and GC, for many years in pharmaceutical laboratories. Thus, they should be established for all types of analytical procedures to generate reportable results. Along with instrument qualification, method validation, and sound laboratory practices, system suitability fills an essential role in establishing that the method and analysis instrument performance is similar to when the method was validated and that the results generated are valid for their intended use.

For the quantitative test, system suitability tests are conducted to confirm the acceptability of the system concerning the accuracy and precision of chromatographic performance. For the qualitative test, system suitability tests are used to confirm the specificity of the chromatographic system.

Table 3.3 Parameters to be included in system suitability tests.

Parameter	HPLC/UPLC	GC	TLC
Precision	X	X	
Resolution[a]	X	X	X
Peak symmetry	X	X	
Retention factor	X	X	X
Check standard	X		

[a] Not applicable for a single peak assay.

The system suitability criteria must be demonstrated at the beginning and throughout the chromatographic run. If the system suitability criteria are not met, the run should be rejected and appropriately documented.

System suitability parameters and some acceptance criteria for chromatographic procedures are described in USP <621> *Chromatography* and similarly in Ph. Eur General Chapter 2.2.46. However, parameters and acceptance criteria in a specific monograph or validated method take priority over the general chapter. Table 3.3 lists parameters that should be verified in the system suitability test for different types of chromatographic methods.

3.4.4.1 Precision

Precision is a measure of the reproducibility of the method. It is evaluated by injecting multiple standard injections at the beginning of the run and throughout the chromatographic run, bracketing the samples.

The precision of the system at the beginning is determined using at least six (6) injections of the target concentrations. This precision is measured by calculating the percent relative standard deviation (%RSD) of the mean peak area of the six standard injections. In rare instances, peak height may also be used.

$$\text{Relative standard deviation}(\text{RSD}) = 100\% \times \text{standard deviation} / \text{mean}$$

The precision of the system throughout the run is determined using the standard injections at the beginning of the run and bracketing standards throughout the run injected in-between every 6 and 10 samples. This requirement confirms that system performance is maintained throughout the run. If the procedure is used to analyze the assay and impurities, then the system suitability must be met for both assay and impurity before samples can be calculated.

Typical levels of precision for five or six replicate standard injections at the beginning of a chromatographic run are ≤1.5% RSD for an API assay, ≤2.0% for a drug product assay, and ≤10% for impurity analysis of typical specifications of 98–102, 90–110, and 0.1%, respectively. However, it is known that modern autosamplers can deliver a tighter precision of 0.5% or better for API assay. The USP has proposed an RSD requirement of 0.73% for API assay with an upper limit of 102.0% for the analysis of small molecules [15].

3.4.4.2 Resolution

Resolution is a calculated factor that quantifies the separation of two peaks in a chromatogram. The USP <621> *Chromatography* defines how the resolution is determined, and modern software algorithms attempt to reproduce this calculation. The resolution calculation uses both retention time and peak width, so in general, the resolution is negatively affected if retention times are close together or peaks are very broad or tailing excessively. Resolution is not always a component of system suitability. It does not apply to a traditional Assay method with only a single peak to quantify, and in some cases, other metrics may provide a more useful measure of method performance. A typical resolution of a critical pair should be greater than 2.0 to ensure baseline separation, thus ensuring accurate measurement of the analytes.

3.4.4.3 Peak Symmetry

Peak symmetry, also known as tailing, evaluates the symmetry of the chromatographic peak in instrumental analysis. In theory, a chromatographic peak should elute as a perfectly symmetrical Gaussian peak. However, in practice, peaks tail slightly in HPLC/UPLC separations due to laminar flow friction inside the HPLC tubing and some separation inefficiency in the HPLC column. Severe tailing can be an indicator of a physical or chemical problem in the separation. For example, as HPLC columns age, the column bed may settle or harsh conditions may dissolve some of the stationary phases, creating a void volume at the head of the column. This event will act as a mixing chamber that spreads the peak out, causing it to tail. Tailing can also be caused by contamination of the stationary phase or a solvent mismatch in the system. Tailing peaks in the separation reduces the sensitivity of impurity methods, negatively affects peak resolution, and reduces the accuracy of integration algorithms. A tailing factor of ≤1.5 is typical for HPLC/UPLC analytical method. GC separations usually have narrower peaks, and tighter tailing specs are not unusual in GC.

3.4.4.4 Column Efficiency

There are a couple of different parameters in the USP that can be used as a measure of the performance and effectiveness of the column. These parameters are discussed in detail in USP <621> *Chromatography*.

3.4.4.5 Theoretical Plates

Theoretical Plates, *N*, is a calculated factor that compares the peak width to the retention time. Narrow peak width and longer retention times indicate that the separation is more efficient, and therefore, the number of theoretical plates will be larger. A typical specification might be, "Not less than 2500 theoretical plates." As HPLC columns age, retention times will shorten due to loss of stationary phase and peaks will broaden as the column settles and the packing geometry changes. A specification for theoretical plates ensures a minimum separation efficiency that might be necessary to separate two closely eluting peaks. Theoretical plates are usually determined on a single standard peak.

3.4.4.6 Retention Factor

The retention factor (k) is also called the capacity factor (k'). It is a ratio of the time an analyte spends in the stationary phase (retention time) versus the time for an unretained analyte to traverse the column, so the higher the retention factor, the more efficient the column. In a chromatographic separation, an unretained peak will elute at the solvent front, A peak that elutes in the solvent front will have a retention factor of zero, while a peak that elutes at exactly twice the retention time of an unretained peak will have a retention factor of 1. One advantage to the retention factor is that it is independent of flow rate or column geometry, whereas theoretical plates are not.

3.4.4.7 Weight Check Standard

Weight checks are not specified in the USP or any regulatory body but have become standard practice in the pharmaceutical laboratory. A weight check standard is a working standard that has been prepared exactly like the quantitation standard in a similar concentration. It is injected at the beginning of the run with the replicate standards and is compared to the quantitation standard using a simple response factor calculation. For example,

$$\frac{\text{Quant.std.response}}{\text{Check std.response}} \times \frac{\text{Check std.weight}}{\text{Quant.std.weight}} \times 100 = 98.0 - 102.0\%$$

The sole purpose of this parameter is to verify that the analytical balance was performing reproducibly when the standards were weighed. The check standard is not usually used in any quantification. As precision acceptance criteria become broader, a check standard becomes less critical to be included in the system suitability. For instance, let us say we have an impurity method that uses a very low concentration standard, and the precision requirement for System Suitability is RSD $\leq$ 10%. If we set a check standard requirement of 98.0–102.0%, it is highly probable that this requirement will not be met due to the imprecision of the method, and a check standard requirement of 90–110% tells us nothing about the performance of the analytical balance. Therefore, when the system suitability precision requirements of the method exceed 3%, a check standard becomes irrelevant. GC analyses are more variable than HPLC/UPLC, with precision specifications of 5–10%, so check standards are not common with GC.

3.4.4.8 Sensitivity

Sensitivity injections are typically used to verify that the system can detect and quantify a minimum concentration. It is important for methods that quantify impurities or degradation products and for quantitative methods that effectively limit tests, such as residual solvents analysis. Instrument-based analytical methods set a signal-to-noise requirement or recovery specification that is based on the original validation of the method and verified similar performance. In TLC, the sensitivity spot on the plate must be visible or of a specified intensity.

3.4.4.9 System Precision Over the Run

It is also necessary to demonstrate that the chromatographic performance remains suitable throughout the run with instrumental systems. It is accomplished by the inclusion of system precision standards or "Standards over the run." A standard solution is injected throughout the run at regular intervals and a precision requirement set. If a certain standard fails the precision requirement, then the analysis will only be considered valid up to the last passing standard. The precision requirement might be a maximum RSD requirement for Peak area, Peak height, or Retention time, or a specified range for percent recovery of the injection. There is no set requirement for how often system precision standards must be run. An analysis with a short acquisition time might specify a system precision injection every 10 or 20 injections, while very long runs might require system precision injections every four or six hours. The decision for how often these injections should be run depends upon the length of acquisition time, the robustness of the method, the equipment, and the type of analysis.

System precision injections may be used as quantitation standards or, in some cases, as check standards. When used as quantitation standards, all standards over the run, the replicates at the beginning of the run, and the system precision standards, are averaged, and the average response used to quantify sample injections. Check standards are used to verify system performance but are not used for quantification. For example, bioanalytical methods, such as a pharmacokinetic analysis, may run for days or weeks using the same standard curve for quantification. In this case, check standards are injected at regular intervals, and if they pass a recovery requirement, the system is considered suitable and the data valid.

Table 3.4 List of examples of injection sequences showing system suitability.

HPLC/UPLC assay	HPLC/UPLC impurities	GC assay
Blank	Blank	Blank
Resolution solution (1×)	Resolution solution (1×)	Resolution solution (1×)
Check Std. (2×)	Sensitivity solution (1×)	Sys. Suit. Std. (6×)
Sys. Suit. Std. (5×)	Sys. Suit. Std. (6×)	Samples (6×)
Samples (6×)	Samples (6×)	Sys. Precision Std. (1×)
Sys. Precision Std. (1×)	Sys. Precision Std. (1×)	Samples (6×)
Samples (6×)	Samples (6×)	Sys. Precision Std. (1×)
Sys. Precision Std. (1×)	Sys. Precision Std. (1×)	Samples (6×). . .etc.
Samples (6×). . .etc.	Samples (6×). . .etc.	
Sys. Precision Std. (1×)	Sys. Precision Std. (1×)	Sys. Precision Std. (1×)

Table 3.4 shows typical injection run sequences to be set up to establish system suitability for different chromatographic methods. The column should be equilibrated before the first standard is injected. The numbers in the parenthesis are the number of injections of such solutions.

Depending on the analysis, system suitability may include the resolution of critical pairs (R), relative retention time (RRT), capacity factor, tailing factor of the peak of interest, number of theoretical plates (N). The acceptance criteria of these components depend on the validation of the analytical procedures. Additional information can also be found from USP <621> *Chromatography*.

3.4.5 Maintenance of HPLC and UHPLC

Regular maintenance of HPLC and UPLC systems and scheduled calibration are required to keep the system in proper operating condition. For new equipment, the installation qualification (IQ) would be required. Preventative maintenance and calibration should be scheduled at least annually and conducted by a qualified, well-trained service provider. Preventative maintenance will consist of visual inspection of components, thorough cleaning of the system, and replacement of consumable items, such as pump seals and inline filters. Replacement of parts recommended by the manufacturer for preventative maintenance at specified scheduled replacement should be done. All items replaced during preventative maintenance will be documented for GMP compliance purposes, and after preventive maintenance, a PQ is performed to ensure that the instrument is functioning properly. An operation qualification (OQ) and PQ are required after any major service or repair, and if an instrument must be sent back to the manufacturer for repair, it is effectively, "uninstalled." Once it returns to the testing facility, a full IQ/OQ/PQ re-qualification must be performed. Failure of any system component to meet acceptance criteria requires documented correction and confirmation by appropriate requalification. Many of the instruments are covered under service contracts, but those companies that maintain a Metrology department will perform routine repairs and qualifications in-house.

In repairs or preventative maintenance of HPLC and UPLC systems, it is preferred to use parts obtained directly from respective manufacturers when possible. Avoid using alternate parts from third parties, as these parts might cause compatibility problems. UPLC systems typically require more frequent preventative maintenance and re-qualifications due to the greater stresses placed upon the instrument from the high pressures.

3.4.6 Gas Chromatography

GC is a separation technique not unlike HPLC, except analytes are in the gas phase rather than the liquid phase. In GC, the sample mixture is vaporized and injected into a stream of inert carrier gas (as nitrogen or helium) moving through a column containing a stationary phase composed of a liquid or particulate solid and is separated into its component compounds according to their affinity for the stationary phase. The separation is typically based upon the interaction between the polarity of the analyte and the polarity of the column, although specialized columns are available. Other factors that influence the separation on the column are the vapor pressure of the analyte, column temperature, carrier gas flow rate, column length, and the amount of sample injected. The signal output is called a chromatogram.

The major components of a GC are the injector, inlet, oven, column, detector, and gases. There are two primary types of injectors, liquid, and gas.

Liquid injectors inject a small amount of liquid typically 1–10 μL, into the inlet, which is heated to a temperature that will instantly vaporize the sample. Headspace injectors inject headspace gasses in a sealed vial directly onto the GC column.

A headspace injector withdraws a known volume of gas from the headspace of a vial containing a small amount of liquid. Headspace samples are heated and agitated to force volatile components out of the liquid phase and into the gas phase. It is important to note that volatile components do not move completely into the headspace, but rather reach equilibrium between the gas and liquid phase. For this reason, headspace vials may only be injected once as removal of gas from the headspace alters the equilibrium.

With a liquid injection, the sample is vaporized in the inlet and introduced onto the column. Liquid injections can be directly injected or split. A direct injection inlet transfers the entire amount of vaporized gas onto the separation column; this is standard when using packed columns and can be performed with wide-bore capillary columns using a modified packed column inlet.

For GC, split injections are performed using a split/splitless inlet and transfer only a portion of the vaporized sample onto the column. Split injection is necessary when the total volume of the vaporized sample exceeds the capacity of the column, and an incorrectly set split-flow can affect both sensitivity and peak shape. It is used for general analysis, and a splitless injection (a direct injection using a split/splitless inlet) is frequently used for trace analysis.

The GC column is where the separation takes place. Separation typically is related to polarity, with polar compounds retained longer on the column. A wide range of stationary phase polarities is available, as are specialized packings, such as chiral packings based upon amino acid derivatives, chiral silanes or cyclodextrins: a complete list of packings (L), phases (G), and supports (S) used in *USP–NF* and *PF, Reagents, Indicators, and Solutions – Chromatographic Columns*. There are two types of GC columns, "Packed" and Capillary. A "Packed" GC column consists of a steel or glass tube, 1/8–¼ inch in diameter, and approximately 1 m long, filled with a stationary phase packing. While packed columns may still be used in a few legacy methods in the USP, they have mainly been eclipsed by capillary columns that provide superior resolution, peak shape, and shorter analysis time. GC capillary columns are made from silica tubing with inner dimensions of 0.18–0.53 mm and lengths up to 105 m. The stationary phase is a thin film on the inside of the tubing, with a range of film thicknesses from 0.10 to 10 μm.

The GC column resides inside an insulated, temperature-controlled oven that can be rapidly heated or cooled. The separation and selectivity of the GC column can be modified by changing the temperature; elevating the temperature of the column will shorten the retention time. Separations can be isothermal, at a fixed temperature, or gradient, in which the oven temperature gradually increases according to a programmed gradient profile. Gradient separations are very useful when the sample contains analytes with a wide range of boiling points. Compounds with a high boiling point have a lower vapor pressure and are likely to spend more time out of the gas phase during the separation. Increasing the temperature during the analysis decreases the retention of the "high boilers" and shortens the overall analysis time.

The sample is transported through the column by the carrier gas. A high flow rate will reduce the retention time but will reduce the interaction of the sample with the stationary phase and result in a less optimal separation. Helium has long been used as a carrier gas in the United States, whereas hydrogen or nitrogen is usually used elsewhere. As helium becomes limited, the use of hydrogen as a carrier gas is becoming more common in the United States.

Once samples leave the column, they pass into the detector. The commonly used detector in the pharmaceutical industry is the flame-ionization detector (FID). As the sample leaves the column, it is mixed with hydrogen and oxygen or air, which passes into a flame. Samples containing carbon are pyrolyzed (burned), converting the carbon atoms to carbon cations. These charged ions are quantified with a highly sensitive electrometer. FIDs are highly sensitive, linear over several orders of magnitude, and can detect most organic molecules. The disadvantages are that they cannot detect inorganic compounds, and the sample is destroyed during pyrolysis.

Other types of detectors are used in pharmaceutical analysis. The most common is the thermal conductivity detector (TCD) and the electron capture detector (ECD). There are many other types of GC detectors that are highly specialized and will not be discussed here.

The TCD measures differences between the thermal conductivity of the column eluent and a reference gas. Because all compounds have a thermal conductivity different from that of helium (carrier gas), the TCD responds to all compounds and is, in theory, a universal detector. However, since an FID is much more sensitive to organic compounds, TCDs are typically used only for the analysis of gases (argon, oxygen, nitrogen, and carbon dioxide) and the occasional rare pharmaceutical compound that does not contain a carbon-hydrogen bond. TCDs are typically qualified for noise and drift only.

The ECD uses a radioactive β-particle source to create an electron beam that collides with the column eluent and analytes. The detector measures the current across the eluent flow that changes when an analyte captures electrons from the beam. ECDs can be 10–1000 times more sensitive than an FID and have an excellent response for halogenated compounds, such as pesticides and CFC's and is highly useful for trace analysis. These detectors are qualified only for noise and drift.

3.4.6.1 Qualification

As a GC is a quantitative instrument, formal, protocol-driven qualification, IQ, OQ, and PQ, are required periodically to assure that the instrument is performing as needed. Typical parameters tested during OQ and PQ are discussed below. However, it is critical to remember that the acceptance criteria are based on the manufacturer's specifications and expected usage of the instrument.

Liquid Injector – Liquid injectors are qualified for injection accuracy and injector precision. Liquid injection volumes are small, 0.1–10 μL, so precision and accuracy are critical parameters.

Headspace Injector – Headspace injectors are considerably more complicated than liquid injectors and, therefore, have a more involved qualification. Headspace injection volumes run from 0.5 to 5 mL and larger; therefore, the linearity of response for different injection volumes and precision of small, medium, and large injection volumes must be qualified. Samples are heated and agitated for a programmed length of time, so temperature accuracy of the sample compartment and transfer line, and accuracy of the timer must also be qualified.

Table 3.5 Qualification parameters.

Qualification test	Description
GC oven	Accuracy and stability of the set temperature at two or more set points. The accuracy of the Ramp rate.
Inlet (split/splitless and purge/packed)	Temperature accuracy Pressure accuracy Pressure decay Gas flow stability
Liquid injector (with split/splitless or purge/packed inlet)	Injection volume accuracy Injection volume precision Injection volume linearity Injection carryover
Headspace injector	Vial heater temperature accuracy. Heated zones temperature accuracy Headspace leak test
FID	Noise and Drift Linearity of response Sensitivity (signal-to-noise)
TCD	Noise and Drift Sensitivity
ECD	Noise and Drift Sensitivity

Inlet – Inlet temperature accuracy is qualified as it is critical to proper vaporization of the sample.
Oven – An accurate oven temperature and precise temperature ramp rate during a gradient analysis are essential for accurate, reproducible analysis.
Detectors – FIDs are qualified for linearity of response over the operating range, accuracy of the temperature settings, sensitivity, noise, and drift. TCDs and ECDs are typically qualified for noise and drift.

Depending on the type of detectors used, different qualification parameters for a GC are used. The difference is in Table 3.5.

3.4.6.2 Residual Solvents

Organic synthesis of API frequently involves the use of solvents; some of the relatively toxic and removal of these solvents to a specified level is part of the validated manufacturing process. Analysis of these residual solvents is the purpose of residual solvents analysis. This analysis is typically performed using headspace

GC. Regulatory Agencies segregate solvents into three classes, where Class 1 solvents have the highest risk of exposure, and Class 3 the least risk. Regulatory requirements for residual solvents are documented in USP <467> *Residual Solvents*, the FDA Guidance for Industry, *Q3C Impurities: Residual Solvents*, and the ICH guidance; *Impurities: Guideline for Residual Solvents, Q3C*.

3.4.6.3 Hyphenated Technologies

Hyphenated technologies, such as gas chromatography-infrared spectroscopy (GC-IR) or gas chromatography–mass spectrometry (GC–MS), pair two or more standalone analytical techniques in a manner that results in a much more advanced analysis. A separation technique (e.g. HPLC or GC), is typically paired with a complementary form of analysis, such as MS or infrared spectrometry. This combination allows for a secondary analysis of multiple components of the sample as they are separated in the front-end analysis. In GC-IR or GC–MS analysis, separation by the gas chromatograph has the benefit that the analyte is introduced into the IR or MS in purified form and the gas phase, eliminating separate sample preparation steps and interferences. Liquid chromatography–mass spectrometry (LC–MS) splits off a small amount of the column eluent into a nebulizer that introduces the sample into the mass spectrometer for analysis. Gas chromatography–tandem mass spectrometry (GC–MS–MS) is a coupling technique of the mass fragments from one mass spectrometer are introduced into a second mass spectrometer, providing increased sensitivity and more detailed structural information.

3.4.7 Thin-Layer Chromatography

Thin-layer chromatography (TLC) is a chromatographic technique used to separate non-volatile mixtures. TLC is performed on a sheet of glass, plastic, or aluminum foil, which is coated with a thin layer of adsorbent material, usually silica gel, aluminum oxide (alumina), or cellulose [16]. Silica gel plates typically have a fluorescent dye incorporated into the gel. When separating organic molecules, the silica gel fluoresced a bright green color, and separated organic molecules show up as dark spots due to the absorbance of UV light by aromatic functional groups and conjugated double bonds. There are also many methods to detect analytes chemically, which use dyes or other compounds that react with the analyte and form colored or UV absorbing complexes. TLC is widely used as a qualitative technique but can be quantitative as a limit test. In the pharmaceutical analysis, regulatory agencies are encouraging companies to replace TLC limit tests due to the subjective nature of the interpretation of the results.

3.4.8 Bio-Pharmaceutical Separations

Electrophoresis is a form of analysis that separates components of a mixture by forcing the migration of analytes through a fluid or gel, using an electric field. The electrophoretic mobility of the analytes depends upon the molecular charge and the viscosity of the eluent. Electrophoresis is used for the separation of macromolecules, such as proteins, peptides, and enzymes, and there are many variants of this type of analysis. Capillary electrophoresis forces the analyte along a silica capillary against an eluent flow in the opposite direction. Detection can be done with an ultraviolet–visible spectrometer (UV–VIS) or a mass spectrometer. Capillary electrophoresis is used for DNA analysis, analysis of antibody-drug conjugates, characterization of proteins, and enables very high-resolution chiral separations.

In slab electrophoresis, the electrical field forces analytes through a gel slab that may be agar, polyacrylamide, or other polymerized material. The proteins are visualized using stains, such as bromophenol blue, Coomassie blue, or Ethidium bromide. This technique is highly useful for the identification of protein mixtures, and individual bands can be cut out of the gel for secondary analysis, such as sequencing.

3.4.8.1 Capillary Zone Electrophoresis

Capillary zone electrophoresis (CZE) separates molecules into bands based upon their electrophoretic mobility.

Capillary Isotachophoresis is similar to CZE except for the high and low mobility ions bracket the sample. It enables the separation, but the bands elute without a buffer in between them.

3.4.8.2 Isoelectric Focusing

Isoelectric focusing is an electrophoresis technique in which amphoteric molecules, molecules with both acidic and basic functional groups (zwitterions), are separated using a pH gradient. These molecules will migrate to the pH where there is no net charge on the molecule. It can be used to identify or purify proteins and peptides using capillary electrophoresis and can purify large amounts using slab gels.

3.4.8.3 Sodium Dodecyl Sulfate–Polyacrylamide Gel Electrophoresis

Sodium Dodecyl Sulfate–Polyacrylamide Gel Electrophoresis (SDS-Page) is a chromatographic technique used in bioanalysis for large molecules. The mobile phase contains SDS, which denatures proteins, unfolding them, so this is called a "denaturing gel," and the stationary phase is a polyacrylamide gel. Samples are loaded onto the gel, and an electric current is applied across the gel, causing charged, denatured proteins to move. Proteins are separated by molecular size, with the smallest proteins traveling the farthest.

3.5 Spectroscopic Sciences

Spectroscopy is a scientific measurement technique. These techniques are used to identify and quantify materials by measuring light that is emitted, absorbed, or scattered by materials. This section gives a summary of the spectroscopic analytical techniques used in pharmaceutical laboratories.

In spectroscopy, scientists extend study from visible light to other regions of gamma rays, X-rays, ultraviolet (UV) radiation, infrared (IR) radiation, microwaves, and radio waves. The associated wavelengths for these regions are illustrated in Table 3.6. For the spectroscopic chemical analysis studies, the regions from 10^{-7} to 10^{-2} are explored; and many useful applications and techniques have been developed like UV–Vis spectrometry, infrared-absorption (IR) [17].

3.5.1 Ultraviolet–Visible

Ultraviolet–visible (UV/Vis) spectroscopy is routinely used in the pharmaceutical industry for qualitative or quantitative analyses. Spectroscopic analysis is commonly carried out in solutions; however, solids and gas materials may also be studied using this application. The principal application of this technique is based on molecular absorption of energy in the form of ultraviolet or visible light to excite electrons from non-bonding to higher anti-bonding molecular orbitals. The development of the UV/Vis spectrophotometer was based on this principle and has been widely applied as a stand-alone spectrometer or as a UV detector used in HPLC. The Beer–Lambert law states that the absorbance of a solution is directly proportional to the concentration of the absorbing components in the solution and the path length [18]. Therefore, the response of the analyte (with a certain statistical degree of confidence) is proportional to the concentration of the

Table 3.6 Wavelengths regions.

Type of radiation	Wavelength range (nanometers)
Gamma rays	Less than 0.01
X-rays	0.01–20
Ultraviolet (UV)	20–400
Visible (Vis)	400–700
Infrared (IR)	1000–1 000 000
Radio	More than 1 000 000

Source: Based on De Caro and Haller [17].

solution. This principle becomes the basis for quantitative analysis for the pharmaceutical industry. To obtain accurate results for an unknown analyte, the response of the instrument is calibrated using a known standard material (e.g. peak area or height of compound of interest of standard material). The analysis of the test sample is carried out, and the quantitation of the test sample is done by comparing the instrument response of the unknown test sample material to the response of the known standard material.

USP <857> *Ultraviolet–Visible Spectroscopy* also provides information regarding validation or verification of UV/Vis procedures as well as details on instrument qualifications [19].

3.5.2 Infrared-Absorption

Infrared spectroscopy is used to identify and study chemical structures and is a powerful technique for qualitative analysis. For a given sample of solid, liquid, or gas, the technique of infrared spectroscopy uses an instrument to generate an infrared spectrum, which is a unique fingerprint of the substance. This IR spectrum can be compared to a library database to identify the material.

A typical laboratory instrument that uses the IR technique is a Fourier transform infrared (FTIR) spectrometer. As a fast, easy, and versatile test, FTIR is an ideal test for identifying unknown substances. It can analyze solids, liquids, gases, and pastes, and it is a powerful technique in identifying counterfeits.

The sophistication of IR instrumentation has been advancing rapidly. A handheld and robust instrument is available at a reasonable cost. Thus this technology has become critical in formulation development as well as process control.

Similar to IR spectroscopy, Raman spectroscopy is a technique that uses a monochromatic beam or laser in the visible, near infrared, or near ultraviolet range of the electromagnetic spectrum instead of the infrared range. This technique has become more popular over the last decade in research as well as clinical laboratories [20].

3.5.3 Mass Spectroscopy

Mass spectrometry (MS) is an analytical technique that is used to identify the amount and type of chemicals present in a sample by measuring the mass-to-charge ratio and abundance of gas-phase ions. A mass spectrum is a plot of the ion signal as a function of the mass-to-charge ratio. MS has become extremely critical in the pharmaceutical industry and is especially useful in quantifying, identifying, and controlling the impurities. Typically, the technique is used with another analytical technique such as HPLC or GC, although it can be used as a standalone instrument for the elucidation of structure.

3.5.4 Atomic Absorption, Inductively Coupled Plasma, Inductively Coupled Plasma/Mass Spectrometry, and Inductively Coupled Plasma/Optical Emission Spectrometry

Atomic absorption (AA) spectroscopy is one of the oldest methods of analyzing metal elements. AA spectrometers measure the absorption of light by free atoms in the gas phase. Samples containing metals are digested to remove organic compounds, then dissolved in a liquid carrier, and injected into a flame source using a variety of techniques. Disadvantages of AA are that typically only one metal may be quantified at a time and not as sensitive as ICP-MS or ICP-OES. Despite these disadvantages, AA is low cost and robust and is widely used in applications that do not require the increased sensitivity of other methods.

Inductively coupled plasma (ICP) spectrometry is a methodology where a gas plasma provides a very high-temperature excitation source for atomic spectrometry. By combining with a mass spectrometer or an optical emission spectrometer, individual isotopes may be identified and quantified [21].

ICP-OES and ICP-MS utilize an ICP, which is plasma that is inductively heated using an electromagnet, to vaporize and ionize the sample. In ICP-OES, the sample is atomized, and the mist delivered into the ICP, after which the concentration of ionized atoms in the sample is detected using an optical emission spectrometer. ICP-OES technique can analyze multiple elements simultaneously and is less affected by matrix effects than of ICP-MS technique. However, with detection limits in the sub-part per million (ppm) range, it is not as sensitive as the ICP-MS technique.

ICP-MS has higher speed, precision, and sensitivity when compared to AA and ICP-OES. It uses ICP to vaporize and ionize the sample for detection by MS. In addition to the advantage of analyzing for several metals concomitantly, ICP-MS has the advantage of an extensive linear range, detection limits in part per trillion (ppt) range, and the ability to quantify isotopes. The disadvantages of ICP-MS are that it is more sensitive to matrix effects than ICP-OES and is much more expensive to purchase and maintain than either AA or ICP-OES.

With the adoption of USP <232> *Elemental Impurities – Limits* and <233> *Elemental Impurities – Procedures*, and worldwide adoption of ICH Q3D, Guideline for Elemental Impurities, ICP-MS and ICP-OES have become the preferred techniques for analysis of potentially toxic metals in pharmaceuticals.

3.5.5 Nuclear Magnetic Resonance Spectroscopy

NMR spectroscopy is an analytical technique using the magnetic properties of specific atomic nuclei to determine the physical and chemical properties of atoms or the molecules in which they are contained [22]. For quantitative analysis, it is used in the R&D or Quality Control laboratories to determine the molecular structures or composition of a mixture containing known compounds. For qualitative analysis, NMR is used to identify the unknown by matching its spectrum against a standard spectrum or interpreting the basic structure [23].

3.5.6 X-ray Absorption and X-ray Emission Spectrometry

For the past decade, X-ray absorption and X-ray emission have gained much popularity in the methodological and instrumental aspects. This analytical methodology was used to quantitatively determine the compositions of different crystal forms of the materials [24].

3.6 Uniformity of Dosage Units

Testing for the uniformity of dosage units, also called content uniformity, is performed to ensure that the content level of drug substance is consistent across the batch. Content uniformity is not to be applied to formulations like gels, suspensions, or emulsions that are intended for cutaneous administration. The official compendia definition of uniformity of dosage units is: *The degree of uniformity in the amount of the drug substance among dosage units,* USP <905> *Uniformity of Dosage Units* [25]. Content uniformity is a metric that is used to control the quality of solid oral dosage forms. Manufacturing processes are dynamic and variable and accepting that some variation in the content of individual tablets or capsules is unavoidable has necessitated a metric that assesses the quality on an individual dose level. It is particularly important for drugs with a narrow therapeutic index, drug/device combinations, and specialized dosage forms, such as inhalation products. The two approaches used to examine the Uniformity of Dosage Units are Content Uniformity and Weight Variation. The choice to use content uniformity or weight variation is dependent on the type of dosage unit and the percent of drug substance in the dosage unit. Chemical testing for content uniformity testing can be used in all cases; however, weight variation is preferred where possible because of the significant cost savings and laboratory time. Content uniformity is also used to assess the precision of dosage for tablets that are scored to split to deliver a one-half dose.

In general, content uniformity testing is performed with 10 dosage units tested individually. Individual results are expressed as a percent label claim, and an average of these 10 units are reported. The number of units to be tested follow USP <905> [26]. It is a good practice that the weight of each dosage unit is taken before the analysis of the units. This information could help understand the variability of the results.

3.6.1 Weight Variation

Weight variation is performed by weighing individual tablets or capsules, or in some cases, liquid dosages. It is only applicable if the individual dosage units contain a large enough active/excipient ratio that any error in the dose from manufacturing imprecision is negligible. The applicability of weight variation and those instances in which it may be applied is clearly defined in USP <905> *Uniformity of Dosage Units* [26].

3.6.2 Acceptance Criteria per USP <905>

The acceptance criteria for either solid dosage units or liquid dosage form is calculated using the following formula:

$$|M - \bar{X}| + ks$$

where

$\bar{X}$ is the mean of the individual contents,
M is the Reference value. The reference value depends on the target content (T) of each dosage unit expressed at the percent label claim. T is 100.0% unless or the manufactured approved target concentration per dosage unit. Table 2 in USP <905> defines the value for M.
k is the acceptability constant, which is 2.4 if $n = 10$ and is 2.0 when $n = 30$ with $n =$ number dosage units tested.
s is the sample standard deviation

3.7 Elemental Analysis

Elemental analysis is the quantification of elemental impurities, defined as those with a high density or atomic weight. In the pharmaceutical industry, however, elemental impurities are those which have a deleterious effect on human health. Elemental impurities originate in nature in minerals and ores, and they find their way into pharmaceutical products. The sources are usually from raw materials (stabilizers, coatings, flavors, colors, etc.) that utilize minerals in their manufacture; from plants that have taken up metals from the soil or air during the growth process; from exposure to metal catalysts, a typical tool used in the synthesis of APIs; from contaminants introduced during the manufacturing process and leachable metallic impurities in packaging materials. Exposure to trace amounts of some metals is essential for human metabolism, but many metals exhibit toxic effects at higher exposures.

The introduction of new guidelines relating to elemental impurities in pharmaceuticals from the International Conference on Harmonization (ICH), U.S. Food and Drug Administration (FDA) and the United States Pharmacopeia (USP), has created new and complex challenges for the pharmaceutical industry. These challenges include replacing the wet chemistry "heavy metals" limit test with new analytical technology, specifically ICP-based technologies and mandated conformance to be new and specific limits for individual elements. In 2013, the ICH adopted Q3D, *Guideline for Elemental Impurities.* The European Medicines Agency (EMA) announced that all European products must comply with the new Q3D guidelines by December 2017, and the FDA and USP have required compliance by 1 January 2018. By January of 2018, most pharmaceutical manufacturers must replace the obsolete wet chemistry method with modern methods utilizing ICP-MS or ICP-OES.

The FDA Guidance for Industry, Q3D Elemental Impurities, has included a list of Elemental Impurities in Drug Product with permissible daily exposures [27].

3.8 Appearance

Physical appearance can be subjective but is one of the most important tests performed upon drug substances and drug products. For a drug product, it fulfills the most basic need to verify that the correct product is being tested. Typical testing involves viewing the material or product against a white background and reporting the color, form, and identifying marks, if appropriate. The color change can be an indication of the presence of degradation or contamination of the products.

3.9 Visual Inspection

Visual inspection testing is used mainly for parenterals. Each sample container is inspected for the presence of visible foreign particulates. The visual inspection process is subjective testing that requires vigorous training to ensure that the product meets quality acceptance criteria. The subjectivity is mainly due to testing performed by human employees with their eyes, often with no additional technologies. The inspector must be qualified before conducting the test and must also be requalified periodically. Annual eyesight exams must also be done to ensure a 20/20 vision. Therefore, proper inspector training is critical to a visual inspection program [28].

In a general overview, the approach of the regulators seems to cover two principles: (i) specification of the inspection method and analyst training to ensure the inspection process is controlled, and (ii) an assessment of the reproducibility of the result. However, it is also important to recognize that there is no particle-free parenteral solution. Therefore, the objective of GMP has been the manufacture of injectable solutions with the lowest possible particulate burden.

Visual inspection ranks as an essential operation in parenteral drug production to ensure the safety of drug products in its containers, such as vial or syringe. There is an increased emphasis by regulators on having a well characterized and robust inspection process [29]. The USP has worked with many experts and published many requirements for injectables in the USP <1> *Injections and Implanted Drug Products (Parenterals) Product Quality Tests* and several other General Chapters in this area [30–34].

3.10 Microbiological Testing

Microbiological testing is an important area of pharmaceutical testing as it plays a significant role in assuring the appropriate quality of drugs. The role of microbiological testing is to provide data indicating the effective controlling or eliminating microbes throughout the manufacturing process of drugs and biologics, especially the manufacturing of vaccines. This area is typically handled by a group of well-trained microbiologists in a specialized microbiological lab. The World Health Organization (WHO) has issued a Worldwide Quality Standard on Microbiological Test Methods in 2004, which details these microbiological methods [35].

Microbiological testing is often classified into two methodology groups: Quantitative and Qualitative Methods. The quantitative methods are analyzed for *Total Aerobic Microbial Count (TAMC), Total Combined Yeasts and Moulds Count (TYMC), Test for bile-tolerant gram-negative bacteria*. The Qualitative methods (also known as tests for specified microorganisms) are tests for *bile-tolerant gram-negative bacteria*, for *Escherichia coli*, for *Salmonella*, for *Pseudomonas aeruginosa*, for *Staphylococcus aureus and Candida albicans*.

3.10.1 Microbial Limits

The microbial limit test measures the antimicrobial activity of the sample to assure that it is maintained at an effective level. There are two general techniques: (i) the cylinder place assay and (ii) the turbidimetric assay [36]. Detailed information can be found in the USP <61> *Microbial Examination of Nonsterile Products: Microbial Enumeration Tests*.

3.10.2 Sterility

Sterility testing is a test to confirm that the batch of drug products is sterile or has been sterilized. The testing indicates that the product is free of viable microorganisms such as bacteria, mold, or yeast. Sterility is a qualitative test that is required during the validation of the sterilization process, in-process testing, or for routine quality control. Sterility testing is very tedious and must be performed by qualified laboratory analysts to detect growth during the incubation period. Also, the number of units needed for testing, the total volume required for the test, and the control of the media to provide accurate results are important factors for sterility testing. Testing detail can be found in the USP <71> *Sterility Tests*. The result is either reported as pass or fail. A satisfactory result indicates that no contamination has been found in the sample under the conditions of the test [37].

3.10.3 Bacterial Endotoxins

Bacterial endotoxins (BE) testing is an in-vitro assay for determining and quantitating bacterial endotoxins, a component of the cell wall of Gram-negative bacteria. The BE test is performed for parenterals and medical devices. This test is also called the Limulus Amebocyte Lysate (LAL) test or a pyrogen test. The Gel Clot test is done by determining the detection of endotoxin in each sample using the combination of LAL reagents with the sample or a dilution of the sample. The LAL reagent will gel in the presence of endotoxin. Another technique is the Kinetic endotoxin test. The change in optical density (e.g. turbidity for kinetic, color for kinetic) of a standard or sample is measured continually over an incubation period until a predetermined optical density is achieved. The endotoxin content of the sample is determined by measuring the optical density and interpolating from the standard curve. USP <85> *Bacterial Endotoxins Test* has more details on the requirements of this test. It is also another microbiological test needs to be performed as part of release testing of pharmaceutical products [38].

3.10.4 Antimicrobial Effectiveness Testing

Antimicrobial effectiveness (AE) or preservative effectiveness test is performed for sterile multi-dose parenteral products. The purpose is to challenge the inhibiting and killing the growth of microorganisms that may have been inadvertently introduced into the container of a parenteral formulation. The AE test is performed during formulation development and stability testing of the multidose drug product. The preservative and its concentration must be carefully determined based on historical data, characteristics of the drug substance and excipients, and or the physicochemical attributes of the product [39].

The two typical techniques used for AE testing are membrane filtration and pour plate. The membrane filtration is done using a single membrane for each enumeration test. A direct aliquot of the test article may be used, or dilution of the sample may be used as determined by validation. The pour plate is a technique using materials that are insoluble or reconstituted as emulsions or suspensions that cannot be analyzed using the membrane filtration technique. It is done by adding an appropriate amount of sterile molten agar to an equal volume of dilution of test preparation in a sterile petri dish. Additional detail testing is described in USP <51> *Antimicrobial Effectiveness Test* [40].

3.11 Summary

This chapter is intended to provide information for some commonly used analytical tests and techniques, which are part of the analytical laboratory testing operations in a CGMP environment at a pharmaceutical company. Additional detailed information is provided to help the reader develop their own set of instructions for their Standard Operating Procedure.

This chapter is nowhere close to the complete list. The readers should consult other analytical references if the testing of your interest is not one of these described or if there are not enough details needed. Also, information on major analytical technologies such as chromatography or spectroscopy is very limited in this volume. If these areas play a major part of your responsibilities, it is advisable to seek an additional book for a complete explanation. Good basic analytical practices are the foundation for advancement in a career in this dynamic, highly regulated industry.

List of Abbreviations

AA	atomic absorption
AE	antimicrobial effectiveness
API	active pharmaceutical ingredient
ASTM	American Society for Testing and Materials
ATC	automatic temperature control
CZE	capillary zone electrophoresis
DSC	differential scanning calorimetry
ECD	electron capture detector
FDA	Food and Drug Administration
FID	flame ionization detector
FTIR	Fourier transform infrared
GC	gas chromatography
GC-IR	gas chromatography–infrared
GC–MS	gas chromatography–mass spectrometry
GMP	Good Manufacturing Practices
HPLC	high-performance liquid chromatography
IC	ion chromatography

ICH	International Council on Harmonization of Technical Requirements for Pharmaceuticals for Drug Product for Human Use
ICP	inductively coupled plasma
ICP-MS	inductively coupled plasma-mass spectrometry
ICP-OES	inductively coupled plasma-optical emission spectrometry
ID	identification
IQ	installation qualification
ISE	ion selective electrode
LAL	limulus amebocyte lysate
LOD	Loss on Drying
MS	mass spectrometry
NMR	nuclear magnetic resonance
OQ	operational qualification
PPM	parts per million
PQ	performance qualification
QC	quality control
qNMR	quantitative nuclear magnetic resonance
R&D	research and development
ROI	residue on ignition/sulfated ash
SDS	sodium dodecyl sulfate
SEC	size exclusion chromatography
SOP	Standard Operating Procedure
TAMC	total aerobic microbial count
TCD	thermal conductivity detector
TGA	thermal gravimetric analysis
TLC	thin layer chromatography
TYMC	total combined yeasts and mold count
UHPLC	ultra-high-pressure liquid chromatography
UPLC	ultra-performance liquid chromatography
USP	United States Pharmacopeia
UV	ultraviolet–visible

References

1 Food and Drug Administration (2011). Code of Federal Regulations. Title 21, Subchapter A, Part 58, Good manufacturing practices. https://www.accessdata.fda.gov/scripts/cdrh/cfdocs/cfcfr/CFRSearch.cfm?fr=58.3 (accessed 7 August 2020).

2 USP <41> (2011). Balances. *USP 39–NF 34*. Rockville, MD: United States Pharmacopeial Convention.

3 USP <1251> (2011). Weighing on an analytical balance. *USP 39–NF 34.* Rockville, MD: United States Pharmacopeial Convention.

4 USP <31> (2011). Volumetric apparatus. *USP 39–NF 34.* Rockville, MD: United States Pharmacopeial Convention.

5 Nernst equation (2011). pH http://Meter.info website. http://www.ph-meter.info/pH-Nernst-equation (accessed 7 August 2020).

6 Agilent Technologies (2011). VK 200 tablet hardness tester operator's manual. https://www.agilent.com/cs/library/usermanuals/public/70-9015h_vk_200_op_man.pdf (accessed 7 August 2020).

7 Mettler Toledo (2011). Karl Fischer titrator instruments, Operator's Manual. https://www.mt.com/us/en/home/products/Laboratory_Analytics_Browse/Product_Family_Browse_titrators_main/Karl_Fischer_Titration.html (accessed 15 August 2020.

8 Nilsen, C.L. (1998). *The QC Laboratory Chemist – Plain and Simple.* Boca Raton, FL: CRC Press.

9 USP <731> (2011). Loss on drying. *USP 39–NF 34.* Rockville, MD: United States Pharmacopeial Convention.

10 Altamimi, M.J., Wolff, K., Martin, G.P., and Royall, P.G. (2017). An investigation of the anomeric stability of lactose powder stored under high-stress conditions. *Pharmaceutical Technology* 41 (3): 36–45.

11 USP <701> (2011). Disintegration. *USP 39–NF 34.* Rockville, MD: United States Pharmacopeial Convention.

12 USP <788> (2011). Particulate matter. *USP 39–NF 34.* Rockville, MD: United States Pharmacopeial Convention.

13 Ettre, L.S. (2003). Tswett and the invention of chromatography. *LC-GC The Magazine of Separation Science* 21 (5): 458–467.

14 Dong, M.W. (2006). *Modern HPLC for Practicing Scientists.* Wiley: Hoboken, NJ.

15 Olsen, B., Parente, E., Daniels, J., et al. System suitability for USP chromatographic procedures – small molecules. *Pharmacopeial Forum* 39 (5).

16 Wikipedia contributors (2020). Thin-layer chromatography. Wikipedia, The Free Encyclopedia, 06:29 UTC. https://en.wikipedia.org/w/index.php?title=Thin-layer_chromatography&oldid=970761109 (accessed 7 August 2020).

17 De Caro, C.A. and Haller, C. (2015). UV-Vis spectrophotometry – fundamentals and applications, Mettler-Toledo Publication. https://www.researchgate.net/publication/321017142_UVVIS_Spectrophotometry_-_Fundamentals_and_Applications (accessed 15 August 2020).

18 Metha, A. Derivation of Beer-Lambert law. PharmaXChange.info website. https://pharmaxchange.info/2012/04/ultraviolet-visible-uv-vis-spectroscopy-%E2%80%93-derivation-of-beer-lambert-law/#:~:text=Lambert%20Derivation&text=Thus%20Lambert's%20Law%20was%20defined,length%20b)%20of%20the%20solution (accessed 8 August 2020).

19 USP <857> (2011). Ultraviolet-visible spectroscopy. *USP 39–NF 34*. Rockville, MD: United States Pharmacopeial Convention.

20 Eberhardt, K., Stiebing, C., Matthaus, C., Schmitt, M., Popp, J. Advantages and limitations of Raman spectroscopy for molecular diagnostics: an update. *US National Library of Medicine.* 2015 Jun;15(6):773–87. doi: https://doi.org/10.1586/14737159.2015.1036744.

21 Kealey, D. and Haines, P.J. (2002). *Analytical Chemistry*, 209–217. Oxford, UK: Bios Scientific Publishers Ltd.

22 Wikipedia contributors (2020). Nuclear magnetic resonance spectroscopy. Wikipedia, The Free Encyclopedia, 20:38 UTC. https://en.wikipedia.org/w/index.php?title=Nuclear_magnetic_resonance_spectroscopy&oldid=967870890 (accessed 7 August 2020).

23 Kealey, D. and Haines, P.J. (2002). *Analytical Chemistry*, 248–269. Oxford, UK: Bios Scientific Publishers Ltd.

24 Kealey, D. and Haines, P.J. (2002). *Analytical Chemistry*, 214–217. Oxford, UK: Bios Scientific Publishers Ltd.

25 Nema, S. and Ludwig, J.D. (ed.) (2010). *Pharmaceutical Dosage Forms: sParenteral Medications, Volume 3: Regulations, Validation, and the Future*, 58–67. New York, NY: Informa Healthcare.

26 USP <905> (2011). Uniformity of dosage forms. *USP 39–NF 34*. Rockville, MD: United States Pharmacopeial Convention.

27 Food and Drug Administration (2015). Q3D elemental impurities. Rockville, MD: Food and Drug Administration. https://www.fda.gov/media/135956/download (accessed 15 August 2020).

28 Barber, T.A. (2000). *Control of Particulate Matter Contamination in Healthcare Manufacturing, Visual Inspection of Injectables and Devices.* Denver, CO: Interpharm Press.

29 Forcino, H. Trends and best practices in visual inspection. *Pharmaceutical Technology* 38 (3): 42–47.

30 USP <1> (2011). Injections, foreign and particulate matter. *USP 39–NF 34*. Rockville, MD: United States Pharmacopeial Convention.

31 USP <790> (2011). Visible particulate matter in injections. *USP 39–NF 34*. Rockville, MD: United States Pharmacopeial Convention.

32 USP <788> (2011). Particulate matter in injections. *USP 39–NF 34*. Rockville, MD: United States Pharmacopeial Convention.

33 USP <1790> (2011). Visual inspection in injections. *USP 39–NF 34*. Rockville, MD: United States Pharmacopeial Convention.

34 Akers, M.J., Larrimore, D.S., and Guazzo, D.M. (2002). Particulate matter testing. In: *Parenteral Quality Control: Sterility, Pyrogen, Particulate, and Package Integrity Testing*, 3ee, 221. Marcel Dekker Inc.

35 Food and Drug Association (1993). Guide to inspections of microbiological pharmaceutical quality control labs. https://www.fda.gov/

inspections-compliance-enforcement-and-criminal-investigations/inspection-guides/microbiological-pharmaceutical-quality-control-labs-793 (accessed 15 August 2020).
36 USP <81> (2011). Microbial limits. *USP 39–NF 34*. Rockville, MD: United States Pharmacopeial Convention.
37 USP <71> (2011). Sterility. *USP 39–NF 34*. Rockville, MD: United States Pharmacopeial Convention.
38 USP <85> (2011). Bacterial endotoxins. *USP 39–NF 34*. Rockville, MD: United States Pharmacopeial Convention.
39 Moser, C.L. and Meyer, B.K. Comparison of compendial antimicrobial effectiveness tests: a review. *AAPS PharmSciTech* 12 (1): doi: 10.1208/s12249-010-9575-9.
40 USP <51> (2011). Antimicrobial effectiveness testing. *USP 39–NF 34*. Rockville, MD: United States Pharmacopeial Convention

4

Control Strategies for Pharmaceutical Development

Judy Lin

Bristol-Myers Squibb, Princeton, NJ, USA

TABLE OF CONTENTS

Analytical Testing for the Pharmaceutical GMP Laboratory, First Edition. Edited by Kim Huynh-Ba.

4.1 Introduction

The quality of pharmaceutical products is critical since it has a direct effect on patient welfare. The entire pharmaceutical development process and the quality assurance activities in the manufacturing of commercial products are an endeavor to ensure the quality, safety, and efficacy of the drug product. In the twenty-first century, the pharmaceutical industry has started to develop a systematic approach to ensure the quality of pharmaceutical products throughout the entire lifecycle, including development, commercial launch, and postapproval changes. The pharmaceutical product quality is not only guaranteed by testing the finished product but also more importantly, it is secured by an integrated quality system. In addition to the implementation of the Current Good Manufacturing Practice (CGMP), the joint quality system includes establishing an effective control strategy of product quality through its lifecycle based on analysis, monitoring, and mitigation of potential risks. This chapter will provide an overview of the control strategies and risk management, as well as commonly used tools for risk assessment.

4.2 Quality-by-Design Concept

The objective of pharmaceutical development is to design a quality drug product with a manufacturing process to consistently meet consumer's needs by delivering drug products with consistent quality and efficacy [1]. Quality-by-design (QbD) concept helps in ensuring this outcome by starting with predefined objectives and emphasizing product and process understanding and process control based on risk management. The quality target product profile (QTPP) forms the basis of all product and process development activities. The following elements are considered during the establishment of QTPP:

- Intended use in a clinical setting, route of administration, dosage form, and delivery systems
- Dosage strength(s)

- Container closure system
- Therapeutic moiety release or delivery and attributes affecting pharmacokinetic characteristics
- Drug product purity, stability, and microbial attributes

Based on the QTPP, critical quality attributes (CQAs) can be determined for the drug product. A CQA is a physical, chemical, biological, or microbial property that should be within a suitable limit, range, or distribution to ensure the product quality. To illustrate this, a hypothetical example of an immediate release oral tablet, manufactured by a wet granulation process, is provided in Sections 4.2–4.5. While some CQAs are common for all dosage forms, dosage-form-specific CQAs should be considered, such as sterility for parenterals. During the progression of product development, the list of CQAs should be reassessed and modified based on the increased understanding of the product and manufacturing process. Table 4.1 lists some typical CQAs of this hypothetical drug product.

Input material attributes and process parameters that can impact the CQAs should be identified at the beginning of development. A conventional approach to linking the material attributes and process parameters to CQAs is the risk assessment. From the risk assessment, studies are designed and performed to identify design space and control space of material attributes and process parameters so that the risks are mitigated and reduced. A comprehensive control strategy is then formulated to ensure that the residual risks are well controlled, and the product quality is consistent. During the product lifecycle, process performance is continuously monitored to ensure that it delivers a drug product with consistent quality. Figure 4.1 illustrates the QbD concept [2, 3].

Table 4.1 Typical CQA of an immediate release oral solid dosage form (tablets).

Critical quality attribute (CQA)	Impact of quality target product profile (QTPP)
Potency	Dosage strength
Impurity profile	Dosage strength; product stability
Dissolution or disintegration	Drug release and bioavailability
Uniformity of dosage units	Consistency in dosing
Water content	Product stability and not to promote microbial growth
Hardness	May impact drug release and tablet appearance
Friability	Tablet appearance
Microbial limit	Ensure no microbial contamination

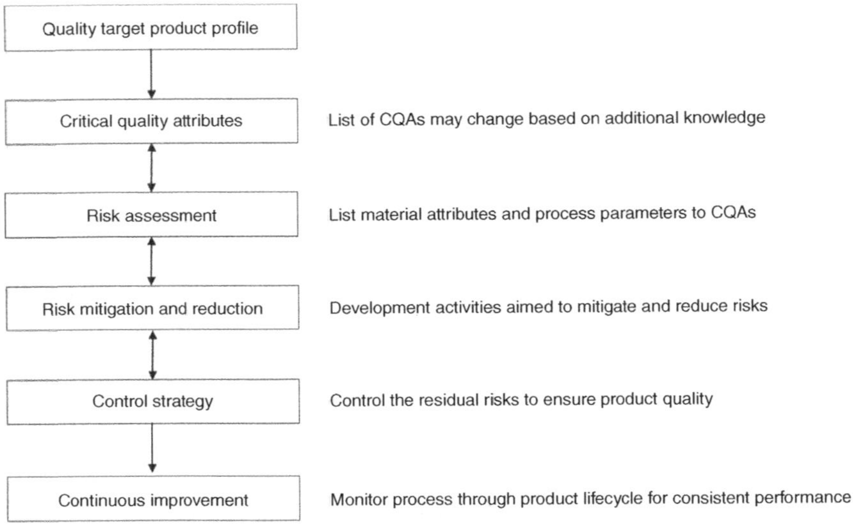

Figure 4.1 Quality-by-design (QbD) concept. *Sources:* Adapted from Food and Drug Administration [2]; Woodcock [3].

4.3 Risk Management

Many potential risk factors can negatively impact the quality of the pharmaceutical product in various degrees. It is unrealistic to eliminate all the risk factors due to multiple factors, such as the physicochemical properties of the active pharmaceutical ingredient (API) and limited large-scale commercial manufacturing experience. Risk management provides a systematic approach to assessing, mitigating, controlling, and monitoring the risks throughout the lifecycle of the pharmaceutical product, from development to commercial manufacturing and post-marketing. Figure 4.2 illustrates the essential elements of the risk management process and their corresponding development stages [2, 3].

4.3.1 Risk Assessment

Risk assessment is the process of identifying risk factors, understanding the probability of the risk, and assessing the risk severity.

4.3.1.1 Risk Identification

In identifying risk factors, material attributes and process parameters link to the product CQAs. A commonly used tool in risk identification is the cause-and-effect

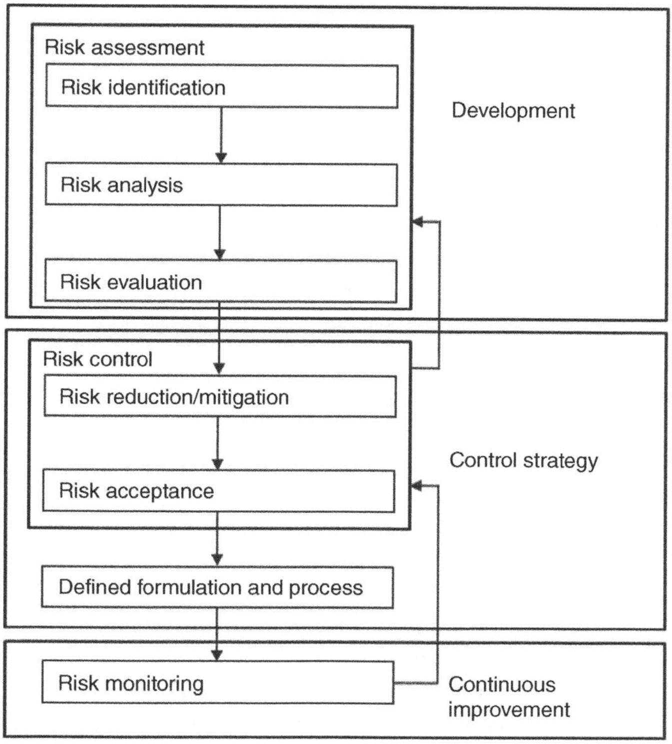

Figure 4.2 Basic elements of the risk management process. *Sources:* Adapted from Food and Drug Administration [2]; Woodcock [3].

diagram, also referred to as the Ishikawa diagram or fishbone diagram as shown in Figure 4.3.

In Figure 4.3, risks of the tablet manufacturing process that can impact the uniformity of dosage, one of the CQAs, are identified. Each unit operation in the manufacturing process is evaluated to determine all the potential risk factors in the assessment regardless of its probability or the severity of its consequence [4–6].

Example The unit operation of raw material addition has an impact on the CQA of uniformity of dosage. The API is the first material added to the container; it may disproportionally be concentrated at the bottom of the container and stick to the container wall, which will negatively impact the uniformity of dosage. Additionally, when the drug loading of the drug product is low and the amount of API is relatively minor compared to the excipients, the method of geometric

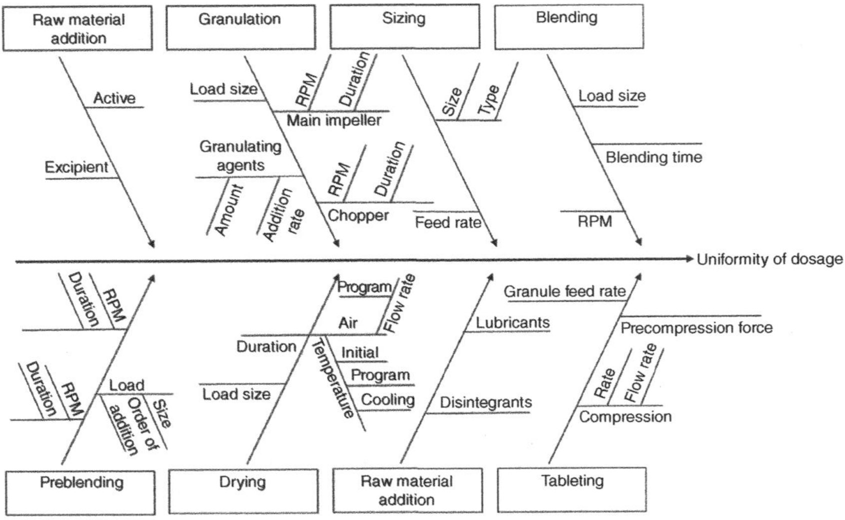

Figure 4.3 Example Ishikawa diagram of tablet manufacturing process.

mixing will dramatically improve the homogeneity of the mixture, thereby having a positive impact on the CQA of uniformity of dosage.

Another example is the unit operation of blending. For this unit operation, load size, mixing time, and speed all have a direct impact on the homogeneity of the mixture. To optimize the unit operation of blending, all these parameters need to be within a specific range to have a uniformed blend at the end. In some instances, when the setting of blending time has exceeded a limit, there could be a phenomenon of de-mixing occurring due to the segregation of different materials with different particle sizes.

4.3.1.2 Risk Analysis

All the identified risk factors are further analyzed to determine strategies for mitigating and controlling the risks. The risk assessment captures the risk severity, probability, and detectability. Failure mode effects analysis (FMEA) is the most common analysis for risk assessment. For each risk factor, scores are assigned to risk severity, probability, and detectability. Most of the methods include scoring all three parameters; however, some methods may separate the detectability from the other two parameters. Based on the values, the risk factors are categorized as low, medium, and high. Low-risk factors are acceptable due to their low probability of happening and minor impact. The high-risk factors caused by their high probability of happening, and the high result must be reduced to a low or medium risk category. For the medium-risk

Table 4.2 Example of an FMEA analysis.

	Severity			
Probability of occurrence	**Irrelevant (5)**	**Moderate (10)**	**Critical (20)**	**Disastrous (40)**
High (5)	25	50	100	200
Probable (4)	20	40	80	160
Moderate (3)	15	30	60	120
Low (2)	10	20	40	80
Remote (1)	5	10	20	40
High risk				
Medium risk				
Low risk				

factors, additional efforts are needed to reduce the risk. However, the risk factors in this category may be acceptable if a strategy can be developed and implemented to control the risks so that product quality can be ensured. In addition to the analysis of probability and severity above, the detectability of the defects caused by the risk factors is an essential element in determining whether a risk is acceptable. If the defect is readily detectable or even better, can be corrected during the manufacturing process, the risk is controllable and is acceptable. In-process control (IPC) tests can be implemented to control/mitigate the risk during the manufacturing process. Appropriate process intervention may be performed based on the IPC results. IPC testing is an essential element of the control strategy. With the control strategy in place, the medium risks are reduced to low risks. Table 4.2 is an example of the FMEA analysis.

FMEA analysis should be repeated at different development stages to capture a further understanding of the product and process. Usually, the study is performed at the inception of the development, after the design of the prototype formulation and process, after the finalization of formulation and process, before process validation, and after any significant change in the formulation and process.

Example Table 4.3 below is an illustration of an initial FMEA analysis on the unit operation of blending for the manufacture of tablets in the hypothetical case.

The risk assessment shows that these studies determine the applicable ranges and optimal conditions of the process parameters identified in Table 4.3 with blend uniformity as the quality attribute. After defining the ranges and optimal conditions, the risk assessment is verified, as illustrated in Table 4.4.

Table 4.3 Initial FMEA analysis on blending.

	Severity	Probability	Risk score	Risk category
Bin fill level too high	20	3	60	High
Blending time too short	20	3	60	High
Blending time too long	20	3	60	High
Bin rotation too fast	20	3	60	High

Table 4.4 Final FMEA analysis on blending.

	Severity	Probability	Risk score	Risk category	Control
Bin fill level too high	20	1	20	Low	Set bin fill level
Blending time too short	20	2	40	Medium	BU test
Blending time too long	20	2	40	Medium	BU test
Bin rotation too fast	20	1	20	Low	Set bin rotation speed

The risks of the process parameters are reduced to medium and low after the control, and the residual risks are acceptable with the traditional control strategy, such as process parameters settings and in-process tests.

Qualitative risk assessment is also a critical approach to the risk assessment. Below is an example of assessing the critical material attributes (CMAs) that start with the evaluation of the drug substance.

Example Based on the hypothetical case study of the immediate-release tablet above, the physicochemical properties of the drug substance should be assessed in correlation with the CQAs of the drug product. Based on the results of the assessment, an initial risk assessment on drug substance attributes can be performed. Table 4.5 is an example of the initial risk assessment on drug substance attributes and the CQAs of the drug product (DP).

From this table, several CMAs of drug substances are identified to be of high risk, such as particle size and solubility on dissolution and flow property on the uniformity of the dosage unit. In this example, assuming the flowability of the drug substance is poor, a wet granulation process is adapted to formulate and manufacture the drug product to mitigate this risk to content uniformity. A forced degradation study on the drug substance is often performed to determine the physicochemical stability so that decisions can be made on the selection of

Table 4.5 Example of risk assessment on drug substance attributes.

	Critical material attributes (CMAs) of drug substance								
DP-CQA	**Solid-state form**	**Particle size**	**Solubility**	**Moisture content**	**Hygroscopicity**	**Residual solvent**	**Process impurities**	**Chemical stability**	**Flow property**
Physical attributes	Low	Low	Low	Low	Low	Low	Low	Low	Low
Assay	Low	Low	Low	Low	Low	Low	Low	High	Low
Related compounds	Low	Low	Low	Low	Low	Low	Low	High	Low
Uniformity of dosage unit	Low	Low	Low	Low	Low	Low	Low	Low	High
Dissolution	Low	High	High	Low	Low	Low	Low	Low	Low

excipients and container closure systems, among others. For example, a drug substance that is susceptible to oxidation should avoid excipients with high levels of oxidizing agents, such as peroxides, and an anti-oxidizing agent may be included in the formulation. Also, if a drug substance is prone to hydrolysis, then some desiccant should be used in the container closure system.

Additionally, if a drug substance has poor solubility, then the physical property of particle size distribution (PSD) should be controlled to mitigate the risk to the CQA of dissolution. To establish proper limits on PSD, batches of a drug substance with various PSDs should be evaluated during the pharmaceutical development process to determine the range of PSD suitable for the drug product via *in vivo* bioavailability and bioequivalence studies. Once appropriate measures, such as wet granulation process, have been adopted, the risk assessment of the drug substance should be repeated to demonstrate that high risks have been reduced to low or medium.

4.3.2 Risk Control

Efforts should be made to mitigate and control risks throughout the process of pharmaceutical development and commercial manufacturing, including formulation and process design, process parameter selection, in-process and finished product control, and formulation and process optimization. An example of risk reduction via formulation design is the excipient selection. During the process of excipient selection, the excipient-drug substance compatibility study, commonly referred to as excipient compatibility study, provides the information on the chemical interactions between the drug substance and different excipients.

To illustrate how an excipient compatibility study is typically performed, the same hypothetical example of the oral solid dosage form of tablets will be used.

Example The test samples are prepared by blending and placing the following materials (typically powders) into the respective vials in duplicates:

- Drug substance (single powder)
- Drug substance + each of the excipients (binary powder blends)
- Drug substance + all excipients (optional, blend of multiple powders)

For the duplicate samples, one set is placed into a stability chamber under 40 °C/75 RH and the other under 25 °C/60 RH as a control. After a period of time, usually one month, both sets of samples are analyzed to determine if there is any significant degradation of the drug substance in the presence of an excipient as compared to the drug substance without the excipient. Higher temperatures, such as 50 °C and in an open container, can be used to accelerate the study. Samples can be withdrawn at various time points from one week to one month, to assess the stability of the drug substance in the mixtures.

For the development of generic drug products, one conventional approach is to select the same excipients as the reference listed drug (RLD), which has been thoroughly evaluated by the RLD manufacturer to be compatible. However, an excipient compatibility study is still needed when developing a generic drug product to demonstrate compatibility as part of the pharmaceutical development process.

By choosing the excipients that have no or limited chemical interaction with the API (i.e. compatible with the API), the risk of product degradation during manufacturing and storage in shelf life is reduced or eliminated. Likewise, a well-designed manufacturing process can also reduce the risk to product quality. For example, for a tablet formulation with a tendency of lumping, adding a sieving and delumping step in the manufacturing process can help improve the blend uniformity and the subsequent uniformity of dosage units.

After the excipients have been selected, to identify those with high risks, a risk assessment should be performed to determine the risk factors for each excipient to the CQAs of the drug product. Going back to the same case study of the immediate-release tablet above, the level of a lubricant magnesium stearate usually has a high risk on the CQA of dissolution. Also, the levels of binding agent and disintegrant have high risks on the CQAs of disintegration and dissolution. Appropriate studies should be planned and performed to discover the design space and control space to mitigate these risks. Refer to Chapters 5 and 6 on how to design such studies and interpret the data; refer to Figure 4.4 below on the concept of design space and control space. Once the studies are completed, with the control and design space determined, the risk assessment can be repeated to confirm the risks are under control.

The selection of process parameters can mitigate and reduce the risks to product quality [7]. For example, the load capacity, rotational speed, and the rotation

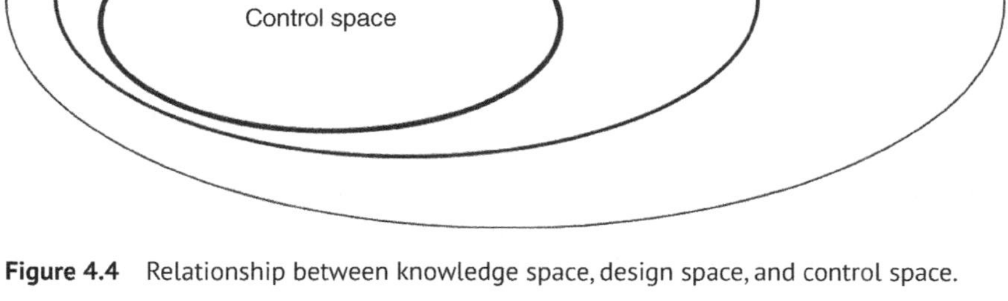

Figure 4.4 Relationship between knowledge space, design space, and control space.

number of the blender in a tablet manufacturing process can impact the blend uniformity and, consequently, the CQA of uniformity of dosage units. Different methodologies can be applied to the unit operation of blending to improve uniformity, such as geometric blending and adding the API in between bulk excipients. Another example is wet granulation, where chopper speed, blade speed, granulation agent level, and speed of addition can impact the CQA of dissolution and disintegration. Process characterization studies are performed to understand the potential impact on product quality from individual process parameters as well as the interaction of multiple process parameters. Design of experiments (DoE) with full factorial or partial factorial design is a commonly used statistical tool for process characterization and is discussed in Section 4.5.

Furthermore, when the process is being scaled up from pilot to production, the critical process parameters (CPPs) evaluated during the pilot-scale should be revisited. For example, load capacity, rotational speed, and the rotation number of the blender for the unit operation of blending, chopper speed, blade speed, and granulation agent level/addition speed for the unit operation of wet granulation. With the knowledge accumulated through the various development phases, including CQAs and its relevant material attributes and process parameters, a control strategy is formulated to ensure products with consistent quality can be delivered from commercial manufacturing to the patient. The main components of a control strategy are:

- Critical material attributes, including drug substance (API) and excipients
- CPP
- In-process monitoring and testing
- Finished product release testing

In-process and finished product release testing, as well as the subsequent stability program, are essential elements of the control strategy. They provide the confirmation of the process performance and the satisfactory product quality at release and throughout the shelf life [6, 8, 9].

The application of Process Analytical Technologies (PAT), a mechanism to design, analyze, and control pharmaceutical manufacturing processes through real-time measurement of quality attributes, has become more prevalent in the manufacturing of drug products. PAT tools provide a real-time picture of how process parameters can impact quality attributes, thus, providing a better understanding of the process [5, 10]. In some cases, a feedback loop can be established to adjust the process parameters in real-time based on the quality attributes results. The implementation of PAT reduces the risk of over-processing, enhances consistency, and minimizes rejects. The analytical technology of near-infrared (NIR) is often deployed to monitor blend uniformity during the blending and moisture level during the fluid bed drying [11–13].

The control strategy needs to be reviewed periodically for continuous improvements in the process performance and product quality throughout the product lifecycle, which is a crucial element of the QbD concept and the risk management paradigm [1, 7]. According to the Food and Drug Administration (FDA) Guidance for Industry ICH Q8: Pharmaceutical Development, the design space is defined as the multidimensional combination and interaction of input variables (e.g. material attributes) and process parameters that have been demonstrated to assure product quality [1]. Working within the design space is not considered a change; only changes out of the design space require a regulatory postapproval change process. The applicant proposes the design space, and it is subject to regulatory assessment and approval. The control space is within the design space and narrower and represents the optimal conditions of the process. The knowledge space encompasses design space; it represents the general knowledge and understanding of the conditions for process parameters.

Figure 4.4 illustrates the relationships between knowledge space, design space, and control space established during the development process.

4.4 Establishing Specifications

4.4.1 What Is the Specification?

The DP is a finished dosage form, such as a tablet, capsule, or solution, which contains a drug substance, generally, in association with one or more other ingredients. The drug substance (DS) is an active ingredient that is intended to furnish pharmacological activity or another direct effect in the diagnosis, cure, mitigation, treatment, or prevention of disease or to affect the structure or any function of the human body but does not include intermediates used in the synthesis of such ingredient. In many publications, the drug substance is also called the API.

The specification is the quality standard to confirm the quality of drug substances, drug products, intermediates, raw materials, reagents, components, in-process materials, container closure systems, and other materials used in the manufacture of a drug substance or drug product. Each specification must include a list of required tests, references to analytical procedures, and appropriate acceptance criteria. According to the International Conference on Harmonisation (ICH), ICH Q6A, and Q6B, the acceptance criteria are the numerical limits, ranges, or other measures for the tests describing the critical attributes of the drug products submitted and justified by the manufacturer and approved by the regulatory agencies. Therefore, the set of specifications is an integral part of a total control strategy to ensure drug product quality and consistency [14, 15]. For drug products and drug substances, the FDA has provided definitions of specifications in Title

21 CFR Section 314.3, while other regulatory agencies, such as the European Medicines Agency (EMA), require a two-tier specification system, which is release and shelf-life specification [16]. The concept of different acceptance criteria for release vs. shelf-life specifications applies to drug products only. The purpose of having separate specifications for release and stability is to place a more stringent acceptable range when the drug product is released so that the levels of assay and impurity remain within the acceptance criteria to the end of the shelf life of a drug product and to allow for an acceptable gradual change in potency and impurity levels.

Synthetic or semi-synthetic drugs of low molecular weight (small molecule drugs) have significantly different manufacturing processes and characterization techniques compared with those of proteins, peptides, or their derivatives (biological drugs). Therefore, the typical tests included in the specifications of these two categories of drugs are discussed separately in this section. More information on this topic can be found in Chapter 3.

4.4.2 Typical Tests Included in the Specification of a Small Molecule Drug

4.4.2.1 Drug Substance

4.4.2.1.1 General Tests The following tests can be found in the specification of a drug substance:

Description/Appearance: A qualitative statement on the physical appearance of the material usually based on visual observation, for example, white to off-white powder with lumps.

Identification: The identification test should be specific to the drug substance and able to discriminate between compounds with similar structures. Typical techniques used in this test include infrared spectroscopy (IR), liquid chromatography/ultraviolet (LC/UV) diode array, liquid chromatography/mass spectrometry (LC/MS), or gas chromatography/mass spectrometry (GC/MS). An identification test solely based on the retention time of one chromatographic method is not sufficient. The FDA requires two tests of orthogonal techniques used in combination for identification.

Assay: The assay test is used to determine the purity of the drug substance. It is preferred to use a specific and stability-indicating method for the analysis such as LC/UV or gas chromatography/flame ionization detection (GC/FID). If a nonspecific method is used, such as titration, it should be combined with a suitable test for impurities for assay value reporting.

Impurities: This category contains many types of impurities: organic impurities, inorganic impurities, and residual solvents [17].

The *organic impurities* are process-related, raw materials, or packaging-related components, such as starting material, by-product, intermediates, reagents, and degradation products during storage. The impurities can be structurally identified or unidentified. Organic impurities are usually quantitated using chromatographic methods, such as LC/UV, LC/MS, GC/FID, or GC/MS [18, 19]. For impurities that are DNA reactive, or the so-called genotoxic impurities, there is a separate framework defined by regulatory authorities applying to the identification, categorization, qualification, and control of these mutagenic impurities to limit potential carcinogenic risk. Limits for genotoxic impurities are commonly set at ppm (parts per million) levels depending on the threshold of toxicological concern (TTC). Very often, a separate specific method needs to be developed and validated to analyze a genotoxic impurity, which is usually a chromatographic method such as LC/MS [17, 20, 21].

Inorganic impurities can be process-related or found in raw materials. They include reagents, heavy metals, and inorganic salts, etc., and are usually known. US Pharmacopeia (USP) test methods have been used for the detection and quantitation of inorganic impurities, such as USP General Chapter <731> *Loss on Drying*, USP <281> *Residue on Ignition*, and inductively coupled plasma-optical emission spectrometry/inductively coupled plasma-mass spectrometry (ICP-OES/ICP-MS) in USP <233> *Elemental Impurities – Procedures* [22–24].

Residual solvents are aqueous or organic liquids used during the manufacturing process. Because some of the solvents used in the process can pose safety concerns, residual solvents in the drug substance should be controlled to ensure safety and quality consistency. Residual solvents are frequently detected and quantitated using GC methods. In some cases, LC or titration methods are utilized [25].

4.4.2.1.2 Specific Tests The following tests may be included in the specification on a case-by-case basis. These tests are determined by the test's effectiveness and appropriateness to examine the quality of the drug substance for its intended use.

Physicochemical properties: The properties include pH in a solution, melting point, optical rotation, and refractive index. USP test methods are usually used for the tests [26–29].

PSD: PSD of the drug substance may have a significant impact on the bioavailability of the drug and thus the content uniformity of dosage units for low-drug loading products. The most common technique is particle size analysis based on laser light diffraction [30].

Polymorphic forms: Drug substances may exist in different crystalline forms. Polymorphism also includes solvation and amorphous forms. Different polymorphic forms may exhibit different physical properties and can potentially impact the processability, bioavailability, or chemical stability of the drug substance. The techniques commonly used to characterize the polymorphic forms are X-ray powder diffraction, thermal analysis (DSC, TGA, and DTA), Raman spectroscopy, and solid-state NMR, or a combination of these techniques [31].

Chirality: When a drug substance is a chiral compound and is developed as a single enantiomer, control of the undesirable enantiomer should be demonstrated. LC, supercritical fluid chromatography (SFC), and capillary electrophoresis (CE) are commonly used techniques for chiral purity determination [32].

Water content: Water content determination is important if the compound is prone to hydrolytic degradation or is hygroscopic. Water content is usually determined using Karl Fischer titration [33].

Microbial Limits: Depending on the manufacturing process and the intended use of the material, the microbial limit test may be needed. For example, if the drug substance will be used in formulating injectable drug products, the microbial limit test is necessary. The testing methods are included in the USP compendium [34, 35].

4.4.2.2 Drug Product

4.4.2.2.1 General Tests The following tests are typically included in the specification of drug products, and the techniques are similar to the ones used for the drug substance.

Description/Appearance: A qualitative statement of the physical appearance of the product usually based on visual observations, such as white film-coated oval tablet with “10” debossed on one side and a plain face on the other side.

Identification: The identification test should be specific to the active ingredient in the drug product and be able to discriminate between compounds with similar structures. Typical techniques used in this test are, infrared spectroscopy (IR), LC/UV photodiode array, LC/MS, or GC/MS. An identification test solely based on the retention time of one chromatographic method is not sufficient.

Assay: The assay test is used to determine the dose strength or the potency of the finished product. It is preferred to use a specific and stability-indicating method for the analysis (e.g. LC/UV or GC/FID). If a nonspecific method is used, such as titration, it should be combined with a suitable test for impurities for assay value reporting. Often, the same method used for drug substance analysis can be used for the drug product analysis with some modifications in the sample preparation procedure.

Impurities: Organic impurities, inorganic impurities, and residual solvents are included in this category. Organic impurities in drug products are usually from process impurities from the API manufacturing process, degradation of the API during the manufacturing of the product, and potential degradation of the product in the shelf life. Acceptance limits are set to control the individual impurities and total impurities in the product at release and through its shelf life. Organic impurities are usually quantitated using chromatographic methods such as LC/UV, LC/MS, GC/FID, or GC/MS. As with the assay test, the method used for the API impurity analysis can often be adapted for the drug product impurity analysis. Similar to the specification of the drug substance above, the same principles on the control of genotoxic impurities apply to drug products when there is a genotoxic impurity as a potential degradation product of the drug product. Unless there is additional solvent introduced during the product manufacturing process, residual solvents and inorganic impurities in the drug product is usually controlled by setting appropriate specifications for the API and excipients, therefore they are not monitored in the drug product.

4.4.2.2.2 Specific Tests The specific tests required for the drug product depend on the dosage form. This section is focused on the most common dosage forms, which include oral solid dosage form (e.g. tablets and hard gelatin capsules), oral liquids (oral liquids and powder are intended to be reconstituted as oral liquids), and injectables.

4.4.2.2.3 Oral Solid Dosage Form Dissolution: This test measures the release rate of the drug substance (API) from the drug product. The release rate can impact the bioavailability of the active ingredient, therefore, affecting the efficacy and safety of the drug product. The dissolution methodology is discussed in detail in Chapter 7.

Disintegration: The disintegration test may be used instead of a dissolution test when the API in the drug product is highly soluble across the physiological range (pH 1.2–pH 6.8). Often, the test results can be correlated with the dissolution results. See Chapter 3 for detailed discussions on disintegration testing.

Hardness/Friability: These two tests are performed to ensure tablet integrity during the subsequent processing steps after compression, such as coating and packaging, as well as the distribution of finished products. Hardness and friability are usually performed as in-process tests. The hardness test is used to measure the force that is needed to cause the breakage of a tablet. The official terminology of this test in USP <1217> is the tablet breaking force, which is the force required to cause tablets to fail (i.e. break) in a specific plane [36]. If the tablet is too hard, it may impact the disintegration or dissolution, consequently, it may affect the bioavailability of the drug. Conversely, if the tablet is too soft, tablet breakage may be seen in subsequent process steps causing inefficiency in the manufacturing process or patient compliance with medication dosing.

The Friability study is a test used to measure the physical strength of uncoated tablets to determine their resistance to chipping and surface abrasion by tumbling them in a rotating cylinder. The percentage weight loss after tumbling is recorded as the friability of the tablets. The equipment configuration and testing parameters of the friability test is described in detail in USP <1217>, and the information is harmonized between the European Pharmacopeia (Ph. Eur) and the Japanese Pharmacopeia (JP) [36].

Uniformity of dosage units: This test ensures that the active ingredient contained in each dosage unit is consistent. USP <905> includes detailed information about the test, including the calculations of results and acceptance criteria. The information in <905> is harmonized with Ph. Eur. and JP [37].

Water Content: This test is performed when the drug product quality can be affected by the level of water content. Karl Fischer titration is the most used technique, which is also a pharmacopeia method (USP <921>) [33].

Microbial limit: For an oral drug product, the microbial limit test is usually included in the specification with the limits on the total count of aerobic microorganisms, the total count of yeasts and mold. Depending on the patient population, the absence of specific objectionable bacteria, such as *Escherichia coli* and Salmonella, may need to be demonstrated. The testing method is detailed in USP <61> and <62>. If the composition, moisture content, and primary packaging do not allow micro growth, then this test may be eliminated, or the testing frequency may be reduced with justification [34, 35].

More information on different analytical testing used in the pharmaceutical lab is discussed in Chapter 3.

4.4.2.2.4 Oral Liquid Dosage Form Uniformity of dosage units: Similar to solid oral dosage, this test is to ensure the patient can receive a consistent amount of the active ingredient from each dose. If the homogeneity of the oral liquid can be demonstrated, this test is not necessary as a routine release test. In this case, the container volume of the liquid may be used to assure the uniformity of the drug product.

pH: Acceptance criteria for pH is necessary when it is applicable. The proposed range should be justified based on development knowledge, batch analysis, and stability data.

Microbial limits testing (MLT): This is usually a required release test for oral liquids.

Antimicrobial preservative content: Antimicrobial preservative is added to the formulation to ensure there is no microbial growth during the storage and use period of the drug product. Because the antimicrobial preservative level may drop during long-term storage and the use period, it should be demonstrated that the preservative is still effective after storage. Antimicrobial effectiveness test (AET) is used to demonstrate the effectiveness of the preservative. USP <51> describes the test in detail [38].

Antioxidant preservative content: Testing for antioxidant content should normally be done at the release stage.

Extractables: Extractables are impurities that can potentially migrate from the container closure system to the drug product due to direct contact with the product. In general, the waiving of this test can be granted where extractables from the container closure systems are demonstrated to be consistently below the acceptable safe levels as indicated by development knowledge and stability data.

Alcohol content: If there is alcohol in the formulation, its content needs to be disclosed on the product label according to pertinent regulations.

Dissolution: This test may be appropriate if the active ingredient is insoluble in the product matrix. USP <711> includes information on the dissolution test for oral liquids. More information can be found in Chapter 7 [39].

PSD: It may be appropriate for oral suspensions to include quantitative acceptance criteria and a procedure for the determination of PSD. Developmental data determine the need for either a dissolution procedure or a PSD procedure for these formulations.

Redispersibility: If the oral liquid, such as oral suspensions, can settle on storage, a dispersibility test is needed to ensure the product can be redispersed appropriately, and the correct dose strength of the product can be dispensed to the patient.

Viscosity: Testing for viscosity is usually needed for relatively viscous solutions (e.g. syrup) or suspensions. This test can be performed with a reduced frequency (skip lot) if justified [40].

Reconstitution time: For a dry powder that is intended for reconstitution to form an oral liquid, reconstitution time is evaluated to ensure that the product performs as intended.

Water content: This test is usually required for dry powder intended for reconstitution or oral liquid formulation that is not water-based.

4.4.2.2.5 Injectable Uniformity of dosage units: Extractable volume testing is usually done to ensure uniformity of the units.

pH: This is a required test for injectables. The proper pH range is based on the physiological requirement from the route of administration and product stability.

Osmolality: Osmolality is an important quality attribute of injectables. It is the concentration of all particles in a fluid. For examples, a thick syrup would have a much higher osmolality than a cup of water with just a teaspoon of sugar. For a particular formulation, its osmolality is consistent; therefore, an osmolality test can verify the correct formulation [41].

Sterility: All injectables must be sterile. USP <71> has a detailed description of the methodology, validation consideration, and requirements of the sterility test. For terminally sterilized drug products, the parametric release may be proposed when justified by data generated during development and validation [42].

Endotoxins/Pyrogens: It is important to control the endotoxins level since it can cause a severe immune response from the patient. USP <85> describes different methodologies and method validation consideration. Ph. Eur. also provides acceptable endotoxin levels in drug products for different routes of administration [43].

Particulate matter: Injectables should have appropriate acceptance criteria for a particulate matter based on the route of administration. Particulate matter includes both visible particle and sub-visible particulates.

Functionality testing of delivery systems: Injectables packaged in prefilled syringes, autoinjector cartridges, or the equivalent should have test procedures and acceptance criteria on the functionality of the delivery system. These may include control of syringeability, pressure and seal integrity (leakage), and parameters such as tip cap removal force, piston release force, piston travel force, and power injector function force. If justified, the tests can be performed as in-process tests.

Other tests, such as water content, antimicrobial preservative content, antioxidant preservative, extractables, reconstitution time, should be considered similar to an oral liquid.

4.4.3 Typical Tests Included in the Specification of Biological Drugs

4.4.3.1 Drug Substance

Physical appearance test: A qualitative statement of the physical appearance of the material is usually based on visual observation or instrument measurements. For most biological drugs, the drug substance is in liquid form; therefore, the color and clarity of the solution are two important characteristics of the appearance/description test.

Identity: The identity of the material should be specific and usually requires a combination of multiple tests, such as peptide mapping, sulfhydryl groups and disulfide bridges, carbohydrate structure, biological activity assessment, and ELISA.

Purity and impurities: The purity of a biological drug is method dependent, requiring a combination of test methods to characterize the drug. The impurities can be process-related, product-related, or from raw materials. Process-related impurities typically include cell culture media, host cell protein, DNA, monoclonal antibodies, solvents, and buffer components. Product-related impurities are molecular variants with properties that differ from those of the desired product, such as truncated forms, deamidated forms, and aggregates. The techniques used in characterizing the impurities include ion exchange chromatography (IEC), size exclusion chromatography (SEC), sodium dodecyl sulfate-polyacrylamide gel electrophoresis (SDS-PAGE), capillary electrophoresis-sodium-dodecyl sulfate (CE-SDS), capillary isoelectric focusing (cIEF), and peptide mapping.

Potency: Testing for potency is a quantitative measure of biological activity based on the attribute of the product that is linked to the relevant biological properties. If an appropriate potency measurement is performed at the product stage, the alternative physicochemical method may be sufficient.

Quantity: Quantity is a measurement of the protein content. The commonly used technique of this analysis is quantitative UV measurement. If the correlation of the results of quantity and the biological assay is established, quantity measurement can be used instead of the biological assay.

4.4.3.2 Drug Product

All the tests listed above under the drug substance section should be included in the drug product specification. Other tests related to the specific dosage forms, such as pH and osmolality, may be included. Microbiological properties of the drug product should be controlled, such as sterility and bacterial endotoxin test (BET).

4.4.4 Considerations of Setting Acceptance Criteria

The specification (or acceptance criteria) is a quality standard for the drug substance and drug product. The acceptance criteria are set based on the drug product's efficacy and safety and demonstrate the consistency of raw material qualities and manufacturing processes. Several aspects as following are often considered when setting the appropriate acceptance criteria.

4.4.4.1 Process Capability

Process capability is a key factor to consider when setting the specification acceptance criteria. For example, the limit of an individual impurity can be set at the level qualified in toxicology studies. If the final process for the commercial product could consistently produce batches with a lower level of impurity, its acceptance criteria in the specification should be set at a lower level than the toxicology qualified limit (i.e. the specification limit is set based on the process capability). Under this principle, it is possible to set tighter acceptance criteria than ICH Q3A and Q3B. This principle is because the release data should be within a much narrower range than shelf-life acceptance criteria when process parameters are within normal operating ranges. When release results fall outside the typical range, it may indicate a major change in the quality of the raw materials or deviation outside of the process operation range. However, for generic drug products, a different principle applies that allows the limit of an individual impurity to be set as per ICH Q3A and Q3B guidelines, unless the limit is listed in the corresponding USP monographs.

4.4.4.2 Impact of Drug Substance to Drug Product Specification

In general, it is not necessary to include the specific attributes of the drug substance in the drug product specification. When setting the acceptance criteria of attributes in the drug product, the impact of the drug substance should be taken into consideration. For example, if an impurity is a process impurity in the drug substance or a degradation product in the drug product, the acceptance criteria of the impurity level in the drug product should be set at a higher level than that of the drug substance specification. By the same principle, if an impurity is a process impurity in the drug substance, but not a degradation product, the limit should remain the same in the drug product specification when it is also listed as a specified impurity. However, process impurities should not be included in the drug product specification if it is not a degradation product.

4.4.4.3 Release Vs. Shelf Life

Different acceptance criteria for release vs. shelf-life specification of the drug product may be implemented in some regions of the world, such as countries regulated by the European Medicines Agency (EMA). Tighter acceptance criteria can be set for the release specification with the understanding that potential chemical/physical changes may occur during the drug product storage up to its shelf life. For example, the level of a degradation product may increase during storage; therefore, the acceptable impurity level in the shelf-life specification is set at a higher level than in the release specification. A similar concept can be used for potency and water content. The differences in the acceptance criteria between release and shelf-life specifications should be justified based on the carefully designed stability studies and the statistical analysis of the stability data that shows trends over the shelf life of the drug product.

4.5 Design of Experiments

The DoE is a commonly used systematic method to determine the impact of process parameters on the CQAs [44].

4.5.1 Common Terms

The following are some of the common terms used in the DoE study.

Factor: This is the input parameter of the experiment, such as drying temperature, drying time, and starting level of water content. While multiple factors can be evaluated simultaneously in a single study, in practice, a maximum of four or five factors are included in a single study before the number of experiments and data analysis become overwhelming.

Level: The discrete values of each factor. Usually, a center point (target) and a range of low and high limits are defined for the study.

Response: The response is the output measure of the experiment. In pharmaceutical process characterization, this is usually the CQA that can potentially be impacted by the process parameters, (i.e. factors included in the study).

Full factorial design: In statistics, a full factorial experiment is an experiment whose design consists of two or more factors, each with discrete levels, and all combinations of these levels across all factors. This study design evaluates not only the effect of each factor but also the interactions between factors on the response. The full factorial design is the most used study design to understand the effects of the process parameters.

Fractional factorial design: In statistics, fractional factorial designs are experimental designs consisting of a carefully chosen subset (fraction) of the experimental runs of a full factorial design. The subset is chosen to understand the effects of individual factors and two-factor interactions. During development, this study design can be used for scouting purposes when there are relatively large numbers of factors to be studied. Based on the outcome of the study, some factors may be eliminated so that a full factorial design study can be employed to thoroughly understand the high impact factors.

4.5.2 Conducting the Study

Before performing the actual experiments, one needs to decide on the number of factors, the target value of each factor (center point), the range of each factor (high and low levels), and the responses. Table 4.6 is an example of factors in the blending process for the hypothetical immediate-release tablet in this chapter. The response of the study is blend uniformity, and the goal of the study is to understand the impact of the blending process parameters on the blend uniformity.

In a typical full factorial design, the number of experiments to be conducted is 2^n, where n is the number of factors. For a three-factor study, eight experiments are needed. The number of experiments can increase rapidly as the number of factors increases. Also, it is common to include replicate experiments with factors set at the center point. The replicate experiments provide information on experimental variability. During data interpretation, it is important to determine if the differences in response are due to the factors under study or they are the results of inherent experiment variability. Table 4.7 presents the study design of the factors from Table 4.6.

Table 4.6 Factors to be studied for the blending step.

Factor	Center point (0)	High level (+1)	Low level (−1)
Blender loading	60%	80%	40%
Rotation speed	10 rpm	14 rpm	6 rpm
Number of revolutions	120	160	80

Table 4.7 The DoE study design of factors from Table 4.6.

Experiment number	Blender loading	Rotation speed	Number of revolution
1	−1	−1	−1
2	+1	−1	−1
3	−1	+1	−1
4	+1	+1	−1
5	−1	−1	+1
6	+1	−1	+1
7	−1	+1	+1
8	−1	+1	+1
9	0	0	0
10	0	0	0
11	0	0	0

When experimenting according to the matrix in Table 4.7, it is extremely important to randomize the experiment order so that there is no systematic bias due to any potential drift in data.

In addition to the example presented above, there are many other experimental designs available, such as randomized design and central composite design. Many off-the-shelf software packages are available for DoE study design and data analysis.

4.5.3 Results Interpretation

The full factorial design experiment can provide information on the effect of each factor, the two-way interaction between the factors, and higher-order interactions, such as three-way interaction. A statistical analysis, hypothesis testing, or p-value approach is used to assess the significance of the effect. Even when a factor does not have a significant effect on the response by itself, it may still have an impact on the response when it is combined with other factors, such that the interactions between factors may be significant.

4.5.4 Summary

DoE is a powerful statistical tool to understand the impact of process parameters and material attributes on CQAs. It not only sheds light on the effect of each factor but also provides knowledge of the interactions between factors and provides the foundation for a holistic control strategy of the process parameters and material attributes.

4.6 Common Statistical Analysis

4.6.1 Mean, Standard Deviation (SD), and Relative Standard Deviation (RSD)

Mean ($\bar{x}$) is the average of a group of numbers or data points. The mean (i.e. average) is obtained by adding all values and dividing by the number of data points available.

$$\bar{x} = \frac{\sum_{i=1}^{n} x_i}{n}$$

Standard deviation, SD or σ, is a measure that is used to quantify the amount of variation or dispersion of a set of data values.

$$\sigma = \sqrt{\frac{\sum_{i=1}^{n} \left(x_i - \bar{x}\right)^2}{n-1}}$$

The relative standard deviation, RSD, is the absolute value of the coefficient of variation. It is often expressed as a percentage.

$$RSD(\%) = \frac{\sigma}{\bar{x}} \times 100$$

An example calculation is presented in Table 4.8

4.6.2 Confidence Interval

The confidence interval is a range of values so defined that there is a specified probability the value of a parameter lies within it. For example, if 10 replicates of a drug product assay analysis were performed, and the confidence interval was calculated based on this group of 10 discrete results with 95% probability, then the confidence interval means that there is a 95% chance that another assay result would fall within the range. The confidence interval is calculated by

$$\bar{x} \pm \frac{t\sigma}{\sqrt{n}}$$

where $\bar{x}$ is the average of the group of the values, σ is the standard deviation of the values, n is the number of the values, and t is a constant determined by the

Table 4.8 Mean, standard deviation, and relative standard deviation calculation.

Measurement	Result
1	98.5
2	97.4
3	99.2
4	98.7
5	99.9
6	98.0
7	97.9
8	98.3
9	99.1
Average ($\bar{x}$)	98.6
SD (σ)	0.77
RSD (%)	0.78

confidence level and the degree of freedom. The degree of freedom is the number of values minus one [5, 6]. Table 4.9 lists the t values.

Example The 95% confidence interval of the data group listed in Table 4.2:

$$\text{Upper boundary} = \bar{x} + \frac{t\sigma}{\sqrt{n}} = 98.6 + \frac{2.306 \times 0.77}{\sqrt{9}} = 99.2$$

$$\text{Lower boundary} = \bar{x} - \frac{t\sigma}{\sqrt{n}} = 98.6 - \frac{2.306 \times 0.77}{\sqrt{9}} = 98.0$$

The t value of 2.306 is obtained from Table 4.9, using the column of 95% confidence level and degree of freedom of 8 (n–1).

4.6.3 Statistical Significance (t-Test)

A t-test's statistical significance indicates whether the difference between the two data groups' averages mostly likely reflects a "real" difference. The t-value is calculated based on the following equations:

$$S_p = \sqrt{\frac{(n_1 - 1)\sigma_1^2 + (n_2 - 1)\sigma_2^2}{(n_1 + n_2 - 2)}}$$

Table 4.9 Table of t-values.

	Confidence level (%)			
	90	95	98	99
	2 tails α =			
Degree of freedom	**0.10**	**0.05**	**0.02**	**0.01**
1	6.314	12.706	31.821	63.657
2	2.920	4.303	6.965	9.925
3	2.353	3.182	4.541	5.841
4	2.132	2.776	3.747	4.604
5	2.015	2.571	3.365	4.032
6	1.943	2.447	3.143	3.707
7	1.895	2.365	2.998	3.500
8	1.860	2.306	2.896	3.355
9	1.833	2.262	2.821	3.250
10	1.812	2.228	2.764	3.169
11	1.796	2.201	2.718	3.106
12	1.782	2.179	2.681	3.055
13	1.771	2.160	2.650	3.012
14	1.761	2.145	2.624	2.977
15	1.753	2.131	2.602	2.947
16	1.746	2.120	2.583	2.921
17	1.740	2.110	2.567	2.898
18	1.734	2.101	2.552	2.878
19	1.729	2.093	2.539	2.861
20	1.725	2.086	2.528	2.845
∞	1.645	1.960	2.326	2.576

Sources: Adapted from Knop and Kleinebudde [5]; Yu [6].

$$SE(\bar{x}_1 - \bar{x}_2) = S_p\sqrt{\frac{1}{n_1} + \frac{1}{n_2}}$$

$$t = \frac{\bar{x}_1 - \bar{x}_2}{SE(\bar{x}_1 - \bar{x}_2)}$$

Table 4.10 Assay results from the uncoated and coated tablets.

	Uncoated tablets	Coated tablets
1	99.5	99.9
2	98.4	100.2
3	99.2	98.9
4		101.2
5		99.1
6		98.7
Average ($\overline{x}$)	99.0	99.7
SD (σ)	0.57	0.95

Where $\overline{x}_1$ and $\overline{x}_2$ are the means of the data in each group, σ_1 and σ_2 are the standard deviations of two data groups, and n_1 and n_2 are the number of values in the two data groups.

When the calculated t value is higher than the t value found in the table for the corresponding confidence interval and degree of freedom, the means of two data groups are a statistically significant difference. The degree of freedom in this test is n_1+n_2-2 [6].

For example, during a process development study, the potency (% Label Claim) of the uncoated tablets were analyzed in triplicates. After the tablets were coated, the potency (% Label Claim) of the coated tablets were analyzed in six replicates. The data are listed in Table 4.10.

$$\sigma = \sqrt{\frac{(n_1-1)\sigma_1^2+(n_2-1)\sigma_2^2}{(n_1+n_2-2)}} = \sqrt{\frac{(3-1)0.57^2+(6-1)0.95^2}{(3+6-2)}} = 0.86$$

$$SE(\overline{x}_1-\overline{x}_2) = \sigma\sqrt{\frac{1}{n_1}+\frac{1}{n_2}} = 0.86\sqrt{\frac{1}{3}+\frac{1}{6}} = 0.61$$

$$t = \frac{|\overline{x}_1-\overline{x}_2|}{SE(\overline{x}_1-\overline{x}_2)} = \frac{|99.0-99.7|}{0.61} = 1.15$$

The t value is 2.365, which is obtained from Table 4.3 with a degree of freedom of 7 (i.e. 3 + 6–2, and 95% confidence level). The calculated t value is less than the one found in the table. Therefore, the uncoated tablet and coated tablet mean assay results have no statistically significant difference.

4.6.4 Outlier Detection

Outlier detection is an important statistical tool to understand the analytical data trend. This tool needs to be explained in a Standard Operating Procedure (SOP) to clearly identify the scope of the outlier analysis and how it is to be used. The outlier detection should be used only by a qualified statistician. If a datum is considered as an outlier based on the statistical analysis discussed below, a thorough lab investigation must be performed to understand the cause of the outlier. One of the commonly used outlier tests is Dixon's Q-test.

$$Q = \frac{Gap}{Range}$$

Where *Gap* is the absolute difference between the outlier data in question and the closest data to it and *Range* is the absolute difference between the large data and the smallest data in the group.

Example Six measurements of a sample with the results as:
99.7, 98.2, 97.6, 99.4, 99.8, 96.7
Is 96.7 an outlier?

$$Gap = |97.6 - 96.7| = 0.9$$

$$Range = |99.8 - 96.7| = 3.1$$

$$Q = \frac{Gap}{Range} = \frac{0.9}{3.1} = 0.290$$

The calculated Q value is less than the Q value in the table at 95% confidence level and 6 measurements, 0.625. Therefore, 96.7 is not an outlier (Table 4.11).

4.7 Summary

QbD concept is a crucial element in pharmaceutical development. It provides a framework to ensure drug product quality. Under the QbD paradigm, the product quality is based on the comprehensive understanding of the drug product and its manufacturing process. The product quality is ensured by a holistic control

Table 4.11 Q value.

	Confidence level (%)		
Number of measurement	**90**	**95**	**99**
3	0.941	0.97	0.994
4	0.765	0.829	0.926
5	0.642	0.710	0.821
6	0.560	0.625	0.740
7	0.507	0.568	0.680
8	0.468	0.526	0.634
9	0.437	0.493	0.598
10	0.412	0.466	0.568

strategy that is derived from the knowledge acquired during the development process, as well as the understanding of risk through the entire lifecycle, from product design to continuous improvement in postapproval commercial manufacturing.

List of Abbreviations

API	Active Pharmaceutical Ingredient
BET	Bacterial Endotoxin Test
CE	Capillary Electrophoresis
CE-SDS	Capillary Electrophoresis-Sodium-Dodecyl Sulfate
CGMP	Current Good Manufacturing Practice
CIEF	Capillary Isoelectric Focusing
CMA	Critical Material Attribute
CPP	Critical Process Parameter
CQA	Critical Quality Attribute
DNA	Deoxyribonucleic Acid
DoE	Design of Experiments
DP	Drug Product
DS	Drug Substance
DSC	Differential Scanning Calorimetry
DTA	Differential Thermal Analysis
ELISA	Enzyme-Linked Immunosorbent Assay

FMEA	Failure Mode Effects Analysis
GC/FID	Gas Chromatography/Flame Ionization Detection
GC/MS	Gas Chromatography/Mass Spectrometry
ICH	International Council for Harmonisation of Technical Requirements for Pharmaceuticals for Human Use
ICP-OES	Inductively Coupled Plasma-Optical Emission Spectrometry
ICP-MS	Inductively Coupled Plasma-Mass Spectrometry
IEC	Ion Exchange Chromatography
IR	Infrared Spectroscopy
IPC	In-Process Control
JP	Japanese Pharmacopeia
LC/UV	Liquid Chromatography/Ultraviolet
LC/MS	Liquid Chromatography/Mass Spectrometry
NIR	Near-Infrared
NMR	Nuclear Magnetic Resonance
PAT	Process Analytical Technologies
Ph. Eur	European Pharmacopeia
PSD	Particle Size Distribution
QbD	Quality-by-Design
QTPP	Quality Target Product Profile
RLD	Reference Listed Drug
RSD	Relative Standard Deviation
SD	Standard Deviation
SDS-PAGE	Sodium Dodecyl Sulfate-Polyacrylamide Gel Electrophoresis
SEC	Size Exclusion Chromatography
TGA	Thermal Gravimetric Analysis
TTC	Threshold of Toxicological Concern
SFC	Supercritical Fluid Chromatography

References

1 Food and Drug Administration (2009). *Guidance for Industry: Q8 (R2) Pharmaceutical Development*. Rockville, MD: Food and Drug Administration https://www.fda.gov/media/71535/download (accessed 8 December 2020).

2 Food and Drug Administration (2009). *Guidance for Industry: Q10 Pharmaceutical Quality System*. Rockville, MD: Food and Drug Administration https://www.fda.gov/media/71553/download (accessed 8 December 2020).

3 Woodcock, J. (2004). The concept of pharmaceutical quality. *American Pharmaceutical Review* 7 (6): 10–15.

4 Aulton, M. and Talyor, K. (2017). *Aulton's Pharmaceutical E-Book: The Design and Manufacturing of Medicines*, 1e, 180. Elsevier Health Science.

5 Knop, K. and Kleinebudde, P. (2013). PAT-tools for process control in pharmaceutical film coating applications. *International Journal of Pharmaceutics* 457 (2): 527–536.

6 Yu, L. (2008). Pharmaceutical quality by design: product and process development, understanding, and control. *Pharmaceutical Research* 25 (4): 781–791.

7 Food and Drug Administration (2006). *Guidance for Industry: Q9 Quality Risk Management*. Rockville, MD: Food and Drug Administration https://www.fda.gov/media/71543/download (accessed 8 December 2020).

8 Burgess, D., Duffy, E., Etzler, F., and Hickey, A.J. (2004). Particle size analysis: AAPS Workshop Report. Cosponsored by the Food and Drug Administration and the United States Pharmacopeia. *The AAPS Journal* 6 (3): Article 20.

9 Qiu, Y., Chen, Y., Zhang, G. et al. (ed.) (2016). *Developing Solid Oral Dosage Forms: Pharmaceutical Theory and Practices*, 2e, 155–169. Academic Press.

10 Food and Drug Administration (2004). *Guidance for Industry: PAT – A Framework for Innovative Pharmaceutical Development, Manufacturing and Quality Assurance*. Rockville, MD: Food and Drug Administration https://www.fda.gov/media/71012/download (accessed 8 December 2020).

11 Badawy, S., Badawy, S., Narang, A. et al. (2012). Mechanistic basis for the effects of process parameters on quality attributes in high shear wet granulation. *International Journal of Pharmaceutics* 439 (1–2): 324–333.

12 Sulub, Y., Wabuyele, B., Gargiulo, P. et al. (2009). Real-time on-line blend uniformity monitoring using near-infrared reflectance spectrometry: a noninvasive off-line calibration approach. *Journal of Pharmaceutical and Biomedical Analysis* 49 (1): 48–54.

13 Frake, P., Greenhalgh, D., Grerson, S.M. et al. (1997). Process control and end-point determination of a fluid bed granulation by application of near infra-red spectroscopy. *International Journal of Pharmaceutics* 151 (1): 75–80.

14 Food and Drug Administration (2000). *Guidance for Industry: Q6A Specifications: Test Procedures and Acceptance Criteria for New Drug Substances and New Drug Products: Chemical Substances*. Rockville, MD: Food and Drug Administration https://www.fda.gov/regulatory-information/search-fda-guidance-documents/q6a-specifications-test-procedures-and-acceptance-criteria-new-drug-substances-and-new-drug-products (accessed 8 December 2020).

15 Food and Drug Administration (1999). *Guidance for Industry: Q6B Specifications: Test Procedures and Acceptance Criteria for Biotechnological/Biological Products*. Rockville, MD: Food and Drug Administration https://www.fda.gov/regulatory-information/search-fda-guidance-documents/q6b-specifications-test-procedures-and-acceptance-criteria-biotechnologicalbiological-products (accessed 8 December 2020).

16 Food and Drug Administration (2019). Application for FDA approval to market a new drug. Code of Federal Regulations. Title 21, Part 314.3. Rockville, MD: Food and Drug Administration. https://www.accessdata.fda.gov/scripts/cdrh/cfdocs/cfcfr/CFRSearch.cfm?fr=314.3 (accessed 8 December 2020).

17 ICH Harmonised Tripartite Guideline (2003). Evaluation for stability data Q1E. https://database.ich.org/sites/default/files/Q1E_Guideline.pdf (accessed 8 December 2020).

18 Food and Drug Administration (2008). *Guidance for Industry. Q3A(R) Impurities in New Drug Substances*. Rockville, MD: Food and Drug Administration https://www.fda.gov/media/71727/download (accessed 9 July 2020).

19 Food and Drug Administration (2006). *Guidance for Industry. Q3B(R2) Impurities in New Drug Products*. Rockville, MD: Food and Drug Administration https://www.fda.gov/media/71733/download (accessed 9 July 2020).

20 Huynh-Ba, K. (ed.) (2009). *Handbook of Stability Testing in Pharmaceutical Development*, 1e. New York: Springer-Verlag Chapter 13.

21 International conference on harmonisation of technical requirements for registration of pharmaceuticals for human use. ICH guideline M7(R1) on assessment and control of DNA reactive (mutagenic) impurities in pharmaceuticals to limit potential carcinogenic risk (August 2015). https://www.ema.europa.eu/en/documents/scientific-guideline/ich-guideline-m7r1-assessment-control-dna-reactive-mutagenic-impurities-pharmaceuticals-limit_en.pdf (accessed 8 December 2020).

22 USP <731> Loss on drying. *USP 43–NF 38*. Rockville, MD: United States Pharmacopeial Convention.

23 USP <281> Residue on ignition. *USP 43–NF 38*. Rockville, MD: United States Pharmacopeial Convention.

24 USP <233> Elemental impurities–procedures. *USP 43–NF 38*. Rockville, MD: United States Pharmacopeial Convention.

25 USP <467> Residual solvents. *USP 43–NF 38*. Rockville, MD: United States Pharmacopeial Convention.

26 USP <791> pH measurement. *USP 43–NF 38*. Rockville, MD: United States Pharmacopeial Convention.

27 USP <741> Melting range. *USP 43–NF 38*. Rockville, MD: United States Pharmacopeial Convention.

28 USP <781> Optical rotation. *USP 43–NF 38*. Rockville, MD: United States Pharmacopeial Convention.

29 USP <831> Refractive index. *USP 43–NF 38*. Rockville, MD: United States Pharmacopeial Convention.

30 USP <788> Particulate matter. *USP 43–NF 38*. Rockville, MD: United States Pharmacopeial Convention.

31 USP <941> Xray diffraction. *USP 43–NF 38*. Rockville, MD: United States Pharmacopeial Convention.

32 USP <1053> Capillary electrophoresis. *USP 43–NF 38*. Rockville, MD: United States Pharmacopeial Convention.

33 USP <921> Water determination. *USP 43–NF 38*. Rockville, MD: United States Pharmacopeial Convention.

34 USP <61> Microbiological examination of nonsterile products. *USP 43–NF 38*. Rockville, MD: United States Pharmacopeial Convention.

35 USP <62> Microbiological examination of nonsterile products: tests for specified microorganisms. *USP 43–NF 38*. Rockville, MD: United States Pharmacopeial Convention.

36 USP <1217> Tablet breaking force. *USP 43–NF 38*. Rockville, MD: United States Pharmacopeial Convention.

37 USP <905> Uniformity of dosage units. *USP 43–NF 38*. Rockville, MD: United States Pharmacopeial Convention.

38 USP <51> Antimicrobial effectiveness test. *USP 43–NF 38*. Rockville, MD: United States Pharmacopeial Convention.

39 USP <711> Dissolution. *USP 43–NF 38*. Rockville, MD: United States Pharmacopeial Convention.

40 USP <911> Viscosity. *USP 43–NF 38*. Rockville, MD: United States Pharmacopeial Convention.

41 USP <785> Osmolality test. *USP 43–NF 38*. Rockville, MD: United States Pharmacopeial Convention.

42 USP <71> Sterility tests. *USP 43–NF 38*. Rockville, MD: United States Pharmacopeial Convention.

43 USP <85> Bacterial endotoxins test. *USP 43–NF 38*. Rockville, MD: United States Pharmacopeial Convention.

44 Politis, S., Colombo, P., Colombo, G., and Rekkas, D. (2017). Design of experiments (DoE) in pharmaceutical development. *Drug Development and Industrial Pharmacy* 43 (6): 889–901.

5

Development and Validation of Analytical Procedures

Linda L. Ng

Fresenius-Kabi, Lake Zurich, IL, USA

TABLE OF CONTENTS

Analytical Testing for the Pharmaceutical GMP Laboratory, First Edition. Edited by Kim Huynh-Ba.

5.1 Introduction

Quality finished pharmaceutical drug products/dosage forms are manufactured from quality active pharmaceutical ingredients (APIs), formulation, well-designed processing, packaging, and labeling. Thoughtful design with a good understanding of the processes at each step of the product production, starting from API synthesis to product labeling, will result in safer medicines for the patients and a reduced risk for recalls. Confirmation of the quality is achieved in the data generated by reliable and validated analytical test procedures on equipment that has been qualified. Data can be generated as in-process, release testing, and stability testing. Analytical tests are selected to address appropriate attributes essential to the product and performed at critical steps of the manufacturing process or the life of the product. This chapter will attempt to place the concept and terminology on the same level playing field of understanding for the readers and discuss the science behind the various studies of development, qualification, validation, and verification. The terms method and procedure are interchangeable in this chapter.

Method development, validation, and verification could be performed in a Current Good Manufacturing Practices (CGMPs) laboratory but are not required if the equipment is qualified to meet CGMP requirements. Test data to support commercial market products are performed in a CGMP laboratory with appropriate documentation [1, 2].

5.2 Method Development

A pharmaceutical product is composed of many components, which include the API (also termed the drug substance) and inactive ingredients or excipients that are formulated into a dosage form or drug product. Dosage forms include types such as capsule, tablet, solution, suspension, emulsion, and transdermal patch. This group is further expanded for tablets: orally disintegrated tablets, immediate-release, delayed-release, and sustained-release tablets. Also, sterile dosage forms will require microbiological input besides the chemical and physical assessments.

This chapter does not cover all possible variations of dosage forms, but it will describe a broad scope to show the challenges of developing a reliable method.

The analyst developing the test procedure must reduce the product sample to a form that can be tested. For example, a tablet must be reduced to a homogenous, preferably aqueous solution for impurities testing if reversed-phase chromatography is used. The active ingredient must be extracted with full recovery from a transdermal patch for the assay or impurities testing. When the API is at low strength (e.g. Latanoprost ophthalmic solution, 0.005% w/v), testing will be challenged by the sensitivity of the method, recovery of the API, and detection capability of the analytical technique. The procedure separates impurities from each other for an optimum quantitation of each. For physical measurement of particle size, the measurement technique must be reliable to obtain multiple data points. In general, physical, chemical, and microbiological test procedures must aim to provide accurate data on different pieces of equipment at different times by different analysts. The method developer has a good understanding of the properties of the analyte, the dosage form, the technology of the technique and equipment, and potential environmental effects to produce a reliable procedure to give accurate data.

5.2.1 Development of Physical, Chemical, and Microbiological Procedures

Many techniques and typical procedures used in a CGMP laboratory are described in Chapter 3 in the GMP Laboratory and Chapter 4. Independent of the type of analysis, the purpose is to be able to analyze the compound of interest, qualitatively or quantitatively, in the matrix. Numerous techniques are available to an analyst; however, in a CGMP laboratory, it is common to use known and tried techniques. It is up to the method developer to design an appropriate sample preparation process for the test procedure. That usually depends on the technique used and the complexity of the sample matrix.

5.3 Qualification, Validation, and Verification

The purpose of qualification, validation, and verification is to generate reliable data for the products tested, resulting in the delivery of quality drug products to the patients. Whether it is an instrumental or a physical procedure, experimental studies are performed to confirm that the data are reliable and trustworthy [3]. All analytical procedures, when developed, must be validated. On the other hand, published validated monographs for use at a CGMP laboratory must be verified, that is, select validation parameters need to be addressed to ensure the procedures for the products evaluated will provide reliable data. All instruments should be qualified for assurance that data generated by these laboratory instruments will be accurate and reliably reproduce every time the instrument is used. Good design using optimum conditions in an analytical procedure with validated parameters and qualified studies for the instrument will provide reliable data.

The discussion here will cover only products for commercialization. It is expected that methods for in-process controls, release, and stability data of the products are performed under CGMP conditions. The US Food and Drug Administration (FDA) inspection observations have been issued for inadequate validation and laboratory controls [4].

5.3.1 Qualification

The test procedure is commonly performed using an instrument; therefore, the instrument must be qualified to ensure that generated data are reliable. Appropriate studies are designed and completed to support the instrument's capability to provide reproducible data. The qualification studies are carried out by the instrument manufacturer as well as the firm that purchased the instrument. The type of studies will depend on the type of instrument. In general, the current instrument qualification practice is divided into four parts: (i) design qualification (DQ), which is traditionally vendor-driven; (ii) installation qualification (IQ), user-vendor driven; (iii) operational qualification (OQ), user-vendor driven; and (iv) performance qualification (PQ), user-driven. Table 5.1, as reproduced from USP <1058> [5], clearly delineates the timing, applicability, and activities of each phase of the qualification. More information can be found in Chapters 3 and 4.

Instruments should be requalified following a predetermined schedule for assurance of sustained performance through their lifecycle. Some firms may contract out the studies to outside laboratories to perform the studies. Independent of who performs the studies, the user firm is responsible for the documentation of the studies and should store the data in-house.

For chromatographic instruments, the PQ and the system suitability tests (SSTs) have common properties. Both fulfill a similar purpose at a different time of the instrument's life. PQ is part of the qualification testing for the instrument, and SST is performed each time samples are analyzed. Both rely on the fact that the electronics of the equipment, analytical operations, and samples for analysis are one integral system. Although not a guarantee for error-free chromatography, the target is the independent and objective evaluation of the critical system components (detector, pump, column, and auto-sampler) together to ensure that the instrument is reliable and will generate reproducible and accurate data. The purpose of the SST is to accommodate variations that should be kept to a minimum. The SST is to confirm that the whole system is operating during use. Based on this principle, the design and selection of the SSTs are critical. The discussion on the SST is further described in Section 5.4.8.

Table 5.1 Phases of analytical instrument qualification.

Design qualification	Installation qualification		Operational qualification		Performance qualification
Timing and applicability					
Prior to purchase of a new model of instrument	At installation of each instrument (new, old, or existing unqualified)		Alter installation or major repair of each instrument		Periodically at specified intervals for each instrument
Activities					
Assurance of manufacturer's DQ	Description	⇆	Fixed parameters		Preventive maintenance and repairs
Assurance of adequate support availability from manufacturer	Instrument delivery				Establish practices to address operation, calibration, maintenance, and change control
Instrument's fitness for use in laboratory	Utilities/Facility	⇆	Environment		
	Assembly and installation				
	Network and data storage	⇆	Secure data storage, backup, and archive		
	Installation verification	⇆	Instrument function tests	⇆	Performance checks

Source: USP <1058> [5].
Activities under each phase are usually performed as given in the table. However, in some cases, it may be more appropriate to perform or combine a given activity with another phase. Such activities spanning more than one qualification phase are shown as connected by double arrows. If an activity listed under a given phase is performed under another phase, it is not necessary to repeat the activity under the phase where the activity is listed. Performing the activity is far more important than the phase under which the activity is performed.

5.3.2 Validation

The validation parameters listed in Key Information below were defined and applications shared in ICH Q2(R1) (Validation of Analytical Procedures: Text and Methodology) in November 2005 [6], previously coded ICH Q2A (Guideline on Text) and ICH Q2B (Guideline on Methodology). The validation parameters are also discussed in FDA guidance [7, 8] and USP Chapter <1225> Validation of Compendial Procedures [9] and <1223> Validation of Alternative [10].

Typical validation parameters
Accuracy
Precision
Repeatability
Intermediate precision
Reproducibility
Specificity
Quantitation and detection limits
Linearity
Range
Robustness

All procedures should be validated as confirmed to provide reliable data when used. The number of validation parameters required, as defined in ICH Q2(R1), will depend on the type, history, and complexity behind the procedure.

5.3.3 Verification

Published validated procedures are designed to cover the products developed by the sponsor. Each published procedure is validated by the developer of the method for a specific product. Title 21 CFR 211.194(a)(2) states that "the suitability of all testing methods used shall be verified under actual conditions of use." Therefore, the user of compendia or published procedures should confirm that the developed procedure works for the user's application [7, 11]. Thus, to adapt these procedures to drug products manufactured with different excipients or by different synthetic, fermentation, or extraction API manufacturing process, the validation parameter of specificity is critical. Sample stability for analysis that takes an extended time to complete the study should evaluate the sample stability for the time of analysis. On the other hand, the concern for published cleaning validation procedures is different. Those should be verified to ensure they can detect the limit claimed by the facilities that use the procedure.

If the procedure is not a public standard, then the transfer of the method would be necessary. Chapter 6 will discuss more details of method transfer and the relationship of validation, verification, and transfer.

5.3.4 Frequency of Study

The continual monitoring of the instrument and the procedure is a lifecycle process. Instruments should be requalified at prescribed intervals to ensure that they are in optimal working condition. Procedures, on the other hand, should be assessed when a change occurs to determine the scope of revalidation or reverification needed to ensure that the change has no impact on the procedure and purpose. Not all validation parameters need to be addressed. For example, when adding an alternate API supplier, the drug substance testing procedure will require a repeat of the specificity evaluation. The impurities profile and resolution of impurity peaks should be studied to confirm that the procedure is reliable and able to quantitate the impurities, especially new impurities if any.

5.4 Validation Parameters

Validation parameters are designed to demonstrate the reliability of analytical procedures. The different parameters cover the different facets to ensure the data collected are the true values, and how the data can be reported as described in ICH Q2(R1) [6]. All parameters are not required to be addressed for all analytical procedures. That depends on the type of procedure it is such as qualitative vs. quantitative, such as a procedure designed "from scratch" vs. established, human (touch, visual) vs. instrumental.

5.4.1 Accuracy

The accuracy is a measure of how close the experimental value is to the true value. Data generated from analytical procedures should target to be free of error within the analytical variation of the procedure and instrument. Figure 5.1 illustrates how values are targeted to the true values.

Figure 5.1 Target true value.

The experimental design to study accuracy should aim to address recovery during sample preparation, interaction with excipients in drug products, and bias, if any, of the instrument. Recovery is a measure of the assayed value from the true theoretical amount in the sample. For a drug product, a known amount of the API(s) is added to the placebo formulation, i.e. all ingredients minus the API of the drug product. For liquid or semi-liquid dosage forms, this could be a true recovery study. However, for solid dosage forms like tablets, transdermal patches, and suppositories, it is not practical to add a known amount of API and create each unit dose product. Hence for such dosage forms, the spiking study will consider only the recovery during sample preparation and the assaying procedure. The objective is always to aim for 100% recovery. Having less of a procedure is a concern as it is uncertain if the missing portion of the API has interacted with the ingredients of the placebo or the container or has degraded to other impurities. On the other hand, having more will indicate issues relating to the procedure or instrumentation. In either case, the root cause should be determined and corrected with an allowable range of method acceptable uncertainty. More information on the uncertainty is discussed in USP Chapter <1010> Analytical Data – Interpretation and Treatment.

5.4.2 Precision

Precision is a measure of how close the experimental values are to each other for measurements under the same analytical conditions, as illustrated in Figure 5.2. The right picture shows low precision, and the left is for high precision.

To address this parameter, ICH Q2(R1) [6] provides a recommendation on how to perform studies and tease out the variability measurements for the samples as:

a) Repeatability – testing by an analyst on the same day and the same instrument using the same operating conditions. This measurement is also called intra-assay precision.
b) Intermediate precision – testing by different analysts using the same prescribed conditions on different days, different instruments, etc., termed intermediate precision.
c) Reproducibility – this measurement ensures the procedure can generate reliable and accurate data among laboratories. Multiple laboratories will participate in the data collection using one or more homogenous samples. This parameter is helpful to a firm with multiple laboratory locations or for standardization of the procedure, for example, in pharmacopeias.

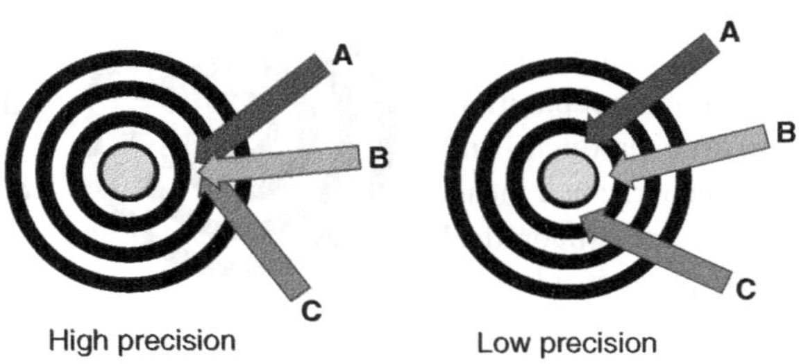

Figure 5.2 High and low precision illustration.

The sensitivity of the procedure, that is, the ability to detect small changes in the amount of analyte being analyzed, is also addressed under this parameter. When all the different facets are covered to give essentially the same value within experimental variation, confidence in the analytical procedure is increased.

Quality is achieved through a combination of accuracy and precision where the values are tight and on target close to the true values, as illustrated in Figure 5.3 (bottom right bullseye).

5.4.3 Specificity

This parameter is for assurance that the analyte(s) of interest has no interference from other compounds in the matrix. In the developmental stage, multiple orthogonal tests can be used. Forced degradation studies should also be designed to ensure the method can separate the analyte from its degradation products. Also, multi-separation techniques such as chromatography with photodiode array detector are often helpful. Peak purity should be established to ensure that there is no interference. Reducing the interference will increase the accuracy of the analyte data.

It is very common to use high-performance liquid chromatography (HLPC) and ultra-high-performance liquid chromatography (UHPLC) with ultraviolet (UV) detection for impurities testing. For an analyte with a chromophore or UV spectrum, the selection of the wavelength for detection will determine the response sensitivity of the procedure. Figure 5.4 illustrates chromatographic data for three compounds – A, B, and C, being monitored at three wavelengths (i.e. 254, 280, and 300 nm). The observed peak is directly proportional to the response factor of the chromophore of the compound. If 305 nm is the detection wavelength of choice, peaks B and C will be observed with B having the best response with no response for peak A. On the other hand, if 254 nm is selected, peaks A and C will be detectable with A having the best response and compound B not detected. Similarly, at 280 nm, peak A and B will be observed and peak C missing. Thus, it is critical to select an optimum detection wavelength for multiple analytes in an assay.

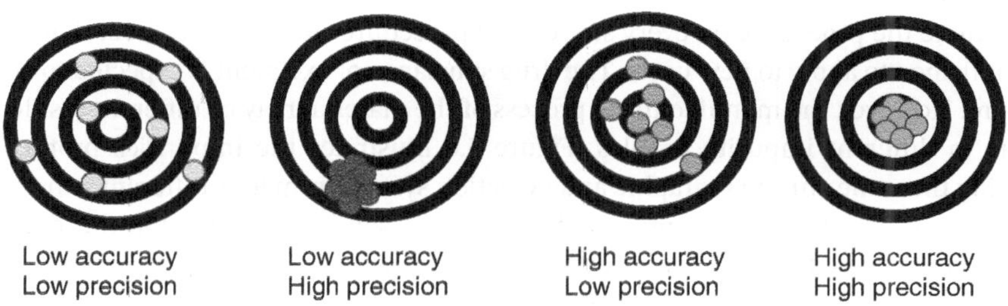

Figure 5.3 Accuracy and precision examples.

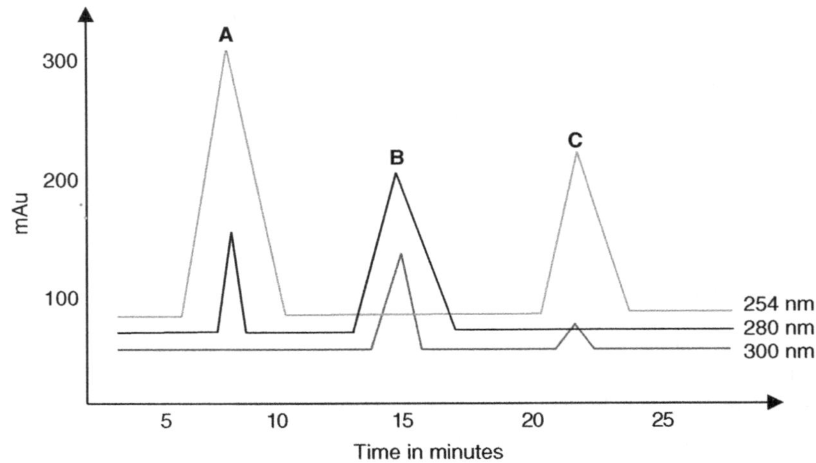

Figure 5.4 Monitoring compounds at various wavelengths.

From the methodology perspective, the availability of columns, detectors, techniques, etc., provide the method developer many choices. Hence, the design of the procedure to ensure a resolution of the various compounds in a combination product or impurities testing will be critical to achieve accurate and reliable data. An example of the separation of three impurities, A, B, and C from the parent, P, using 7 HPLC chromatographic conditions, is illustrated in Figure 5.5 [12, 13].

An examination of the seven chromatograms will show that some conditions provide better separation than others. The ideal objective is to aim to baseline resolve all impurities, and the parent peaks from each other under the shortest chromatographic run time. Commercial software programs are available to assist the developer in optimizing conditions.

Specificity is critical for stability-indicating procedures. Pharmacopeial procedures are generic, that is, not specific to one source, for the drug substance, excipient, or the drug product. The original procedure, when developed by a method designer, is from one source with one profile. The procedure after validation could be accepted and adopted by a pharmacopeia. Since, the original designer cannot evaluate the procedure on every source, the procedure should be verified, that is, confirmed to apply to new users. The drug substance or excipient is expected to be pure. However, the manufacturing process of the material may be different resulting in different impurity profiles. Figure 5.6 illustrates the impurities profiles using the same chromatographic and detection system from four different suppliers, A, B, C, and D with presumably different synthetic routes for the drug substance fluoxetine hydrochloride [14].

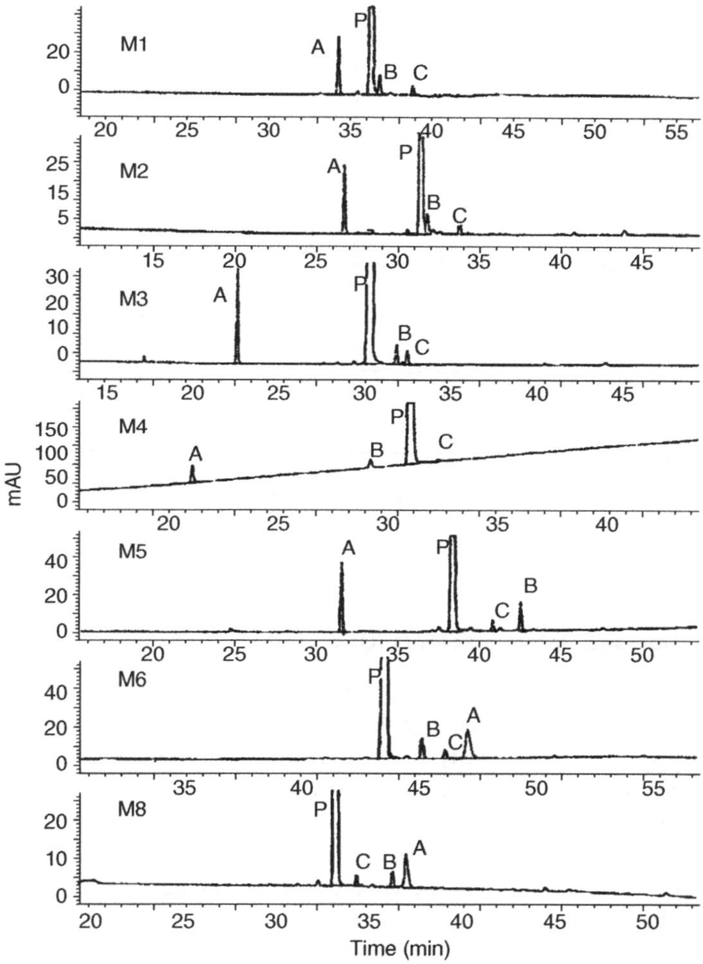

Figure 5.5 Separation of impurities A, B, and C from the peak of interest (P) using seven different HPLC systems. *Sources:* Adapted from Xue et al. [12]; Ng et al. [13].

The type and level of each impurity could be different, resulting in different impurity profiles and total impurities. Note that peaks 1 and 3 are observed in supplier C and D but absent in suppliers A and B. Peak 4 is only observed in supplier C. Supplier A with 0.10% total impurities has the least amount of observable impurities. This example illustrates why specificity should be addressed when published analytical procedures are adopted. Similarly, the drug products manufactured by different firms, that is, Rx and generic products are likely not to have the same type or level of excipients and thus should be verified for specificity.

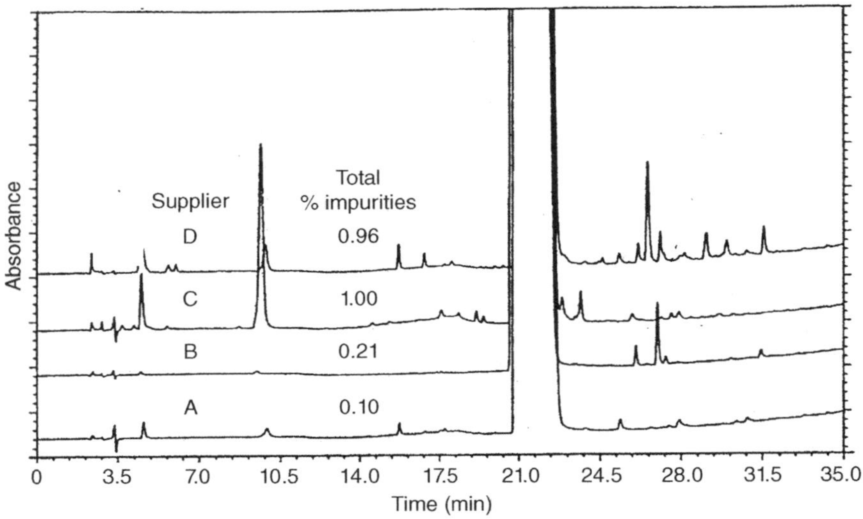

Figure 5.6 HPLC chromatograms of samples of fluoxetine hydrochloride from four different suppliers. *Source:* Wirth et al. [14].

5.4.4 Quantitation and Detection Limits (QL and DL)

QL and DL are not commonly used in qualitative tests. If the sensitivity of the test is critical and contributes to product quality, these parameters should be validated, e.g., Container/Closure system Integrity Test (CCIT) and environmental assay. In addition, it is critical for quantitative tests, where a low level of the analyte(s) is evaluated. For example, if the assay procedure is to quantitate the amount of analyte in a sample, it is not critical how sensitive the assay procedure is. Often, the concentration of the drug substance in the sample solution can be adjusted to the desired concentration for quantitation.

ICH Q3A(R2) [15] and ICHQ3B(R2) [16] defines the quantitation limits for the API and the drug product, respectively. The values are indicated as reporting thresholds for chemically synthesized compounds, as shown in Table 5.2.

Most of the chromatographic methods are capable of lower limits. However, it is critical that the quantified amount is fully recovered from the matrix and that the results collected are accurate. On the other hand, for impurities procedure, quantitation and detection limits are important as that will support the premise that the procedure is sensitive at the low end and the data obtained are reliable.

5.4.5 Linearity

This parameter is not needed for qualitative analysis, where the presence or absence of the analyte is determined. However, for quantitative analysis, the test

Table 5.2 ICH Reporting thresholds for drug substance (API) and finished drug product.

	Maximum daily dose[a]	**Reporting threshold**[b,c]
Drug substance (API)	≤2 g/day	0.05%
	>2 g/day	0.03%
	Maximum daily dose[a]	Reporting threshold[b,d]
Finished drug product	≤1 g/day	0.1%
	>1 g/day	0.05%

[a] The amount of drug substance administered per day.
[b] Higher reporting thresholds should be scientifically justified.
[c] Lower thresholds can be appropriate if the impurity is unusually toxic.
[d] Thresholds for degradation products are expressed either as a percentage of the drug substance or as total daily intake (TDI) of the degradation product. Lower thresholds can be appropriate if the degradation product is unusually toxic.

values will indicate how much of the analytes are present. In this study, a known weight of an API is diluted with a known amount of solvent. Dilutions of the solution are prepared to cover above and below the target concentration of interest. Each of the solutions is assayed and plotted by concentration vs. test values. Linearity will be confirmed when the test values are directly proportional to the concentration of the analyte in the sample following Beer's law (see Figure 5.7) [8].

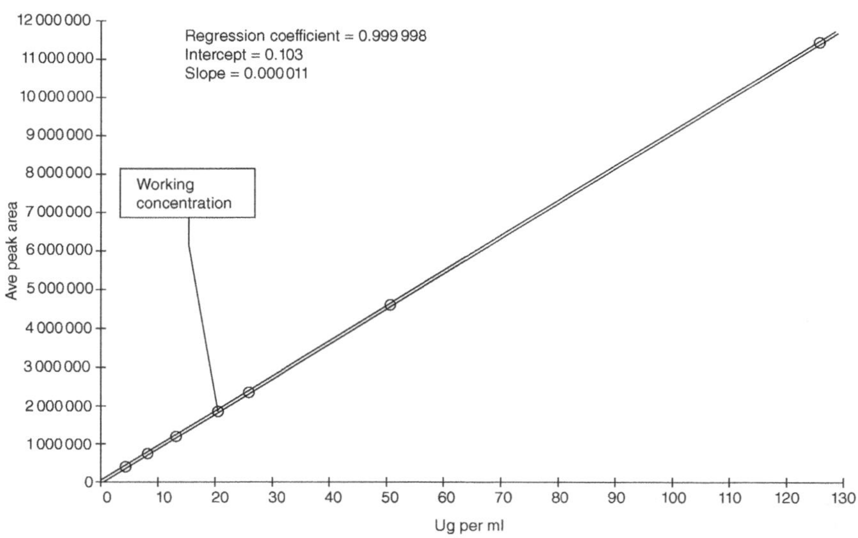

Figure 5.7 The linearity of standard solutions in chromatographic analysis. *Source:* Food and Drug Administration [8]. Public Domain.

For procedures that do not follow Beer's Law, e.g. near infrared (IR), a derivative of the raw data may be used to reproduce readings that become directly proportional to actual values.

5.4.6 Range

Quantitative testing can be performed with external or internal standards. The external standard method for chromatographic procedures is desirable to avoid concern that unknown impurity could be present and missed being detected in the product being analyzed. For the external standard method, a reference standard of known concentration is prepared in addition to samples for analysis. Testing should be performed in the section within the range of linearity values with quantitation methods. It is possible to lose linearity at high concentrations when using chromatographic procedures with UV detection. See Figure 5.8 [8].

It is critical to perform the testing within the range since it is not practical to have standards at the exact concentration of each analyte(s). For example, if the API has degraded with time on stability studies, the preparation of the reference standard cannot be adjusted to reflect the concentration of the analyte since that is unknown before the analysis is completed.

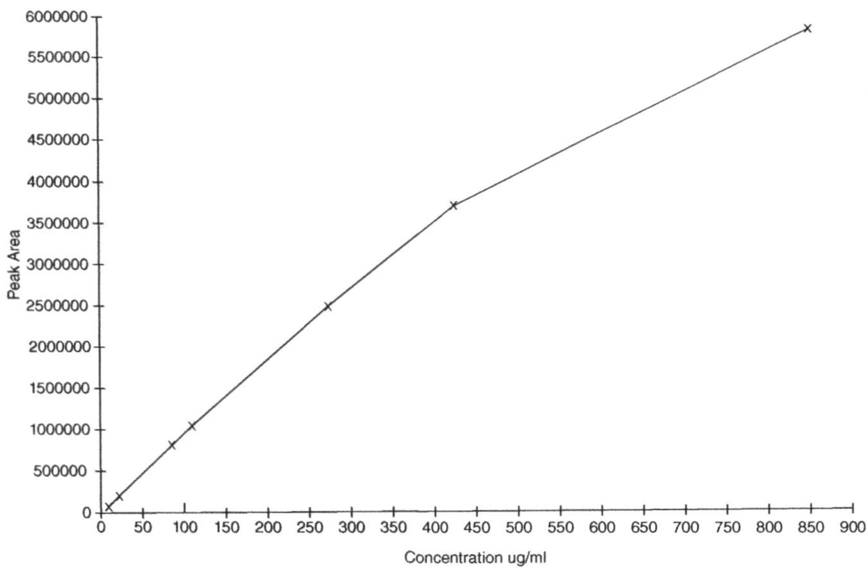

Figure 5.8 Nonlinearity of standard solutions in chromatographic analysis. *Source:* Food and Drug Administration [8]. Public Domain.

5.4.7 Robustness

Robustness is a measure of steadiness or sturdiness of the procedure that is unaffected by small changes or variations in analytical conditions to generate accurate data. The objective is to achieve reproducible data independent of slight changes in analytical conditions or environmental conditions. This parameter should be addressed for methods that could potentially be affected by changes in the operating conditions.

5.4.8 System Suitability Tests (SST)

SSTs [6–8, 17], described in the Key Information section below, **are** to confirm that the properties of the chromatography analysis are adequate for reliable analysis. In general, the SSTs provide results to comprise all parts of the analysis procedure, such as instruments, analytical operations, and sample, as one integral system. These SSTs are performed immediately before the sample analysis to confirm that the system is functioning correctly. The relative standard deviation (RSD) test is six replicate injections, according to USP chapter <621> – Chromatography – System Suitability [6, 17]. It should be noted that RSD shows the repeatability of the analytical procedure used at the time of use. Be aware that variations due to sample preparation and manufacturing are not evaluated.

System suitability tests

- Capacity factor is a measure of the peak of interest with respect to the void volume
- Relative retention time is to show the relative position of another peak to the compound of interest
- Resolution is to show how well the separation is between two closest compounds
- Tailing to indicate the asymmetry or lack of for the chromatographic peak
- Relative Standard Deviation is a measure of injection repeatability to indicate the performance of the chromatographic system, i.e. plumbing, column, and environmental conditions
- Theoretical plates for the measure of column efficiency
- Sensitivity is a measure of assurance that the compounds of interest are detected and quantitated. Normally, a reference standard or an impurity standard at the quantitation limit concentration of the procedure is utilized.

System suitability testing could also be interspersed among samples and at the end of samples depending on type of samples, length of analysis run, and so forth.

The type of tests will depend on the purpose of the analytical procedure. For example, an assay of a single API for a drug substance or a drug substance in a drug product usually covers capacity factor, tailing, and RSD. For analysis of multiple compounds, a resolution test should be included to ensure that the closest pair of compounds that elute is adequately separated from each other to provide reliable values. For impurity or residual procedures, additional tests are recommended to include resolution and sensitivity tests. The sensitivity test includes a standard at the quantitation limit to ensure that impurities at that level will be detectable and quantified during the time of analysis. Additional information may also be found in Chapters 3 and 4.

5.4.9 Stability of Samples During Analysis

When the sample is analyzed instantaneously as online readings, sample stability is not an issue. However, when there is a lapse of time between sample preparation and analysis, the stability of the sample, especially in a different form from its original state, should be evaluated. For example, evaluation of the solution stability is critical when there are samples prepared in solution waiting to be analyzed or many samples waiting in the queue, each analysis would take extended run time, and degradation of the analyte during the lapse time will not provide the correct picture of the original sample. However, it could show a picture of extra degradants and parent analyte if a stability-indicating procedure is used. In any CGMP laboratory, the stability study of the sample is expected to ensure the samples are stable during analysis.

5.4.10 Tie the Pieces Together

Even though each parameter is discussed in separate sections for ease of understanding, the collection of data for the validation study is frequently merged.

For example, to validate a drug product stability-indicating assay procedure by reversed-phase chromatography, the procedure is designed, developed, and optimized to obtain baseline resolution of the parent peak and no interference from excipients and impurities peaks. Once the analyst is confident that a reliable procedure has been developed, experiments to validate the procedure are then performed according to the firm's validation protocol. The validation protocol should be writing clearly and concisely so instructions can be followed and versions tracked. The Key Information section below illustrates the typical sections needed in a validation protocol; please review carefully.

Sections to be included in a typical validation protocol
• Title – including numbering system, version, and implementation date • Purpose/Objective – including types of tests for types of products • Materials, instrumentation, personnel, facilities – including precautions • Design of experiments – including how (instructions), who (analyst A and B), and where (instrument A, B; facility A and B) performed • Acceptance criteria for the validation parameters • References including appropriate pharmacopeia or guidance and definitions, so users have a level of comprehension • Protocol amendment and Deviation – describing when, under what conditions, and who are the responsible parties • Sign off by appropriate parties (names, job titles/department, signatures) with dates

The parameters, such as accuracy, precision, and linearity, can be merged into one experiment. An API can be spiked in duplicate or triplicate in placebo. The formulated product is prepared as a solution, and dilution made to reflect the testing range as described in ICH Q2 [6]. Data are collected and calculated. Once all experiments are completed, the method validation report to support the test procedure is compiled and signed off by the quality group.

5.5 Validation for Physical, Chemical, Biotechnological, and Microbiological Procedures

The same concept should apply for the validation of various types of procedures. That is, equipment should be qualified; validation and verification of procedures will depend on whether the procedure is established vs. developed "from scratch."

Table 5.3 is abstracted from ICH Q2(R1) and the highlights of the draft FDA guidance published in the Federal Register in 2000 [6, 18].

The table illustrates the parameters that should be addressed in four of the most common analytical procedures and specific tests for drug analysis. It does not cover all tests, but the principle of parameters selected can apply.

5.6 Validation of In-process, Environmental, Release, and Stability Procedures

A CGMP laboratory is described in a broad sense. All physical areas where such testing is performed fall under CGMP regulations. This testing could be at the

Table 5.3 Four most common analytical procedures and specific tests for drug analysis.

Type of tests or characteristics	Identification	Quantitative testing for impurities	Limit testing for impurities	Assay dissolution (measurement only), content–potency	Specific tests
Accuracy	–	+	–	+	+†
Precision–repeatability	–	+	–	+	+†
Precision–intermediate precision	–	+‡	–	+‡	+†
Specificity	+§	+	+	+‖	+†
Detection limit	–	–#	+	–	–
Quantitation limit	–	+	–	–	–
Linearity	–	+	–	+	–
Range	–	+	–	+	–
Robustness	–	+	–#	+	+†

Sources: Adapted from Brown et al. [18]; ICH Harmonised Tripartite Guideline [6].

– Signifies that this characteristic normally is not evaluated.

+ Signifies that this characteristic normally is evaluated.

† May not be needed in some cases.

‡ In cases where reproducibility has been performed, intermediate precision is not necessary.

§ Lack of specificity for an analytical procedure may be compensated for by the addition of a second analytical procedure.

‖ Lack of specificity for an assay for release may be compensated for by impurity testing.

May be needed in some cases.

laboratory space in the room where a unit manufacturing operation occurs or in a formal quality laboratory. All physical areas must meet CGMP requirements as per Title 21 CFR 211.160. Good documentation practices [19] also apply. In this respect, all test procedures should either be validated or verified.

5.6.1 In-process Procedures

During the manufacturing of the drug substance and the drug product, tests that are critical for assurance that the manufacturing process is proceeding as written should be included. These tests, when performed as in-process tests and described in Table 5.4, should be validated, or verified.

Independent of type, validation parameters, as described in Section 5.4 may be used. The technique or type of method should be considered when selecting the parameters. In general, similar validation parameters should be selected for the tests whether the tests are performed online, or in-line compared to offline. Consideration for on-line or in-line procedures should include evaluation of positioning of the sensor or probe, the frequency of data collection, calculation formula, and any condition that can affect the reliability of the data collected.

5.6.2 Environmental Procedures

It is common to use established procedures or devices to monitor cleaning surfaces, air, water, and unit operation equipment used in the manufacturing areas. The vendor may supply validated procedures. Thus, it is likely that the user does not validate such procedures. But they should be verified at the site of use. The critical parameter for these procedures is the sensitivity of the procedure (i.e. detection limit and quantitation limit).

Table 5.4 Types of in-process tests.

Type	Category[a]	Data collection speed	Typical tests
Online	Nondestructive	Instantaneous by analyzer	Near IR
In-line	Nondestructive	Instantaneous by probe/ measuring device	pH, temperature, pressure, density, flow
At-line	Destructive	Manual sampling; faster; performed at production area	Hardness, Friability, Weight
Off-line	Destructive	Manual sampling; time-consuming performed in laboratory	Almost all API and finished product tests

[a] Destructive to mean that samples will not be returned to manufacturing batch.

5.6.3 Release and Stability Procedures

All products for release to the commercial market are tested to confirm that the quality is met according to the products' specifications. These tests, which consist of various controls, are to characterize the dosage form. For a sustained release tablet, a dissolution profile test will ensure that the sustained release capability of the tablet is maintained. For biotechnological or biological products, appropriate tests to cover physicochemical properties; biological activity; immunochemical properties; and purity, impurities and contaminants are expected to maintain the quality of such products as described in ICH Q6B [20]. For a pulmonary inhaler, the amount of active delivered by the inhaler using the cascade impactor test will confirm that the particle size distribution is correct and the patient will benefit from the dose. The design and acceptance criterion set for each test is finalized between the manufacturer and the agency that approved the product in the respective countries.

According to ICH Q6B and Q6A, the assay procedure of the API in the finished dosage form does not need to be stability-indicating at release. It assumes that the drug substance has not degraded at the time of release. However, an impurity procedure is still required for the finished dosage form. A stability-indicating procedure aims to resolve compounds from each other, so that impurities, excipients are resolved from the parent compound in the finished dosage form as addressed by the specificity parameter during validation of the procedure [20, 21].

During the development and through the life cycle of the product, studies are performed by the manufacturer to ensure the product is stable during storage or under the condition of use. The stability storage conditions for study in ICH Q1A(R2) are 25 °C/60% RH; and accelerated at 40 °C/75% RH for most finished products. For semi-permeable containers, the storage conditions are 25 °C/40% RH and accelerated at 40 °C/<25% RH to stress the permeability of the container wall for migration of volatiles like moisture and low molecular weights compounds. For API, the storage conditions are also described in ICH Q1A(R2). All stability studies require stability-indicating procedures to differentiate the impurities from the parent compound [22].

All release and stability tests must be validated or verified; the level of validation will depend on the type of tests. More details of stability testing and global conditions to be studied can be found in Chapter 9.

5.7 Other Procedures

5.7.1 Process Analytical Technology (PAT)

Process Analytical Technology (PAT) is a system for designing, analyzing, and controlling the manufacturing process through timely measurements (i.e. during processing) of critical quality and performance attributes of raw and in-process materials and processes to ensure final product quality [23]. These types of measurements, described in the Key Information section that follows, cover biological, physical, and chemical attributes of the materials being processed.

Types of PAT measurements
At-line: Measurement where the sample is removed, isolated from, and analyzed near the process stream
Online: Measurement where the sample is diverted from the manufacturing process and may be returned to the process stream.
In-line: Measurement where the sample is not removed from the process stream and can be invasive or noninvasive

Process analyzers typically generate large volumes of data. For example, data points are collected throughout the process and are monitored for trend, drift, and statistically evaluated. Any method used should be validated according to the appropriate parameters, as described in ICH Q2 [6]. Further, the probes should be qualified and the position of the probes, sensitivity, and the statistical calculations should be considered and confirmed to support the quality claim.

5.7.2 Parametric Release and Real-time Release

It is recognized that a comprehensive set of in-process tests and controls may provide greater assurance of the finished product meeting specification than finished product testing [24]. The parametric release is defined as a sterility assurance release program where demonstrated control of the terminal sterilization process enables a firm to use defined critical process controls, in place of the sterility test, to fulfill the intent of 21 CFR 211.165(a), and 211.167(a) [25]. Terminally sterilized products are treated in a lethal microbial process and must have at least a Probability of Non-Sterility (PNS) of 10^{-6}, which means the probability of product bioburden surviving the sterilization process in any single unit product is less than one in one million. This process allows the elimination of sterility testing for releasing the batch. However, assurance is required to meet the criteria of sterility during stability testing or post-marketing investigations. The parametric release is only permitted after approval by the regulatory authority. All other tests in the drug product specification remain the same. No new procedure is needed for parametric release.

Real-time release (RTR) has a similar concept to monitor production for quality during manufacturing and allows for the elimination of appropriate finished product test(s). It is expected that the facility to have an in-depth understanding of the manufacturing process, robust control strategy, critical in-process controls, and data to support quality during manufacturing. The RTR is approved by the regulatory authority before implementation.

5.8 Validation of Procedures in Continuous and Batch Manufacturing

A batch is a specific quantity of a drug or other material that is intended to have uniform character and quality, within specified limits and is produced according to a single manufacturing order during the same cycle of manufacture as specified in 21 CFR 210.3. Each batch is manufactured independently with recorded testing data for the release of the batch. However, manufacturing technology is continually improving, and as more efficiency is targeted, the movement toward continuous manufacturing is being achieved by more firms for high volume drug products. With that advent of more challenges for the engineer to design the process, the analytical chemist to develop test techniques, the statistician to evaluate the volume of analytical data, and the operator to keep the process running; all with the objective of a quality market product that meets specification or preset test criteria and avoids failures. Tests should be selected to detect changes in the important properties of the product. In-process testing on-line, in-line, and at-line are performed at strategic steps of the manufacturing process to minimize changes. It is too late if a critical change occurs during the manufacturing process and is not detected by the time the final finished dosage form manufacturing step is completed. Release tests are performed only to confirm that the finished product meets the acceptance criteria for each of the listed tests in the product specification. As expected, all tests should be validated or verified accordingly.

5.9 Summary

The importance of validating or verifying a test procedure becomes apparent whether the manufacturing process is by batch or continuous. The quality of the product is determined by how accurate and reliable the procedure is with generating the data.

The main point is the understanding of the design built into the analytical procedures, instruments, data collection systems, general flow, and documentation of the laboratory operations. Also, a life cycle approach for the procedure should be adopted. If changes are made to the procedure, depending on the effect, a repeat of the appropriate validation parameter(s) should be performed. Similarly, the qualification of the instrument should be confirmed periodically to ensure the instrument is functioning properly during use. The purpose is always to have a procedure that will generate reliable and reproducible data to support the quality of products for the consumers.

List of Abbreviations

APIs	Active Pharmaceutical Ingredients
CCIT	Container and Closure System Integrity Test
CFR	Code of Federal Regulations
CGMPs	Current Good Manufacturing Practices
DQ	Design Qualification
FDA	Food and Drug Administration
HPLC	High-Performance Liquid Chromatography
ICH	International Council of Harmonisation of Technical Requirements of Registration of Pharmaceuticals for Human Use
IQ	Installation Qualification
OQ	Operational Qualification
PAT	Process Analytical Technology
PNS	Probability of Non-Sterility
PQ	Performance Qualification
RSD	Relative Standard Deviation
RTR	Real-Time Release
SST	System Suitability Test
UHPLC	Ultra-High-Performance Liquid Chromatography
USP	United States Pharmacopeia
UV	Ultraviolet

References

1 Food and Drug Administration (1978). Current good manufacturing practice for finished pharmaceuticals. Code of Federal Regulations. Title 21, Part 211. Rockville, MD: Food and Drug Administration. https://www.ecfr.gov/cgi-bin/text-idx?SID=a3df941b23a5efab57746e7fb497bb5b&mc=true&node=pt21.4.211&rgn=div5 (accessed 13 July 2020).

2 USP <1078> Good manufacturing practices for bulk pharmaceutical excipients. *USP 40–NF 35*. Rockville, MD: United States Pharmacopeial Convention.

3 Ng, L. (2009). Chapter 14 – the science behind the pharmaceutical regulatory chromatographic procedures. In: *Quantification in LC and GC: A Practical Guide to Good Chromatographic Data* (ed. H.-J. Kuss and S. Kromidas). Wiley.

4 Food and Drug Administration (2017). *Inspection Observations*. Rockville, MD: Food and Drug Administration http://www.fda.gov/ICECI/inspections/ucm250720.htm (accessed 9 July 2020).

5 USP <1058> Analytical instrument qualification. *USP 40–NF 35*. Rockville, MD: United States Pharmacopeial Convention.

6 ICH Harmonised Tripartite Guideline (2005). Validation of analytical procedures: text and methodology Q2(R1). https://database.ich.org/sites/default/files/Q2_R1__Guideline.pdf (accessed 9 July 2020).

7 Food and Drug Administration (2015). *Guidance for Industry. Analytical Procedures and Methods Validation for Drugs and Biologics*. Rockville, MD: Food and Drug Administration Pharmaceutical Quality/CMC. https://www.fda.gov/media/87801/download (accessed 9 July 2020).

8 Food and Drug Administration (1994). *Reviewer Guidance, Validation of Chromatographic Methods*. Rockville, MD: Food and Drug Administration CMC 3. https://www.fda.gov/media/75643/download (accessed 9 July 2020).

9 USP <1225> Validation of compendial procedures. *USP 40–NF 35*. Rockville, MD: United States Pharmacopeial Convention.

10 USP <1223> Validation of alternative microbiological methods. *USP 40–NF 35*. Rockville, MD: United States Pharmacopeial Convention.

11 USP <1226> Verification of compendial procedures. *USP 40–NF 35*. Rockville, MD: United States Pharmacopeial Convention.

12 Xue, G., Bendick, A.D., Chen, R., and Sekulic, S.S. (2004). *Journal of Chromatography A* 1050: 159–171.

13 Ng, L., Lunn, G., and Fastino, P. (2007). *Organic Impurities in Drug Substance: Origin, Control, and Measurement, in Analysis of Drug Impurities*. Oxford: Blackwell Publishing.

14 Wirth, D.D., Olsen, B.A., Hallenbeck, D.K. et al. (1997). Screening methods for impurities in multi-sourced fluoxetine hydrochloride drug substances and formulations. *Chromatographia* 46: 511–523.

15 Food and Drug Administration (2008). *Guidance for Industry. Q3A(R) Impurities in New Drug Substances*. Rockville, MD: Food and Drug Administration https://www.fda.gov/media/71727/download (accessed 9 July 2020).

16 Food and Drug Administration (2006). *Guidance for Industry. Q3B(R2) Impurities in New Drug Products*. Rockville, MD: Food and Drug Administration https://www.fda.gov/media/71733/download (accessed 9 July 2020).

17 USP *<621>* Chromatography. *USP 40–NF 35*. Rockville, MD: United States Pharmacopeial Convention.

18 Brown, R., Caphart, M., Faustino, P. et al. (2001). Analytical procedures and method validation: highlights of the FDA's draft guidance. *LCGC* 19 (1): 74. http://files.alfresco.mjh.group/alfresco_images/pharma//2014/08/22/f924b0b8-18c0-4dd4-a98e-3a03f4bd3505/article-3266.pdf.

19 USP <1029> Good documentation practices. *USP 40–NF 35*. Rockville, MD: United States Pharmacopeial Convention.

20 Food and Drug Administration (1999). *Guidance for Industry. Q6B Specifications: Test Procedures and Acceptance Criteria for Biotechnological/Biological Products*. Rockville, MD: Food and Drug Administration https://www.fda.gov/media/71510/download (accessed 9 July 2020).

21 Food and Drug Administration (2020). *Guidance for Industry. Q6A Specifications: Test Procedures and Acceptance Criteria for New Drug Substances and New Drug Products: Chemical Substances*. Rockville, MD: Food and Drug Administration https://www.fda.gov/regulatory-information/search-fda-guidance-documents/q6a-specifications-test-procedures-and-acceptance-criteria-new-drug-substances-and-new-drug-products (accessed 9 July 2020).

22 Food and Drug Administration (2020). *Guidance for Industry. Q1A(R2) Stability Testing of New Drug Substances and Products*. Rockville, MD: Food and Drug Administration https://www.fda.gov/media/71707/download (accessed 9 July 2020).

23 Food and Drug Administration (2004). *Guidance for Industry. PAT – A Framework for Innovative Pharmaceutical Development, Manufacturing, and Quality Assurance*. Rockville, MD: Food and Drug Administration Pharmaceutical CGMPs. https://www.fda.gov/media/71012/download (accessed 9 July 2020).

24 European Commission (2001). Final version of Annex 17 to the EU guide to good manufacturing practice. http://ec.europa.eu/health/files/eudralex/vol-4/pdfs-en/v4an17_en.pdf (accessed 9 July 2020).

25 Food and Drug Administration (2010). *Guidance for Industry. Submission of Documentation in Applications for Parametric Release of Human and Veterinary Drug Products Terminally Sterilized by Moist Heat Processes*. Rockville, MD: Food and Drug Administration CMC. https://www.fda.gov/media/71461/download (accessed 9 July 2020).

6

Transfer of Analytical Procedures

Kim Huynh-Ba

Pharmalytik LLC, Newark, DE, USA

TABLE OF CONTENTS

Analytical Testing for the Pharmaceutical GMP Laboratory, First Edition. Edited by Kim Huynh-Ba.

6.1 Introduction

Analytical methods are essential tools developed to qualify a laboratory to monitor the quality and integrity of active drug substances or drug products. Analytical procedures in the pharmaceutical industry are typically developed and validated in an analytical laboratory and then transferred to another laboratory during the lifecycle of the drug substance and drug product. This activity is necessary throughout the development and commercialization phases of existing or new drug products to ensure the laboratory can perform the analytical method based on the intended use of the method, and the analytical results obtained are as reliable as demonstrated in the validation. The process of qualifying a laboratory by demonstrating the method performance at a different laboratory is called the *Transfer of Analytical Procedures*.

While Chapter 5, discusses the validation and verification of the analytical procedure, this chapter will discuss the transfer process, the documentation, and important factors influencing the method transfer. Practical process and typical criteria are recommended as a guide to successfully qualify a laboratory. Readers should consider the variability of the procedure, the difficulty level of the procedure, and the capabilities of the laboratories. Statisticians should be consulted, if necessary.

6.2 Purpose of Method Transfer

The purpose of the method transfer is to qualify a laboratory to conduct the testing procedures and demonstrate the reproducibility of the procedures within the predefined acceptance criteria. The source lab, where the method has been demonstrated to deliver results with accuracy, precision, linearity (whether through method validation process or comparison testing), and where the method is obtained from is called the *Transferring Lab*. The lab, where the method is received and evaluated for equivalency, is called the *Receiving Lab*. The goal of the method transfer is to demonstrate the interlaboratory reproducibility. It confirms the receiving lab qualified to apply the analytical procedures, by demonstrating that the resulting data are equivalent to the data generated from the transferring lab. Method transfer also provides additional data for the robustness study of the method validation and could identify additional factors that would influence the performance of the method.

Limited regulatory guidelines are available to describe how this activity should be conducted at the time of publication. Any analytical procedure must be shown to be fit for its intended purpose by ensuring a smooth transition of the analytical method knowledge between the development and manufacturing sites. The United States Pharmacopeia (USP) published General Chapter <1224> for this purpose [1]. Method transfer activity must be documented carefully to show that the receiving lab is qualified to perform the method, and the generated data are as reliable as those generated by the transferring site. Even after validation, certain factors that are not explicitly described in a method may affect the performance of the method in different laboratories. Those factors can be discovered during Method Transfer [2–5].

6.3 Transfer Options

Several options can be used to qualify a new laboratory to perform the analytical procedures through method transfer.

6.3.1 Method Transfer Plan

The success of the Method Transfer depends on the communication between the transferring and receiving sites. It is recommended to have a Method Transfer Plan in place to guide both the transferring and receiving laboratories. The Method Transfer Plan provides detailed information for the receiving lab to prepare adequate resources for the Method Transfer activities. Figure 6.1 shows four options to transfer an analytical procedure from one lab to another, depending on several factors, such as the stage of product development, the stability of the samples, geographic location, shipping, and handling procedures. Figure 6.2 introduces a flow chart on how these options could be used to facilitate the transfer of analytical procedures.

As the transfer team discusses the plan, both the transferring and receiving sites need to agree with the strategy, option, and materials used. This agreement is documented in an Analytical Transfer Plan, and both labs are committed to executing it. This document also explains the justification of the selected option.

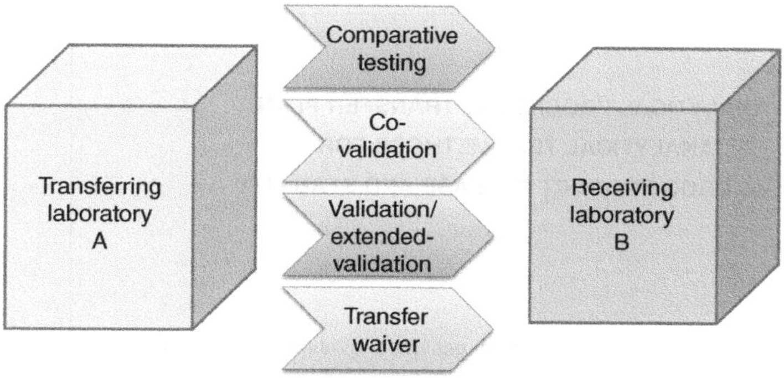

Figure 6.1 Summary of transfer options of analytical procedure.

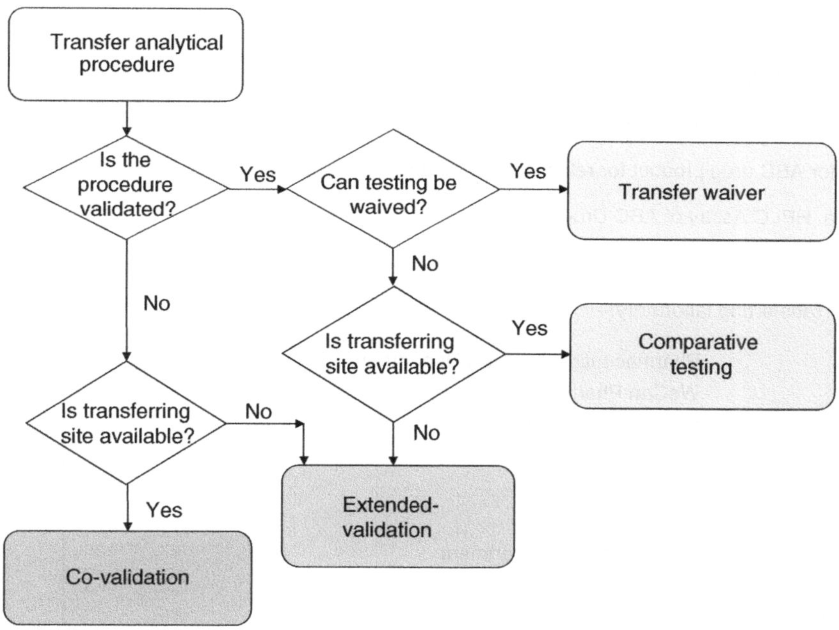

Figure 6.2 Decision tree for transfer options.

An example of the Analytical Method Transfer Plan can be found at the end of this chapter (Figure 6.3).

6.3.2 Comparative Testing

Comparative testing is a study designed to directly evaluate the equivalence of test results between two or more labs through testing duplicate sets of samples. It is a

ANALYTICAL PROCEDURE TRANSFER PLAN
ANALYTICAL TEST METHODS FOR
ABC DRUG PRODUCT RELEASE AND STABILITY

1. PURPOSE

This plan serves as guidance to govern all activities involved in the technology transfer of ABC drug product analytical test methods as defined in the Scope section of this document in accordance with current corporate and site policies and procedures.

2. SCOPE

This document applies to the transfer of the following analytical test methods for ABC drug product for release and/or stability evaluation:

a. HPLC Assay of ABC Drug Product

b. HPLC Assay for preservative in ABC drug product

Transferring laboratory:

Pharmaceutical Analysis Department
WeCan Pharma
City, State, Country

Receiving laboratory:

Quality Control Department
GoodDay Pharmaceuticals
City, State, Country

Figure 6.3 Example of an analytical method transfer plan.

3. BACKGROUND

Analytical technology transfer for ABC drug product will be made to the receiving laboratory, which has experience in performing highly similar methods for WeCan Pharma drug products. The methods were developed and validated at the transferring lab and their analysts have been qualified in the procedures listed above. These analysts will transfer the methods to the receiving lab involved in commercial product of this product.

This plan is prepared based on SOP 124. This document will guide all ongoing and upcoming analytical technology transfer activities for ABC Drug Product analytical test methods.

The Analytical Transfer Team will manage the entire analytical technology transfer and qualification process. Membership of the Analytical Transfer Team consists of the following:

Supervisor –Transferring lab

Analyst –Transferring lab

QA Specialist –Transferring lab

Analyst A – Receiving lab

Analyst B – Receiving lab

QA Specialist – Receiving lab

4. ANALYTICAL TRANSFER PACKAGE

Analytical Transfer Package consists of ABC drug product test methods, validation plans, validation reports, any applicable raw data requested, and specifications for release and stability evaluation from the transferring laboratory to the receiving laboratory. Technical guidance and an informal inventory checklist will be provided by the transferring laboratory during the procurement of critical analytical equipment, chemicals, and other lab supplies if necessary.

Figure 6.3 (Continued)

5. TIMELINES

- Preparation of the Analytical Qualification Protocol by a member or designee is targeted to be completed – Due on
- Review and approval of the Analytical Qualification Protocol by management from both labs to be completed – Due on
- Method familiarization and demonstration provided by trainers from the transferring laboratory to the trainees from the receiving laboratory, according to an approved Training Program – Due on
- Preparation and Approval of the Analytical Qualification Report on completion of work – Due on

6. REQUIREMENTS

- Analytical methods to be transferred must be validated
- Analytical equipment/instruments must be calibrated and qualified
- An approved Qualification Protocol with predetermined acceptance criteria must be available with detailed procedures for qualification experiments
- All personnel from the receiving laboratory involved in the qualification must be trained on the test method prior to performing qualification experiments
- All qualification experiments must be performed in accordance with the pre-approved qualification protocol
- Requirements on raw data recording, deviation investigation, and analytical qualification report preparation will also be provided in the protocol.

Figure 6.3 (Continued)

7. RESPONSIBILITIES

It is the responsibility of the transferring lab to provide test methods, reports, any relevant documents, materials to the receiving lab. The analyst of the transferring lab must also be available to provide instruction and guidance to the analyst of the receiving lab, when needed.

It is the responsibility of the receiving lab to ensure that equipment and analysts are qualified to perform the transfer. Each analyst of the receiving site is responsible to conduct the lab training and execute the transfer protocol as written.

8. ACTION PLAN AND PROCEDURES

8.1 Preparation and approval of Training activities

Training experiments will be designed to ensure that the analysts of the receiving site can perform the methods as stated. A training protocol will be prepared to detail this training exercise.

8.2 Preparation of Analytical Qualification Protocol

- The following qualification experiments will be completed:
- Comparative testing will be the qualification option chosen for all procedures transferred under this protocol.
- Trained analyst(s) from the receiving laboratory will demonstrate linearity, specificity and quantitation limit using spiked placebo samples. Reproducibility will be demonstrated at the receiving laboratory using one batch of ABC drug product that is past expiration.
- The batches will be identified in the protocol.

Figure 6.3 (Continued)

- Qualification of the receiving laboratory will be achieved by demonstrating that all pre-determined acceptance criteria identified in the qualification protocol are met.

8.3 Execution of the Analytical Qualification Protool

- Following successful completion of laboratory demonstration trials, both groups at their respective sites will execute the analytical qualification protocol.
- Test plan and procedures specified in the qualification protocol must be followed.
- All raw data and test results must be recorded and reviewed according to site policies and procedures.

8.4 Deviations and Investigations

Any deviations to the qualification protocol and/or investigations involved must be fully investigated and approved by management from both sites in accordance with site procedures. All deviations and investigations must be resolved and approved prior to approval of the analytical qualification report.

8.5 Preparation and Approval of Analytical Qualification Report

A written report must be prepared to summarize all activities and data generated from execution of the qualification protocol. This includes an assessment of the results against the predetermined acceptance criteria and a summary of all deviations observed (if any), and a conclusion as to the qualification status of the receiving laboratory.

8.6 Analytical Transfer File

The receiving site will maintain a copy of the Analytical Transfer and

Figure 6.3 (Continued)

Qualification File according to the site procedures throughout the lifecycle of the product.

9. REFERENCES

1. SOP 123, "Analytical Technology Transfer and Qualification," Version 1.0.

10. SIGNATORIES

Written by: ______________________________ Date: __________

Project Leader: ______________________________ Date: __________

Transferring Lab:

Approved by: ______________________________ Date: __________

QA Approved by: ______________________________ Date : __________

Receiving Lab:

Approved by: ______________________________ Date: __________

QA Approved by: ______________________________ Date: __________

Figure 6.3 (Continued)

commonly used option when method validation of the analytical procedure is completed. This option ensures the harmonization of the procedure at both labs; however, it does not challenge all validation parameters of the method.

6.3.3 Co-validation

Co-validation is an option when the analytical procedure is not yet fully validated. In this case, the receiving lab participates in and is a part of the validation of the method. Both the transferring and receiving labs identify the critical validation parameters of the method to be conducted. Data generated from the receiving lab and the transferring lab are included in the validation report. The advantages of the co-validation approach are the Receiving lab can contribute to the method before submission and is a part of the validation of the method. Thus, no additional requirement is needed to demonstrate equivalence when the validation results meet the acceptance criteria.

6.3.4 Extended Validation or Partial Validation

Extended validation (revalidation) or partial validation is an option often taken when comparative testing cannot be done. The choice of extended-validation or partial validation will depend on the type of method transferred. In this case, the receiving lab will perform part of the validation or repeat a complete validation. This option is used when:

- the comparative testing is not efficient to determine the discrepancy in the results obtained from an identical set of samples
- when the transferring site is not available
- when the validation was done a long time ago that equivalency is difficult to establish

For example, this strategy choice may be a better option for the content uniformity method than comparative testing, as the procedure would depend heavily on the precision of the method. However, this option is time-consuming, and it may be more difficult to observe the differences that may exist concerning method operation and instrumentation. It does not ensure the equivalency of the test method performed at different labs.

6.3.5 Transfer Waiver

When using transfer waiver, it is not necessary to perform the formal transfer experiments. The waiver is considered if the receiving lab has documentation to qualify it to perform the method. Here are a few examples:

- The new product composition is comparable to that of an existing product, and it uses the same analytical method. The receiving lab already has experience with the method.
- The analytical procedure of the method transfer is described in the United States Pharmacopeia and the National Formulary NF (USP-NF) and is unchanged. Verification is being done using USP <1226> *Verification of Analytical Procedures*.
- The analytical method is the same as, or similar to, a procedure already in use.
- The personnel in charge of the development, validation, or routine analysis of the product at the transferring unit is/are responsible personnel at the receiving unit, and all instruments used at the receiving unit are similar to those at the transferring unit.

6.4 Method Transfer Process

Transfer activities must be carefully selected and agreed upon between the transferring and receiving labs. Issues that arise during a method transfer can be costly and cumbersome. Additionally, if the testing material(s) used for the method transfer happens to be the current lot of material(s) that is (are) related to a pending registration of products, they can also cause investigations, if the discrepancy in results are found. The lab investigation for errors can jeopardize and delay the registration approval of the product.

For a well-planned method transfer, the first requirement is the ability to validate the analytical procedure. Resources must be discussed in advance, to ensure that equipment and analysts are qualified and available for the method transfer activities. If needed, new equipment should be purchased and qualified before the transfer is initiated. The list of equipment is evaluated, and the analysts involved in the transfer must be trained with the techniques. A consistent process for carrying out the transfer of an analytical procedure would help analysts understand their role. The transfer process goes through four major phases:

6.4.1 Preparation Phase

The preparation of each transfer should be planned methodologically. It may be more time consuming for the first transfer between two sites; however, once this process is established, the follow-up transfer would be more straightforward to execute, and the process can be completed consistently.

6.4.1.1 Select a Method Transfer Team

A transfer team is necessary to properly execute the transfer of the analytical procedure from one site to another. As with any team, a team leader provides project leadership and has the overall responsibility for the project. For a large project, a project manager may be needed to communicate the planning, timelines, and facilitation of activities. This person is accountable for the adherence to the project plans and standards, as well as to provide additional resources as needed.

Core Team – The core team consists of members from transferring and receiving sites. At a minimum, two people from each site are included in this team: one representing the laboratory or testing area (quality control) and others representing the quality area (quality assurance). These individuals are responsible for scheduling and conducting activities at their sites and for working with the team leader to direct the project at their sites.

Ad hoc members – Ad hoc team members represent areas such as regulatory, statistics, production, technical services, and provide support to the transfer team as needed.

6.4.1.2 Method Transfer Package

The success of the transfer process depends on a thorough understanding of the method and its limits; the transferring team has more experience with the method through method development and validation data. This knowledge should be communicated with the receiving lab often and early with an information package, of which the receiving site will perform a gap analysis after reviewing. The Key Information section below lists examples of documents included in a Method Transfer Package.

Meetings between the transferring and receiving units are necessary to clarify any issues and answer questions regarding the transfer process and the analytical procedures transferred. The technology should be well established in the receiving lab. If the receiving lab does not have experience in the technique or some aspect of the procedure to be transferred, then training must be done before the qualification experiments start.

Content of a method transfer package

- Analytical methods to be transferred.
- Validation reports for these methods.
- Any subsequent supplemental validation.
- Applicable specifications for release and stability.
- Data generated using the method to be transferred.
- Equipment compatibility checklist.
- Materials availability checklist.
- Equipment inventory checklist.

A review of the validation of the analytical procedure is important. The team will rely on this validation, as well as the database of the samples tested with this procedure, to determine the appropriate acceptance criteria for the transfer. History data of the method is also important and helpful to understand the changes made to the method.

A checklist of the required equipment is critical for the receiving site to evaluate their readiness. This list should include all equipment and all parts, including the software that is currently used. Recommendations and information about suggested vendors for appropriate materials may also be helpful. A checklist of materials such as reagents, salts, samples, reference standards, syringes, filters, vials, etc. are important for the receiving site to conduct their evaluation. The team should discuss if equivalent materials are used in the transfer. Figure 6.4 is an example of a transfer gap analysis checklist.

Gap Analysis for ABC Methods

1. General Section

1.1 Analysts

Any analyst from the transferring lab involved in the transfer of a method will have been trained on that method. The training will be documented on the analyst's training record.

The analysts at receiving lab involved in the transfer will be proficient in the technique that is being transferred such as high performance liquid chromatography (HPLC), dissolution, gas chromatography, etc.

Analysts receive training in basic laboratory procedures and CGMP procedures upon joining the company and on a yearly basis.

1.2 Solvents

HPLC-grade solvents (including HPLC grade water) will be used for the preparation of sample and standard solutions.

HPLC-grade solvents may be used for the gas chromatography work.

Water from laboratory purification systems (e.g. Millipore Milli-Q) may be used for the rinsing of glassware, and other non-quantitative uses.

1.3 Standard Operating Procedures

The analysts will be required to follow the current version of all applicable Standard Operating Procedures, Standard Instrument Procedures, Standard Equipment Procedures and Standard Safety Procedures that are in place.

1.4 Standard Laboratory Equipment

Specific equipment items are shown in the tables that follow.

All relevant pieces of equipment must meet current site specific metrology requirements.

All pipets are glass, volumetric, Class A, "to contain" (TC) pipets.

All flasks are glass, volumetric, and Class A, "to contain" (TC).

Mobile phase may be degassed by sonication under vacuum or by helium sparging.

Figure 6.4 Example of a transfer gap analysis checklist.

1.5 Instrumentation

The required analytical instrumentation is summarized in the tables that follow. Additional details can be found in the test methods themselves.

All instrumentation used in the transfer will be calibrated and in proper working order.

All instrumentation must meet site-specific metrology requirements.

2. SPECIFIC SECTION

2.1 Assay Method – Procedure no.123

List of Equipment

List of Reagents

List of Reference Standards

List of Test Articles required per Analysis

2.2 Residual Solvent –Procedure no. 456

List of Equipment

List of Reagents

List of Reference Standards

List of Test Articles required per Analysis

3. COMPLETION OF GAP ANALYSIS

In signing this document, the receiving lab confirms that the gap analysis has been performed. All the issues have been addressed and documented. The receiving lab is ready to initiate the training and transfer qualification activities of these methods.

4. SIGNATORY

Performed by: ______________________________ Date: __________

QA Approved by: ______________________________ Date: __________

Figure 6.4 (Continued)

After reviewing the information package, the core team will decide and agree on which option to use based on the method, sites, and personnel involved with the transfer. The materials will be selected, including reference standards, samples, placeboes, and excipients.

6.4.2 Gap Analysis

The receiving site reviews the transferring package and evaluates the readiness of their lab. Analytical methods to be transferred must be validated and approved. Four different areas are considered in the gap analysis.

6.4.2.1 Equipment Capabilities

The receiving site should decide on which equipment to be used in the transfer. The types of analytical equipment transferred must be qualified and calibrated for the duration of the transfer. The difference in instrumentation between the two sites should be understood. As an example, all C18 columns may not give the same separation due to their packings. Different high-performance liquid chromatography (HPLC) systems could exhibit different impurity profiles; thus, column source and maintenance should be established. Reagents should be commercially available so that the transferring site does not have to provide supplies to the receiving lab. Data systems must be equivalent and qualified, and the Laboratory Information Management System (LIMS) or Excel spreadsheets should be qualified if used.

6.4.2.2 Facilities Readiness

Facilities where the laboratory is located are very critical. Different systems such as water, sewage, gas lines, power source must be critically evaluated, especially if the transfer involves laboratories residing in different regions of the world. Climatic conditions and handling procedures are evaluated for samples that are not known to be stable.

6.4.2.3 Analyst Qualifications

All personnel from the receiving laboratory involved in the qualification must be trained on the test method prior to their involvement in the execution of transfer experiments. Qualification of analysts should be reviewed according to the methodology and technology used in the transfer. Training records of all analysts involved must be available and completed up to date. Analysts should go through the training phase to ensure that the method is feasible before performing the transfer experiment.

6.4.2.4 Materials Used in the Transfer

All materials needed for the transfer must be readily available at the receiving site. Extra samples or materials should be included and accessible in case an investigation arises. Appropriate storage conditions must exist to ensure that quality and integrity are retained for the samples tested. Reagents and placebo(es) must be equivalent. It is preferred to use commercially reference standards. Depending on the stability of the materials, reference standards and reagents may need to be shipped from the transferring site to the receiving site.

The sample lot to be used in the transfer must be carefully selected. It is recommended that an expired lot be used in the transfer to avoid compliance issues if data generated does not meet acceptance criteria used in the transfer; however, expired materials may not be available for the early development stage at the time of the transfer.

The objective of the transfer activity is to qualify a different lab to perform the same analytical procedure. The quality and uniformity of the sample lot may affect the transfer. Information such as product variability, number of batches available, stage of development, and several strengths should be considered when the samples are selected. A statistician should be involved to determine the number of lots and samples used in the transfer.

After the receiving site has performed a gap analysis, a plan should be developed to address the differences. This plan will include timelines and personnel. After these gaps are addressed, the receiving site can initiate the transfer activities. The readiness of the receiving site is documented and retained as a part of the transfer report.

Typical questions when performing a gap analysis

- Does the receiving lab have the needed instrument? If not, is there a budget to acquire and qualify it?
- Does the receiving lab have adequate facilities and services to support the equipment performance? If not, is there any modification needed?
- Does the receiving lab have the necessary reagents? Are they the same grade listed in the procedure?
- Does the receiving lab have personnel who possess the skills necessary to perform the procedure?
- Which lot of samples will be used in the transfer? How are these samples shipped to the receiving lab?
- What are the possible differences in the equipment used in the procedure?
- Is instrumentation compatible between the transferring lab and the receiving lab?

- Is the procedure fully validated according to ICH guidelines (Q2 R1)?
- Was the procedure previously transferred?
- If yes, what were the acceptance criteria?
- Were these criteria similar to validation acceptance criteria?
- Were there any exceptions to the method transfer protocol?
- Where will the receiving lab source their materials (reagents, standards, controls, and reference materials) after the transfer?
- What is the projected usage of reference standards at the receiving laboratory? How will this impact the transferring lab's supply?

6.4.3 Method Training Phase

The training period is a crucial phase in the transfer process. Most transfer failures can be avoided if issues are identified and addressed during the training. A training protocol with predetermined criteria can be used to familiarize the analyst with the procedures. The training program should cover the procedures in the transfer and any applicable training acceptance criteria. Approval of the training documentation and any issues identified during training should be addressed before executing the analytical qualification experiments. Issues observed during training, their cause, and action steps taken must be documented on the training form. The training program will consist of two main goals.

6.4.3.1 Method Familiarization

The training analyst at the receiving lab should review and discuss the test methods, equipment, supplies, and safety procedures with an experienced analyst from the transferring lab. The training analyst will observe and perform all laboratory techniques required to perform the test methods, including any safety procedures and method precautions. If needed, the analyst at the receiving lab can practice the laboratory techniques needed to run the procedure under the supervision of the analyst from the transferring lab. This training will be conducted at the transferring lab. If travel is not feasible, this can be done via telephone or using any other form of technology such as video, webcast, etc. This activity can be waived if the analyst from the receiving lab is experienced with the method or technology. Upon completion, this activity should be documented.

6.4.3.2 Method Demonstration

Once the analyst from the receiving lab completes the method familiarization step, he/she proceeds to demonstrate this by repeating the procedure at the receiving lab using the equipment of the receiving lab. The analyst from the transferring lab should be available for guidance as needed by the training protocol. This step allows the receiving analyst to work out all the final details of the method. This activity is sometimes used as a dry run before the formal qualification step. Upon completion, training data is evaluated against acceptance criteria and documented as a part of the analyst qualification. Additional details or issues learned during the training phase should be addressed before executing the qualification exercise.

6.4.4 Method Qualification Phase

Once the analyst at the receiving lab is trained to perform the procedure, the qualification phase can begin. A method transfer protocol is prepared including predetermined acceptance criteria and a representative sample for the transfer is selected. Depending on the strategy selected by the transferring team, the qualification is executed. Qualification data are evaluated against the pre-determined acceptance criteria and reported as proof of site qualification.

6.5 Transfer Protocol

6.5.1 Content of a Transfer Protocol

A transfer protocol is necessary to document the plan that was discussed and agreed upon with pre-determined acceptance criteria. The qualification approach, detailed procedures on qualification experiments, and reference test articles to be used during the qualification are documented in the protocol. The following sections are necessary for a transfer protocol:

Content of a method transfer protocol

- Objectives/scope
- Roles and responsibilities
- Assessment of the receiving lab
- Materials, facilities, and instrumentation
- Analysts training
- Qualification procedure
- Acceptance criteria
- Protocol amendment and deviation (if any)

Figure 6.5 gives an example of an Analytical Transfer Protocol. This protocol includes the sites to be transferred as well as those who are responsible for the activities. It also includes the detailed procedure and materials used for the transfer.

ANALYTICAL PROCEDURE TRANSFER PROTOCOL

ANALYTICAL TEST METHODS FOR

ABC DRUG PRODUCT RELEASE AND STABILITY

Written by: ______________________________ Date: __________

Project Leader: ______________________________ Date: __________

Transferring Lab:

Approved by: ______________________________ Date: __________

QA Approved by: ______________________________ Date: __________

Receiving Lab:

Approved by: ______________________________Date __________

QA Approved by: ______________________________Date: __________

1. ANALYTICAL PROCEDURES

This protocol applies to the transfer of the following analytical test methods for ABC drug product for release and/or stability evaluation:

- HPLC Assay of ABC Drug Product
- HPLC Assay for preservative in ABC drug product using HPLC
- Cleaning Verification of ABC drug product manufacturing equipment by HPLC

Figure 6.5 Example of an analytical qualification protocol.

2. SITES INVOLVED

Transferring laboratory:

Pharmaceutical Analysis Department

WeCan Pharma

City, State, Country

Receiving laboratory:

Quality Control Department

GoodDay Pharmaceuticals

City, State, Country

3. OBJECTIVES

- This protocol includes experiments that serve to demonstrate equivalence between the transferring lab and the receiving lab.
- It provides training information for the analysts. Successful training will be demonstrated by meeting the acceptance criteria asdescribed in the training protocol (see attached). This training will be approved by management of the receiving lab.
- It provides qualification experiments that both labs execute and report on the report forms (see attached). Successful qualification will be demonstrated by meeting the pre-determined acceptance criteria, as specified in this protocol. This qualification will be approved by the transferring lab.
- An analytical technology transfer report based on the results of the studies described in this method transfer protocol will be issued upon completion of these experiments. This document will serve as the qualification of the receiving lab.

Figure 6.5 (Continued)

4. REFERENCES

The following procedures will be covered within this protocol:

- HPLC Assay of ABC Drug Product
- HPLC Assay for preservative in ABC drug product using HPLC
- Cleaning Verification of ABC drug product manufacturing equipment by HPLC
- SOP 123, "Analytical Technology Transfer and Qualification," Version 1.0.

5. ANALYTICAL TRANSFER TEAM

Supervisor –Transferring lab

Analyst –Transferring lab

QA Specialist –Transferring lab

Analyst A –Receiving lab

Analyst B –Receiving lab

QA Specialist –Receiving lab

6. PROCEDURES FOR ANALYTICAL QUALIFICATION

6.1 The qualified analyst from the transferring lab will train the analysts from the receiving lab per the training protocol. The laboratory demonstration trials must be successfully completed and approved prior to the start of the method transfer and qualification experiments

6.2 All raw data generated during the qualification phase will be recorded, calculated, and reported by each analyst and reviewed/approved by each site's management according to site procedures.

6.3 Both labs will issue a summary report, including associated data and the original result sheets, to the Transfer Team Leader for incorporation into the Analytical Transfer Report.

Figure 6.5 (Continued)

6.4 All reported results will be evaluated against the acceptance criteria. All results should be reported to the same number of significant figures as required by the acceptance criteria of this protocol.

6.5. All raw data and associated calculations will be audited according to site procedures. The Analytical Transfer Report will be audited against the reported data from both sites.

6.6 Any laboratory investigations or deviations will be conducted according to site procedures (SOPs). Such reports, if any, will be provided to the Transfer Team Leader for inclusion into the Analytical Transfer Report.

6.7 The receiving lab is not qualified to run the transferred test methods until all qualification acceptance criteria are met and the Analytical Qualification Report has been completed and approved.

7. MATERIALS

- ABC Drug Product, Batch no. 235345
- ABC Reference Standard, Lot no. 23425645
- Preservatives, Lot no. 5456

8. QUALIFICATION EXPERIMENTS

8.1. QUALIFICATION OPTION

The comparative testing option, as approved in the Analytical Transfer Plan, will be employed.

8.2. PROCEDURE -HPLC Assay of ABC Drug Product

8.2.1 Demonstration of Equipment Performance in the Receiving Laboratory. The analysts trained will perform this testing independently in their laboratory. System suitability is to be established, according to the procedure, prior to each analysis.

Figure 6.5 (Continued)

8.2.2 Demonstration of Qualification Experiments

Detailed procedures including preparation of standard and test samples

8.2.3 HPLC analysis

Samples are analyzed using bracketing standards according to the procedure.

8.2.4 Data Evaluation

8.2.5 Acceptance Criteria

8.2.6 Documentation of results from site comparisons found on the appropriate qualification forms (see attached).

8.3. PROCEDURE -HPLC Assay for preservative in ABC drug product

8.3.1Demonstration of Equipment Performance in the Receiving Laboratory.

The analysts trained will perform this testing independently in their laboratory. System suitability is to be established, according to the procedure, prior to each analysis.

8.3.2 Demonstration of Qualification Experiments

Detailed procedures including preparation of standard and test samples

8.3.3 HPLC analysis

Samples are analyzed using bracketing standards according to the procedure.

8.3.4 Data Evaluation

8.3.5 Acceptance Criteria

8.3.6 Documentation of results from site comparisons found on the appropriate qualification forms (see attached).

9. DEVIATIONS AND INVESTIGATIONS

9.1 If any result fails to meet the acceptance criteria, an investigation will be

Figure 6.5 (Continued)

conducted to determine the reason. Site SOP's on the investigation of analytical results must be followed.

9.2 Any deviations to the qualification protocol and/or investigations involved with the transfer experiments must be fully investigated, justified, and documented by the Analytical Transfer Team and approved by management from both sites in accordance with site procedures.

10. ANALYTICAL TECHNOLOGY TRANSFER REPORT

- Original raw data generated at both the receiving and transferring site should be documented and archived according to site procedures. All data will be sent to the Analytical Transfer Team Leader. The Leader will review the data and evaluate the results against the acceptance criteria listed in this protocol.

- The Analytical Transfer Team Leader or designee will prepare the qualification report for the transfer. It will include an assessment of the data from the transferring and receiving lab using the acceptance criteria listed in this protocol.

- The report documents and certifies, by approval signatures, that the method transfer was performed according to the qualification protocol, data has been audited, and that all data in the report are accurately presented.

11. ATTACHMENTS

- Analytical Transfer Plan for the ABC Drug Product methods
- Qualification test result sheets.

Figure 6.5 (Continued)

6.5.2 Objectives/Scope

Objective(s) include information such as the method(s) to be transferred, the sites that are involved in the activity, types of samples, and any other relevant information. The scope should identify the method(s) being transferred and the sites involved. ICH guidelines indicate that "the objective of validation of an analytical procedure is to demonstrate that it is suitable for its intended purpose" [1, 6].

6.5.3 Roles and Responsibilities

Roles and responsibilities are described for all parties involved. All parties should agree on information such as names of the transferring and receiving labs, the names of analysts involved in the transfer, planned activities, and timelines. The data generated by the labs should be fully audited by the Quality Assurance department of the corresponding labs.

6.5.3.1 Transferring Laboratory

The responsibilities of the transferring laboratory include, as appropriate, but are not limited to:

- Providing approved test methods, development and validation reports, and current specifications of the product to the receiving laboratory.
- Providing technical support to the receiving laboratory during the early analytical transfer package review.
- Preparing and approving the Analytical Qualification Protocol.
- Providing reference test articles for use during the transfer and qualification process.
- Providing training to the receiving laboratory prior to the execution of the qualification experiments and supervision during the actual qualification, if necessary.
- Conducting the receiving laboratory readiness assessment to ensure the receiving laboratory is ready for the qualification regarding people, equipment, safety, and facility.
- Documenting, retaining, and archiving all laboratory-generated raw data at the site according to the qualification protocol as well as site policies and procedures.
- Handling of deviations encountered by the qualified analyst during qualification according to site policies and procedures.
- Maintaining, storing, and archiving the original analytical technology transfer and qualification file and subsequent transfer of documents.
- Preparing and approving the Analytical Qualification Report.

6.5.3.2 Receiving Laboratory

The responsibilities of the receiving laboratory include, as appropriate, but are not limited to:

- Ensuring that all analytical equipment used during the transfer and qualification process is qualified and calibrated.
- Executing the qualification according to the approved qualification protocol and applicable corporate policies and standards, site policies, and procedures.
- Documenting, retaining, and archiving all laboratory-generated raw data at the site according to the qualification protocol and site policies and procedures.
- Handling deviations encountered by the analysts during qualification according to site policies and procedures.
- Maintaining, storing, and archiving a copy of the analytical technology transfer and qualification file at the receiving site throughout the lifecycle of the product.
- Documenting analyst qualification for each test procedure, which is transferred according to site policies.
- Approving Analytical Qualification Protocol and the Analytical Qualification Report.

6.5.4 Assessment of Receiving Lab

The receiving laboratory performs a readiness assessment before the execution of the method training and qualification. The receiving lab reviews the inventory list of equipment, chemicals, and other critical materials to be used for laboratory demonstration trials and qualification experiments in comparison with those specified in the test methods.

The analysts must review the analytical procedure(s) and any corresponding document numbers. The complete report (or partial validation, depending on circumstances) should be available to the receiving unit, along with any technical details required to perform the test procedure. Databases available for the products and additional instruction notes, if any, should also be made available to the receiving lab.

6.5.5 Materials, Facilities, and Instrumentation

As indicated, the list of materials should include the reagents, reference standards, samples used in the transfer depending on the type of analytical procedures. The receiving unit should also inspect the suitability of laboratory environmental conditions to perform the tests. Laboratory infrastructure is part of Current Good Manufacturing Practice (CGMP) and is also important for the transfer activities. The laboratory systems should comply with applicable regulations and in-house general laboratory procedures. The U.S. Pharmacopeia recommends that only one lot of samples should be involved in the transfer; however, this should depend on the method variability, process variability, and quantity of available samples. A statistician should be consulted if there are concerns regarding the number of lots needed for the transfer [1, 6].

6.5.6 Analyst Training

Names of the analysts involved in the transfer from both the transferring site and the receiving sites must be listed in the transfer protocol. The receiving lab should assess the skills and experience of the personnel involved in the transfer and qualification experiments. If possible, the transfer should include two analysts from the receiving site. For methods such as identification or residual solvent, one analyst may suffice. The analyst's training record should be included, with a statement to indicate that the analyst is trained to carry out the qualification experiment from the receiving lab.

6.5.7 Qualification Procedure

Transfer activity must be specific based on the phase of the product development, type of the methods transferred, and any additional information on the sample preparation. Details of the protocol should be discussed and agreed upon among the involved parties. The Key Information section below lists some examples of information needed in a transfer protocol.

Key considerations for method transfer

- How many analyses will be transferred?
- How many analysts/instruments will be involved?
- What are the acceptance criteria?
- Can a transfer waiver be used?
- How will investigations of transfer failures be handled?

When a method is transferred, the knowledge of the method is being transferred to the receiving site. Therefore, the receiving site must understand the mechanism, the reliability as well as the variability of the procedure. As part of knowledge management of the analytical procedure, the procedure should be written with sufficient detail and explicit instructions so that a trained analyst can execute it without difficulty. The number of replicates, injection sequences, number of injections, and system suitability in the case of

Table 6.1 Typical number of replicates recommended for different analytical methods.

Analytical procedure	Number (N) of samples
Assay/potency	$N = 6$ for one batch $N = 2$ for multiple batches (e.g. ≥ 3)
Impurity/degradation products	$N = 4$ for one batch $N = 2$ for multiple batches
Content uniformity	$N = 10$
Dissolution	$N = 12$
Karl Fischer titration	$N = 6$ for one batch $N = 2$ for multiple batches
pH	$N = 6$ for one batch $N = 2$ for multiple batches
Residual solvents	$N = 2$ for one batch $N = 1$ for multiple batches
Identification	$N = 1$

chromatography should be written. All qualification experiments must be described in a pre-approved qualification protocol. The experiments depend on the experience of the receiving site and the complexity of the method. Different strategies can be taken to qualify for the receiving lab. Table 6.1 suggests the typical number of replicates for different types of analytical procedures depending on the batches available; however, the number of samples should depend on the variability of the method and the batch. Each analyst should perform the analytical procedure independently, including all preparations of reagents, standards, and working samples. The geographic location of the labs should be considered when selecting the transfer options to make sure the materials can be made available to the receiving site. The transferred procedure must include system suitability, quantitation limit (for impurity methods), batch analysis, and other measures of precision/accuracy. A specific timeframe to complete the transfer testing should be included in the protocol depending on the strategies used.

Validation characteristics listed in the Key Information section should be considered to assure that the receiving site can perform the procedure as accurately and reliably as the transferring site. Qualification experiments are selected based on the difficulty of the procedure, the intended use of the methods, and the experience of the sites involved.

Validation characteristics
• Precision/intermediate precision • Recovery/accuracy • Linearity • Limit of detection • Limit of quantitation • Batch analyses

When an impurity method is being transferred, an actual sample containing spiked impurities should be used in the transfer study. Limit of quantitation (LOQ) and limit of detection (LOD) depend on the detector used; therefore, the receiving lab must confirm these characteristics. If authentic substances of relevant impurities are not available for the impurity study, then a stressed sample can be used to ensure appropriate identification and quantitation of all impurities.

6.5.8 Acceptance Criteria

Acceptance criteria must be predetermined and documented in the protocol. These values depend on method performance as established via validation, product specifications, quality of released batches, stability profile of the product, and variability of the analytical procedure. A statistician can assist in the setting of the acceptance criteria for high-risk drug products such as biologics or special dosage forms, or more complex methods as needed. Typically, relative standard deviation and an average of test results are set to ensure the precision and accuracy of the method Tables 6.2 and 6.3 list examples of acceptance criteria for typical analytical procedures.

It should be noted that the acceptance criteria in Tables 6.2 and 6.3 do not exert any statistical basis. Proper choice of acceptance criteria will impact the ability to control the product quality at the receiving site. Process variability should also be considered when setting acceptance criteria for uniformity testing.

Table 6.2 Suggested acceptance criteria used for accuracy.

Analytical procedure	Typical acceptance criteria
Assay/potency	99.0–101.0% for drug substance 98.0–102.0% for drug product
Moisture testing	95–105% of specification
Impurity/degradation products	70–130% of either target impurity specification or expected level of impurity to be measured

Table 6.3 Suggested acceptance criteria used for precision.

Analytical procedure	Typical acceptance criteria
Assay/potency	2.0%
Dissolution	5% at Q value Or F2 evaluation to compare profiles
Impurity/degradation products	Varies depending on the impurity level and LOD Examples: 15–20% for 0.1–0.3% level 10% for 0.3–1.0% level 5% for 1.0% and above

If acceptance criteria are not met, an investigation should be conducted to determine the cause of failure. The investigation may provide guidance on the nature and extent of the remedial steps, which may bring changes to further analyst training and clarification for more complex approaches depending on the procedure.

6.5.9 Protocol Amendment and Deviation

A protocol amendment or protocol deviation is required to document the event when there is a change made to the originally signed protocol of the method transfer. A justification should be documented with any impact on the original plan. These changes should also be included in the method transfer report. These amendments or deviations must be reviewed and approved by all the members who approved the original protocol (or designees, as applicable). All deviations and investigations must be resolved with appropriate corrective action and preventive action (CAPA), before the approval of the method transfer report. If these changes cause changes of method, then they should be incorporated into the method revision.

6.6 Method Transfer Report

6.6.1 Objectives

Upon completion of all activities of the method transfer plan, a written report is prepared to summarize all activities and report the generated data to certify that the execution of the qualification protocol is completed. The protocol should have specified which site is responsible for assembling the final transfer report.

The method transfer report is evidence that the receiving lab is qualified to carry out the testing procedures and reporting the results as accurately and reliably as the transferring lab. The method transfer report should include an assessment of the results against the predetermined acceptance criteria, and a summary of all deviations observed, if any. The report provides a conclusion of the qualification status of the receiving laboratory. Both transferring and receiving labs and the analysts who participated in the transfer are listed in the report. Results should be reported consistently to the corporate procedures and policies. Data of the transfer activities must be verified by quality assurance for accuracy and compliance. The conclusion should also include any change to the method, or any issue discovered during the transfer exercise. It should be noted that this exercise is to qualify a laboratory rather than a particular individual; therefore, the receiving laboratory must have sufficient training programs to ensure additional analysts can perform the procedure as stated.

6.6.2 Data Evaluation

All laboratory raw data are retained and archived at the site where they were generated according to site procedures. Generated data are tabulated and compared against predetermined acceptance criteria. Data from both labs should be reported consistently (e.g. the same significant figures for numeric data; the same terminology describes the testing sample). Example spectra or chromatograms should be included to support the identification of the components of interest. Any spreadsheet used in the evaluation of data must be verified and documented according to established procedures for that company or site.

Variability and mean values should align across labs and be consistent with past experiments. Data of samples and standards are evaluated for evidence of bias, whether by site, analyst, or instrument. If data do not meet acceptance criteria or if a bias is observed, an investigation should be done to understand the differences between the datasets and the method variability. For the chromatographic method, system suitability data should also be evaluated. If the source of bias is identified, additional steps should be taken to control the source. A statistician should be involved if additional testing is necessary or to analyze the equivalency of different datasets.

6.6.3 Conclusion of Transfer Report

The final report is written and approved to show that the receiving lab is qualified to carry out the procedures for their intended purpose (e.g. for release and stability testing, or to support clinical studies). All relevant documentation such as methods, validation reports, amendments, deviations, sample analyses, chromatograms, or spectra is included as part of the transfer report. Any investigations conducted are also included, or referenced, with the appropriate decision.

Typically, the receiving site maintains the original copy of the transfer report. If the receiving site is a contract lab, then the transferring site may want to retain the original copy and provide a duplicate copy to the contract lab.

The receiving laboratory is considered a qualified lab after the Analytical Qualification Report is approved. The test procedures for commercial products should go through appropriate change control to issue an official quality control procedure before use.

6.6.4 Analytical Transfer File

Because the purpose of the method transfer is qualifying an analytical lab to perform a Good Manufacturing Practice (GMP) procedure, a comprehensive file should be maintained at the working site to support the lifecycle of the analytical method. The analytical technology transfer and qualification file consists of original documentation relating to, but not limited to:

- Any early review of the relevant test method, inventory of essential chemicals and lab supplies by the Transferring site,
- The assessment of the receiving lab,
- Training document of the analysts of receiving lab,
- Analytical transfer and qualification plan,
- Analytical Qualification Protocol with all attachments including the Training Program,
- Completed training forms generated from the execution of the Training Program,
- Analytical Qualification Report with deviation reports, if applicable.

As indicated earlier, the receiving site should maintain the original copy of the Analytical Transfer File and provide a copy to the transferring site. All raw data generated during the transfer and qualification is retained and archived at the site where it is qualified.

6.7 Related Documents

The transfer protocol and report should include references to the SOPs that the labs follow for the testing as well as for reporting the data. It should also include the procedures to follow in case an investigation arose. Additional laboratory experiments beyond the scope of the transfer protocol should also be listed. The following Key Information lists the SOPs that may be needed when working on the transfer of an analytical procedure.

List of SOPs that may be needed during the method transfer

- Reporting analytical data
- Validation of an analytical method
- Transfer of analytical method
- Investigation of out-of-specification results
- Handling out-of-trend of analytical results
- Qualification of analytical instrumentation
- System suitability testing
- Review and approval of analytical results
- Setting acceptance criteria
- Averaging and rounding analytical results

6.8 Handling Transfer Failures

Even with all the preparations and precautions, transfer failure does happen; therefore, it is important to have a pre-approved plan to handle method transfer failures. The transferring and receiving sites should agree on which SOP to follow and how the investigation should be conducted in the event of failure. The investigation may discover details that were not included as a part of the method, and method robustness or validation may need to be revisited. Additional changes to the analytical procedure may be necessary to accommodate the new laboratory. It is important to keep in mind that the intended purpose of transfer activity is to qualify a new lab to perform the analytical procedure; therefore, transfer data are not to be used to release a batch.

Qualification results obtained by the trainee in the receiving laboratory are not related to the quality of the batch. If transfer data fails to meet the qualification acceptance criteria, an investigation, directed by the receiving laboratory management, should be conducted to define, and document the cause of failure. Upon investigation, if the reason for the failure is due to a misinterpretation of the procedure or operational error, then it does not affect the performance of the procedure, the findings are documented according to site procedures and then continued with the transfer experiments. Upon investigation, if the reason for the failure is due to a major issue regarding the procedures of the method, the analytical transfer team leader will be contacted to determine if any additional work is deemed appropriate before continuing with the qualification experiments.

During the qualification experiments, if the data generated by a qualified analyst in the transferring laboratory fails to meet the qualification acceptance criteria, then an investigation directed by the transferring laboratory management, must be conducted to define, and document the cause of failure.

All deviations encountered during the qualification process must be fully investigated, justified, documented, and approved before the Analytical Qualification Report is approved.

A statistician may be needed to determine if any additional testing or sampling is necessary. It may be necessary to visit the transferring site or the receiving site to troubleshoot the failure.

The following Key Information lists possible reasons for transfer failure. Method transfer is very cumbersome and an expensive way to develop an analytical procedure; therefore, analytical issues should be addressed during method development and validation before the transfer is considered.

Possible reasons for transfer failure
• Inadequate discussion between two labs before the initiation of the transfer • The analytical procedure is lacking adequate instructions and details • The analytical procedure is not rugged or reproducible • Resistance to the analytical method by the receiving lab • The protocol has poorly designed acceptance criteria • Variation of samples used in the testing • Materials and reagents are not equivalent between sites • Analytical instrumentation is not equivalent between sites

6.9 Transfer to a Contract Lab

Many companies are using contract laboratories as part of their testing laboratories; therefore, transfers to contract labs are performed quite often. Before any contract research organization (CRO) can receive a test, it is recommended that a technical evaluation and quality audit is performed to verify the suitability of the facility to handle the procedure. Contractual and quality agreements should be established.

When working with the CRO, it is important to understand their infrastructure and their standard Operating Procedures (SOPs). The key SOPs that one should be familiar with is data handling, the training procedure for additional analysts, retest procedure, out-of-specification notification procedure, and investigation procedure.

6.10 Transfer to an International Site

When the transfer of analytical procedure is involved with an analytical lab at an international site, local custom regulation requirements should be considered. Face-to-face training is important to ensure analytical practices are understood without the barrier of languages or other limitations. Methods and protocols may need to be translated into local languages before transfer, as well as future operations that are accurate and free of ambiguity.

One of the issues for transferring to an international site is the equivalence of materials being used in the transferred exercise. The integrity of the samples should be known to ensure the shipping of materials does not affect the sample stability. Materials such as reagents, reference standards, and solvents should be commercially available so that it does not drain the supply of the transferring site which would result in additional costs to ship the material.

6.11 Summary

Method transfer is proof to demonstrate the ability of the receiving lab to execute the method with accuracy and reproducibility as its intended use for testing the drug substance or drug products. It is important to spend time during the transfer process to ensure that the qualification of the receiving lab is successful, and it is expected that issues will arise as method transfer is part of the precision study of the method validation. Therefore, all issues should be resolved to satisfy both the parties and the analytical procedure should be updated to include all relevant details. Modifications should be incorporated; this may lead to additional revalidation of the method in some cases.

The transferring site and receiving site partner together in deciding on the transfer strategy, communication style, and people involved in the process. A transfer protocol with predetermined acceptance criteria is necessary and a procedure should be in place to handle transfer failures [7–24].

List of Abbreviations

CAPA	Corrective Action and Preventive Action
CRO	Contract Research Organization
EMEA	European Medicines Agency
GMPs	Good Manufacturing Practices
NF	National Formulary
HPLC	High-Performance Liquid Chromatography
LIMS	Laboratory Information Management System
LOD	Limit of Detection
LOQ	Limit of Quantitation
SOP	Standard Operating Procedure
TC Pipet	To Contain Pipet
TD Pipet	To Deliver Pipet
USP	United States Pharmacopeia

References

1 USP <1224>(2019). Transfer of analytical procedures. *USP 43-NF 38*. Rockville, MD: United States Pharmacopeial Convention.

2 Scypinski, W., Roberts, D., Oates, M., and Etse, J. (2001). Pharmaceutical research and manufacturers association acceptable analytical practice for analytical method transfer. *Pharmaceutical Technology* 26 (3): 84–89.

3 Swartz, M. and Krull, I. (2006). Analytical method transfer. *LCGC North America* 24 (11): 1204–1214.

4 Food and Drug Administration (2003). Guidance for industry. Q1 A (R2) Stability testing of new drug substances and products. Rockville, MD: Food and Drug Administration. https://www.fda.gov/media/71707/download (accessed 27 August 2020).

5 ICH Harmonised Tripartite Guideline (1996). Stability testing: photo stability testing of new drug substance and products Q1B. https://database.ich.org/sites/default/files/Q1B%20Guideline.pdf (accessed 27 August 2020).

6 ICH Harmonised Tripartite Guideline (1996). Valuation of analytical procedures: text and methodology Q2 (R1). https://database.ich.org/sites/default/files/Q2_R1__Guideline.pdf (accessed 27 August 2020).

7 USP <1225> (2019). Validation of Compendial procedures. *USP 43-NF 38*. Rockville, MD: United States Pharmacopeial Convention.

8 Food and Drug Administration (1994). Guidance for industry. Reviewer guidance, validation of chromatographic methods. Rockville, MD: Food and Drug Administration. https://www.fda.gov/media/75643/download (accessed 27 August 2020).

9 Miller, J. and Crowther, J. (ed.) (2000). *Analytical Chemistry in a GMP Environment, a Practical Guide*. Hoboken, NJ: Wiley.

10 Swartz, M. and Krull, I. (2009). Analytical method validation: back to basics, part I. *LCGC North America* 27 (11): 989–995.

11 Swartz, M. and Krull, I. (2011). Analytical method validation: back to basics, part II. *LCGC North America* 29 (1): 44–52.

12 Chambers, D.G., Kelly, G.L., Lister, A. et al. (2005). Analytical method equivalency: an acceptable analytical practice. *Pharmaceutical Technology* 29: 68–80.

13 Miller, J.M. and Miller, J.C. (2010). *Statistics and Chemometrics for Analytical Chemistry*. Pearson Education Limited: London, UK.

14 Huynh-Ba, K., Wong, A., and Dong, M.W. (2020). Validation of stability indicating HPLC methods for pharmaceuticals: overview, methodologies and case studies. *LCGC North America* 38 (11): 606–618. Stability Studies and Testing of Pharmaceuticals: An Overview (chromatographyonline.com). https://www.chromatographyonline.com/view/stability-studies-and-testing-pharmaceuticals-overview-0.

15 Anderson, R.L. (1987). *Practical Statistics for Analytical Chemists*. Van Nostrand Reinhold Company: New York, NY.

16 USP <621> (2019). Chromatography. *USP 43-NF 38*. Rockville, MD: United States Pharmacopeial Convention.

17 Ermer, J. and Miller, J. (ed.) (2005). *Method Validation in Pharmaceutical Analysis. A Guide to Best Practice*. Wiley-VCH Verlag GmbH & Co: Weinheim, Germany.

18 Thatcher, S.R., Mansfield, R.K., Miller, R.B. et al. (2001). Pharmaceutical photostability: a technical guide and practical interpretation of the ICH guideline and its application to pharmaceutical stability – part I. *Pharmaceutical Technology* 25 (3): 98–110.

19 Thatcher, S.R., Mansfield, R.K., Miller, R.B. et al. (2001). Pharmaceutical photostability: a technical guide and practical interpretation of the ICH guideline and its application to pharmaceutical stability – part II. *Pharmaceutical Technology* 25 (4): 50–62.

20 Huynh-Ba, K. and Dong, M.W. (2020). Stability studies and testing of pharmaceutical: an overview. *LCGC Europe* 33 (6): 289–302. Stability Studies and Testing of Pharmaceuticals: An Overview (chromatographyonline.com). https://www.chromatographyonline.com/view/stability-studies-and-testing-pharmaceuticals-overview-0.

21 Dolan, J.W. (2002). Stability-indicating assay. *LC-GC North America* 20 (4): 346–349.

22 Lung KR, Gorko MA, Llewelyn J, Wiggins N. Statistical method for the determination of equivalence of automated test procedures. *Journal of Automated Methods & Management in Chemistry*; 2003; 25:123–127. doi:https://doi.org/10.1155/S146392460300021X.

23 Huynh-Ba, K. (ed.) (2008). *Handbook of Stability Testing for Pharmaceutical Products: Regulations, Methodologies, and Best Practices*. New York, NY: Springer Science & Business Media.

24 Huynh-Ba, K. (ed.) (2010). *Pharmaceutical Stability Testing to Support Global Markets*. New York, NY: Springer Science & Business Media.

7

Dissolution Testing in the Pharmaceutical Laboratory

Vivian A. Gray

V.A. Gray Consulting, Inc., Hockessin, DE, USA

TABLE OF CONTENTS

Analytical Testing for the Pharmaceutical GMP Laboratory, First Edition. Edited by Kim Huynh-Ba.

7.1 Introduction

The dissolution test is an in vitro test typically performed on oral dosage forms to measure the performance of the drug products. More recently, non-oral dosage forms (e.g. mucosal, parenteral, and topical dosage forms) also have *in vitro* testing methods. Since a drug product must dissolve to be absorbed and provide a therapeutic effect for the patient, this test also relates to drug efficacy, given that it determines the drug's ability to dissolve in a physiologically relevant media with gentle agitation and the speed of the dissolution itself. In an ideal situation, the dissolution rate correlates to the plasma level of the active drug substance; however, this is not easily achieved. Dissolution testing and the relevant specifications are required by the Food and Drug Administration (FDA) for all solid dosage forms, especially the conventional tablet and capsule products. However, dissolution testing is not required for *true solutions*. A true solution is defined as a solution that has no undissolved particles. The United States Pharmacopeia (USP) and FDA documents related to dissolution testing will be provided in this chapter, as well as a description of the specialized equipment and operation used in dissolution testing. Method development, setting specifications, harmonization, and validation will be explored. Because drug products are manufactured and tested according to Good Manufacturing Practices (GMP), a section is devoted to the unique GMP aspects that apply to dissolution testing.

7.2 Regulatory and Compendial Role in Dissolution Testing

The USP became the first organization to show an interest in dissolution testing by creating a USP-National Formulary Panel to examine bioavailability and ways to test release mechanisms to provide some, albeit *in vitro*, assurance for drug effectiveness. This panel recommended placing the dissolution test, using the rotating basket apparatus, in USP monographs. During the 1970s, there were 12 official dissolution tests using baskets in the USP monographs. The paddle apparatus followed shortly thereafter. In the early 1980s, the USP Expert Committee on Biopharmaceutics proposed a single-point method: 75%Q in 45 minutes with water as a medium. This specification was, in retrospect, mainly for the Biopharmaceutics Classification System (BCS) Class 1 (Highly Soluble/Highly Permeable) and Class 3 (Highly Soluble/Poorly Permeable) compounds [1].

In 1997, testing using profiles came into prominence through an FDA guidance document [2] where the FDA first started to require dissolution profiles rather than just a single time point test. These profiles showed the dissolution rate over the entire time period of the test. Typically, the measured time points are 15, 30, 45, and 60 minutes for immediate release dosage forms such as tablets or capsules. The profile time points were not used as specifications but as information about the product's release mechanism. The typical single time point specification of 75% dissolution in 45 minutes evolved into the tighter 80% dissolution in 30 minutes and has continued to be used as a manufacturing control step. Dissolution profiles also became a necessary tool to compare performance between two products, such as the innovator products versus the generic-equivalent products. The most common comparison tool is the *f*2 similarity factor, described in many FDA guidance documents [1–4]. The FDA guidance on extended-release [5] requires product specifications of three time points. Instead of relying only on the *f*2 similarity factor, *in vitro and in vivo* correlations are encouraged by this guidance and USP General Chapter <1088> *In vitro and In vivo Evaluation of Dosage Forms* [6].

The following Key Information sections are essential documents related to dissolution testing and, in some cases, will be referred to again during discussions in the later sections of this chapter.

FDA Guidance Documents relevant to dissolution testing

1) Dissolution testing of immediate release solid oral dosage forms extended-release oral dosage forms: development, evaluation, and application of *in vitro/in vivo* correlations
2) The use of mechanical calibration of dissolution Apparatus 1 and 2 – Current GMP
3) SUPAC-IR: immediate-release solid oral dosage forms: scale-up and post-approval changes: chemistry, manufacturing, and controls, *in vitro* dissolution testing, and *in vivo* bioequivalence documentation
4) SUPAC-MR: modified-release solid oral dosage forms: scale-up and post-approval changes: chemistry, manufacturing, and controls; *in vitro* dissolution testing and *in vivo* bioequivalence documentation
5) SUPAC-SS: nonsterile semisolid dosage forms: scale-up and post-approval changes: chemistry, manufacturing, and controls, *in vitro* release testing and *in vivo* bioequivalence documentation
6) Waiver of *in vivo* bioavailability and bioequivalence studies for immediate-release solid oral dosage forms based on biopharmaceutics classification system
7) Bioavailability and bioequivalence studies for orally administered drug products – general considerations
8) Orally disintegrating tablets
9) Quality attributes for chewable tablets
10) Dissolution testing and specification criteria for immediate-release solid oral dosage forms containing biopharmaceutics classification system class 1 and 3 drugs

USP general chapters relevant to dissolution testing

- <711> Dissolution
- <724> Drug release
- <1004> Mucosal drug products-performance tests

- <1088> *In Vitro and In Vivo* evaluation of dosage forms
- <1090> Assessment of drug product performance-bioavailability, bioequivalence, and dissolution
- <1092> The dissolution procedure: development and validation
- <1094> Capsules-dissolution testing and related quality attributes
- <1225> Validation of compendial methods
- <1724> Semisolid drug products-performance tests
- <2040> Disintegration and dissolution of dietary supplements

Chapters for other Pharmacopeia relevant to dissolution testing

- *European Pharmacopoeia, (Ph. Eur.)* 2.9.3 Dissolution test for solid dosage forms
- *Ph. Eur.* 2.9.42 Dissolution test for lipophilic solid dosage forms
- *Ph. Eur.* 2.9.4 Dissolution testing for transdermal patches
- *Ph. Eur.*2.9.25 Dissolution test for medicated chewing gums
- *Ph. Eur.*5.17.1 Recommendations on dissolution testing
- *Japanese Pharmacopeia, JP* 6.10 Dissolution test
- *International Pharmacopeia*, Dissolution test for solid oral dosage forms

Other international guidance documents relevant to dissolution testing

- European Medicines Agency (EMA) Guideline on quality of oral modified-release products
- EMA Guideline on quality of transdermal patches
- EMA Guideline on the investigation of bioequivalence
- EMA Reflection paper on the dissolution specification for generic solid oral immediate-release products with systemic action
- Japanese Guideline for the design and evaluation of oral prolonged-release dosage forms
- Japanese Guideline for bioequivalence studies of generic products
- Japanese Guideline for bioequivalence studies for different strengths of oral solid dosage forms
- Japanese Guideline for bioequivalence studies for formulation changes of oral solid dosage forms.
- Japanese Guideline for bioequivalence studies for different oral solid dosage forms
- FIP Guidelines for dissolution testing of solid oral products

- FIP/AAPS Guidelines for dissolution/*in vitro* release testing of novel/special dosage forms
- Specifications for pharmaceutical preparations (Forty-Sixth Report), WHO Technical Report Series 970. Annex 7 Multisource (generic) pharmaceutical products: guidelines on registration requirements to establish interchangeability. Annex 8 Proposal to waive *in vivo* bioequivalence requirement for the WHO model list of essential medicines immediate-release, solid oral dosage forms.
- The International Conference on Harmonisation (ICH) document Q6A Decision Trees 7: setting acceptance criteria for drug products dissolution
- WHO, dissolution test for solid oral dosage forms. The International Pharmacopeia, Draft February 2018.

7.3 Theory

Dissolution testing is a three-stage process. The disintegration of a gross immediate-release tablet into particles of various sizes usually happens first. It can be measured by the USP <701> *Disintegration* [7]. For capsule shell dosage forms, this stage also includes the rupturing of the capsule shell. The next step, de-aggregation, is a breakdown of the dosage form into discrete particles that increase the surface area, providing a solid–liquid interface and the beginning of dissolution. The dissolution process continues, and the dissolution test measures the rate.

The dissolution rate is represented mathematically by the Modified Noyes and Whitney Equation [8].

$$\text{Rate} = kDS/vh\left(C_s - C_t\right)$$

where D is the diffusion rate constant, S is the surface area of the solid material, v is the volume of the dissolution media, h is the thickness of the saturated layer, C_S is the saturation concentration, k is the dissolution rate constant, and C_t is the concentration of the bulk solution.

The thickness of the saturated layer is influenced by the paddle or basket speed on the dosage unit boundary layer. To perform the dissolution testing, sink conditions should be obtained in the dissolution medium chosen. The sink condition, defined as the quantity of medium used, is not less than three times the value that is required to form a saturated solution of the drug substance, according to the USP <1092> *The Dissolution Procedure: Development and Validation* [9].

A factor in the equation is the drug substance surface area; therefore, the particle size is very important in the dissolution rate.

Dissolution testing measures the rate at which a drug substance dissolves from the dosage unit; thus, the term "*in vitro* release" is more appropriate in the case of an extended-release pharmaceutical product since the drug substance is released from a matrix then dissolved in the media. The dissolution rate may be defined as the amount of active ingredient in a solid dosage form dissolved in unit time under standardized conditions of temperature, agitation, and media composition. The dissolution results are typically expressed as a cumulative percent dissolved of the label claim, Q (Quantity), over the time intervals until at least 80% dissolution, or an asymptote is reached.

There are three important aspects when discussing the dissolution of a drug product: the solubility of drug substance, the dynamic process of the dissolution rate, and the effect of excipients, aging, and the manufacturing process. Dissolution testing can be discriminative for a specific formulation, especially the manufacturing process, as these factors can increase or decrease the dissolution rate.

7.4 Equipment Operation and Sources of Error

The main principle of the dissolution equipment is that it should provide gentle mixing that eventually leads to complete or near-complete dissolution. The dosage form should be observed during the process of dissolution. The mechanics of the equipment can be a source of error if the equipment is not operating properly.

The typical major components of the most common dissolution equipment are shown in Figure 7.1 and consist of the dissolution tester "drive-head" containing the drive belt, spindle assemblies, and electronics for the mechanical parts of the equipment. The vessels are positioned in a water bath equipped with a water circulator and heater. In some cases, the equipment does not have a water bath but jacketed vessels that are thermally heated. There is also a top plate containing insert holes to house the vessels. The stirring devices are shafts inserted into the spindle assemblies.

These shafts either have a basket attached (Figure 7.2), called the "basket"/USP Apparatus 1, or have a paddle-stirring device (Figure 7.3) called the "paddle"/USP Apparatus 2. The vessels, once inserted into the water bath, are filled with dissolution medium. The USP <711> *Dissolution* [10] gives details on the apparatus dimensions and other parameters of the test. Tolerances are associated with many parameters for the operation of the dissolution tester.

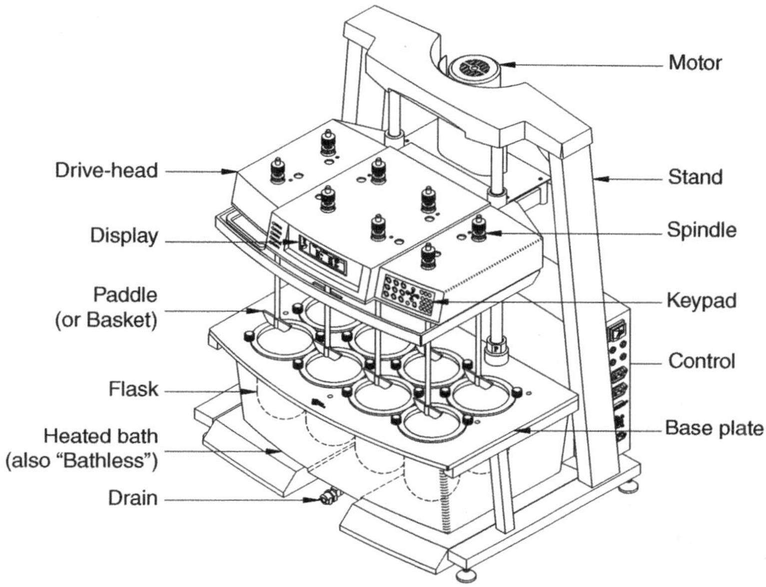

Figure 7.1 Example of modern dissolution test equipment.

Figure 7.2 USP Apparatus 1: basket. *Source:* Based on USP <711> [10].

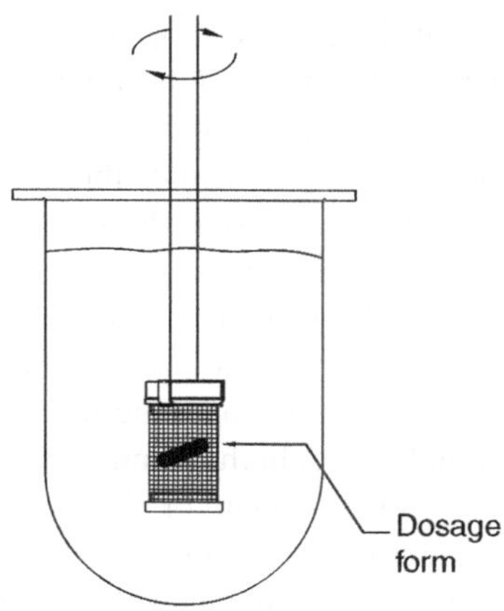

As a regulatory test, the dissolution method must be accurate and practical. The test should have low variability and generate a dissolution profile. The essentials of the test are the accuracy of the results and the robustness of the method. Aberrant and unexpected results do occur, and the analyst should be well trained to evaluate all aspects of the dissolution test and monitor the equipment in operation.

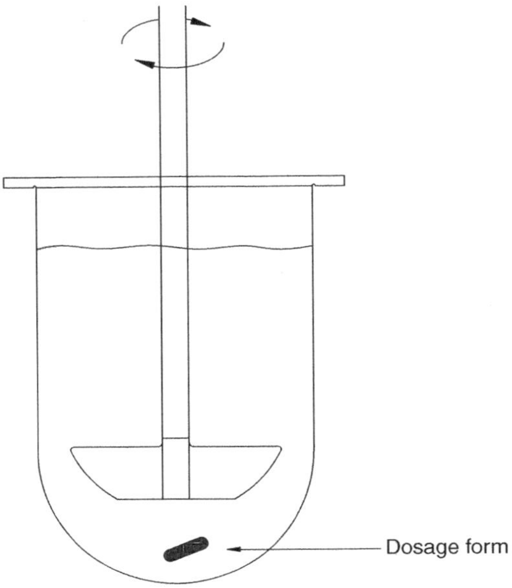

Figure 7.3 USP Apparatus 2: paddle. *Source:* Based on USP <711> [10].

When performing dissolution testing, many factors may affect the test results [11]. The testing equipment and its environment, sample handling, formulation, in situ reactions, automation, and analytical techniques may be the cause of errors and variability. Certain aspects of the equipment performance verification and mechanical calibration process, as well as a close visual observation of the test, may reveal these errors. The physical dissolution of the dosage form should be always unencumbered.

Knowledge of drug properties, especially solubility as a function of pH and/or in surfactants is essential. The precipitation of the drug can occur as the pH of the solution changes or as the amount of drug increases. Complete dissolution of poorly soluble drugs in the standard solution may be difficult; therefore, it is acceptable to use a small amount of alcohol to dissolve the standard completely. Comparing the standard solution absorptivity with the historical data for the typical absorptivity range of that standard can determine if the standard has been prepared properly.

Highly variable results may indicate that the method is not robust, and this can cause difficulty in identifying trends and the effects of various formulation changes. Highly variable has been defined as no more than 20% RSD for data from timepoints of 10 minutes or less and no more than 10% RSD for timepoints of over 10 minutes [1]. Three major factors influence variability: Equipment variables (see Sections 7.4.1 and 7.5), dissolution method conditions, and the dosage form. Variability can arise from the dissolution conditions chosen; thus, visual observations of the dissolution progress of the product are important to help identify any possible anomalies or problems. An apparatus or speed change may be necessary to remediate the problem.

Potential formulation causes of variability are poor content uniformity, the sensitivity of the drug to pH environment in the medium, polymorph changes, and reactions or degradation may be occurring in situ. The film coating may cause the sample to stick to the vessel walls and, thus, impact the dissolution rate. Upon aging, capsule shells are known to form pellicles, and tablets may become harder or softer (depending upon the excipients and drug's interaction with moisture), affecting the dissolution and disintegration rate.

7.4.1 Equipment Variables

The major components of dissolution equipment are the tester, water bath, paddles, baskets and shafts, vessels, samplers, and analyzers. According to USP <711> *Dissolution*, the determination of the suitability of a dissolution test assembly to perform dissolution testing must include conformance to the dimensions and tolerances of the apparatus as given in the chapter. Also, critical test parameters that must be monitored periodically during use include the volume and temperature of the dissolution medium, rotation speed (Apparatus 1 and Apparatus 2), dip rate (Apparatus 3), and flow rate of the medium (Apparatus 4). Sections 7.5.4–7.5.6 give an overview of other apparatus.

Mechanical aspects such as media temperature, paddle or basket speed, shaft centering and wobble, and vibration can significantly impact the dissolution rate of the product. Mechanical and chemical calibration should be conducted periodically, usually every six months, to ensure that the equipment is working properly [2].

The Apparatus Suitability section of USP <711> *Dissolution* requires a Performance Verification Test (PVT) using USP Prednisone Reference Standard Tablets for Apparatus 1 and 2. The reference tablets come with certificates identifying appropriate ranges. The PVT is designed to detect sources of error associated with the improper operation and inadequate equipment conditions [12, 13]. There is an FDA Guidance for Industry (Use of Mechanical Calibration), which states that an enhanced mechanical calibration procedure can be used as an alternative to the Apparatus Suitability procedure.

7.4.2 Media Deaeration

Deaeration of the medium is important because the presence of bubbles caused by dissolved oxygen and other gasses can interfere with dissolution test results and should be avoided [10]. Dissolved air can slow down dissolution by creating a barrier; these bubbles may adhere either to the tablet surface or to basket screens, or particles can cling to bubbles on the glass surface of the vessel or shafts. The test should be performed immediately after deaeration. It is best not to have the paddle rotating before adding the tablet since the paddle movement aerates the medium.

The dissolution testing of the USP Prednisone Reference Standard Tablets uses deaerated water as the medium. There are numerous methods for the deaeration of a medium [14–16], including automated methods. USP <711> *Dissolution* describes heat, filtration, and vacuum as a deaeration method. Helium sparging is also a typical method for deaeration, but this is not considered ideal because it is not environmentally friendly given helium is a nonrenewable resource. Intermittent helium shortages have been reported, and helium can be very expensive outside of the United States.

It is possible to evaluate the impact of deaeration on dissolution results by running controlled experiments (e.g. using deaerated medium vs. non-deaerated medium). In these experiments, it can be valuable to monitor the level of dissolved gasses in the medium. If the experiments confirm that deaeration has an impact on dissolution results, this should be added to the dissolution procedure as part of the analytical control strategy.

7.4.3 Vibration

Vibration can interfere with dissolution testing [17–20]. The bearings of the spindle assembly can become worn, leading to shaft vibration, and wobble. Wear and debris on the drive belts can also cause vibration. The tension of the belt can be adjusted slightly for smooth operation. It is necessary to level top plates and lids of the dissolution bath to avoid vibration. If vessels are not locked in place, slight movement can occur due to the flow of water in the bath, thus influencing the dissolution process.

Benchtop equipment such as shakers, centrifuges, or sonicators can introduce external sources of vibration. A common, but often overlooked source of vibration can be local construction–either in the area or within the lab building itself. The testers themselves should not be near hoods or significant airflow sources. Heavy foot traffic and door slamming should be avoided to help minimize the risk of vibration interference.

Both internal and external vibration sources can be hard to measure. Presently, there are vibration meters that work well, but there have not been standard measurement specifications developed to date.

7.4.4 Water Bath of Dissolution Equipment

Water baths have been optimally designed to eliminate noisy circulators near the bath, and, therefore, the water bath itself is not usually a source of vibration. The temperature of the media should be measured in all the vessels, rather than just one, to assure that the temperature is uniform throughout the water bath. The bathwater level should always be maintained at the fill line of the vessels to ensure uniform heating of the medium. Lastly, the water bath should contain clean, clear water, so observations of the dissolution test can be performed accurately.

7.4.5 Glass Vessels

Vessels can become sources of error. Typically, the method of manufacturing the glass is proprietary. These vessels are often manufactured from large glass tubing, where the vessel bottom is hand blown and molded. Depending upon techniques of the molding process, irregular surfaces can occur, and the uniformity of vessel bottom roundness can vary. Cheaply made vessels are notorious for this problem. There have been extensive studies on the effects of the vessel shape on dissolution results [21–24]. Close examination of newly purchased vessels is very important since surface irregularity can cause significant differences in the dissolution results.

Another common problem with vessels is residue buildup, either from oily products or sticky excipients. An insoluble product that is not rinsed well from the previous testing can cause contamination. Vessels that become scratched and etched after repeated use and washing should be discarded. Lids need to be in place to prevent evaporation. Vessels should be locked down to avoid rocking and vibration.

Because of the wide variation in dimensions of vessels available in the market, it is advisable to use vessels with the same diameter/dimensions to avoid variability in the dissolution results within a run [21].

7.5 Common Errors of Dissolution Apparatus

7.5.1 USP Apparatus 1 and 2

The basket and paddle can be sources of error and should be closely inspected before using. Dimensions should remain as specified. The shafts must be straight. Stainless steel or inert coatings are the most common materials for paddles, and they can become nicked or dented; this can adversely affect dissolution hydrodynamics and be a source of contamination. Thorough cleaning of the paddles following each use is important to eliminate carry-over of the drug, oily matrices, or viscous media.

The baskets need special care and examination as they can become frayed, misshapen, or warped with use. Screen wire can become thinner over time – especially when used with acidic medium – changing the mesh size. There are different designs for attaching baskets to shafts. The attachment can be done using clips, O-rings, or magnets. Since the type of attachment can sometimes affect results, this should be evaluated and documented in the method [18, 25]. Baskets are also prone to gelatin or excipient buildup if not cleaned immediately after use [26]. Off-center shafts can cause increased variability in results.

7.5.2 USP Apparatus 3

The reciprocating cylinder (Figure 7.4) is often used as a research tool where a change of pH in the media is necessary. The dosage unit can be moved from row to row, and in each row, the vessels may contain media of different pH or components. The equipment has a special configuration for beaded products; the beads are contained by the screens in the upper and lower parts of the cell, yet the reciprocating motion allows for good mixing [27–29]. The typical media volume is 200–250 mL.

Sources of error when using this apparatus are mainly associated with the smaller volume because the loss of media through evaporation can be significant, and there may be difficulty in reaching sink conditions when the drug is poorly soluble. The mesh size of the screens is critical – the sample should not adhere to the screen, and the dipping rate should remain constant. For example, in the Liothyronine Sodium Tablets Dissolution test [30], Apparatus 3 is required using a 20-mesh screen on the top and a 40-mesh screen on the bottom of the glass reciprocating cylinder.

When using a surfactant, there can be considerable foaming, but this can be remediated with simethicone drops. However, if you introduce simethicone drops, it is important to demonstrate that these do not affect the dissolution results.

7.5.3 USP Apparatus 4

USP Apparatus 4 is also known as the flow-through cell. The drug product is positioned in a cell where the dissolution medium is constantly flowing and hence dissolving the dosage unit. The liquid passes through a filter at the top of the cell and is then collected in a reservoir. Because of the constant flow of media, an extended-release product matrix can be monitored, or a poorly soluble product can continually be in a sink environment. A diagram of a unique flow through the cell is shown in Figure 7.5.

Sources of error when using this apparatus are the pump operation, flow rate reliability, and the clogging of the filters. Other considerations are related to the flow of liquid through the cell, such as the position of the tablet holders, the quantity and size of glass beads used, and tubing lengths, material, and diameters [31–33].

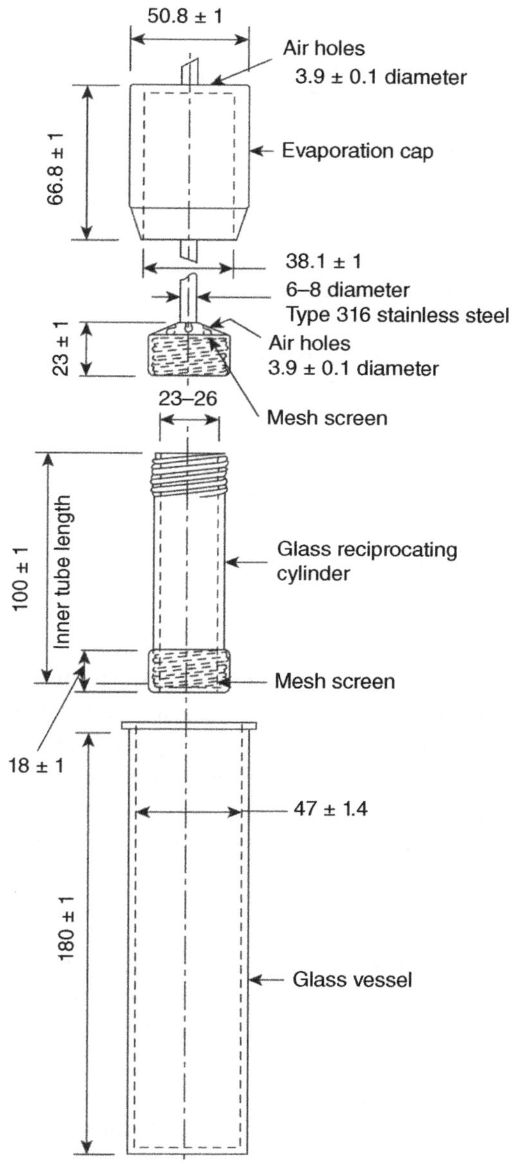

Figure 7.4 USP Apparatus 3: reciprocating cylinder.

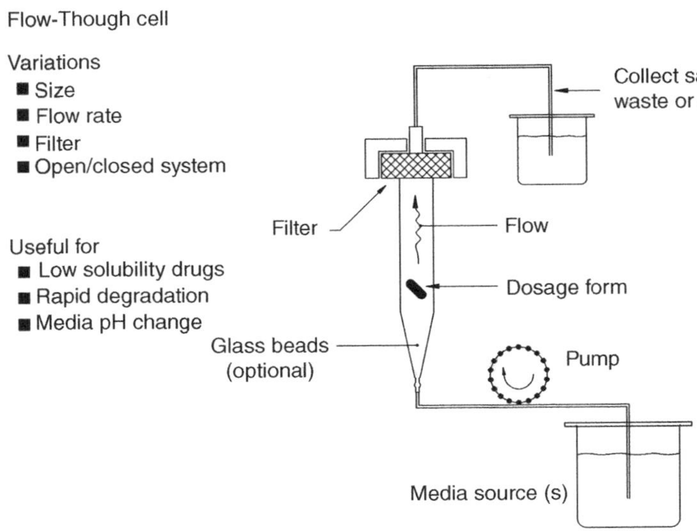

Figure 7.5 USP Apparatus 4: flow-through cell.

7.5.4 USP Apparatus 5

The USP <724> *Drug Release* [34] discusses Apparatus 5, 6, and 7. Apparatus 5 is commonly known as the Paddle over Disk and is used for transdermal patches. There are two patch-holding designs: the watch or sandwich glass assembly (shown in Figure 7.6) and the screen disk. Sources of error for this apparatus would be similar to those mentioned earlier with Apparatus 2 (see Section 7.5.1), and the positioning and attachment of the patch to the device chosen are critical.

7.5.5 USP Apparatus 6

As with Apparatus 5, this apparatus is exclusively used for transdermal patches. As shown in Figure 7.7, the patch adheres to the cylinder in such a way that the "active" side of the patch (the layer where the drug substance permeates out through the membrane) is facing the medium.

Sources of error for this equipment are the same attributes as for Apparatus 2, with the straightness of the shaft is most important along with the proper and firm adherence of the patch to the surface.

7.5.6 USP Apparatus 7

Apparatus 7 is commonly known as the Reciprocating Holder. This apparatus has five designs (Figure 7.8). It operates in a reciprocating motion (as seen in

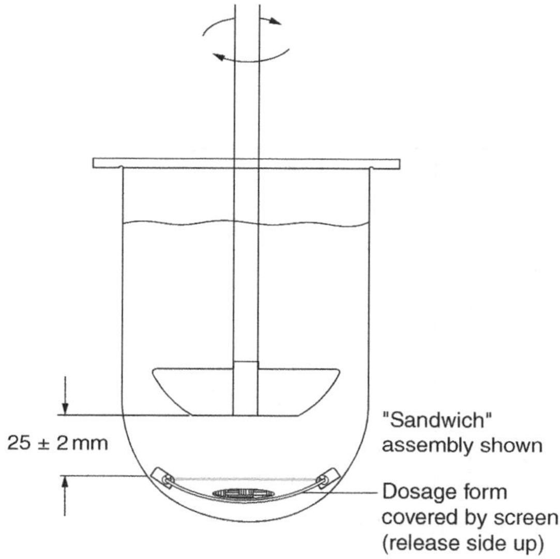

Figure 7.6 USP Apparatus 5: paddle over a disk with "Sandwich or Watch Glass" assembly shown. *Source:* Based on USP <724> [34].

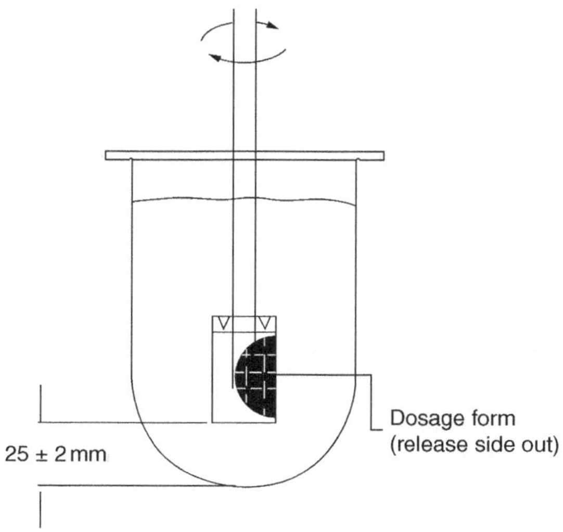

Figure 7.7 USP Apparatus 6: rotating cylinder. *Source:* Based on USP <724> [34].

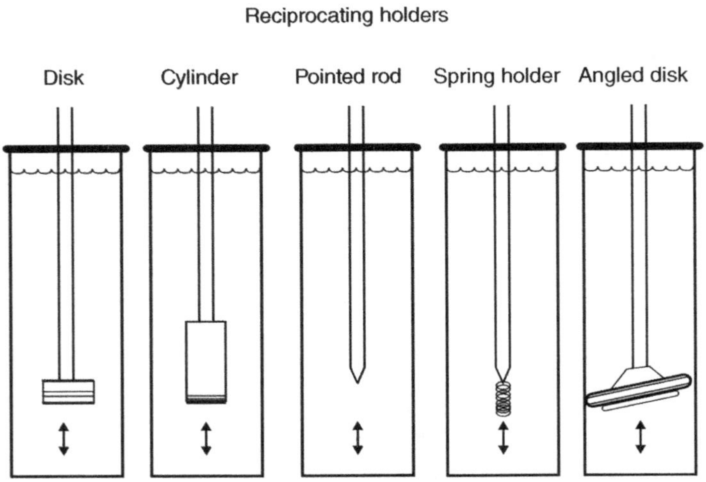

Figure 7.8 USP Apparatus 7: five designs of reciprocating holder. *Source:* Based on USP <724> [34].

Apparatus 3) and also goes from one beaker/vessel to another. There are three designs for use with transdermal patches (disk, cylinder, and angled disk); the other two designs are for specially designed tablets as the osmotic pump (pointed rod and spring holder). These tablets usually have a laser hole where there is a push/pull effect of the drug from a polymeric matrix. The hole must be uniformly exposed to the medium; hence, the design is a rod-like shaft where the dosage form is glued to the tip of the rod. Another variation is a spring-like cage at the end of the rod that houses the dosage unit.

Sources of error are similar to Apparatus 3, where reciprocation is the agitation principle. The accuracy of the indexer is also a critical parameter.

7.6 Dissolution Method Considerations

Method development is important for the dissolution method to avoid testing errors and to prevent aberrant data. The following considerations are particularly important during the development and validation phases of a dissolution procedure.

7.6.1 Sample Introduction

The sample introduction can be non-reproducible and uncontrollable, and the position that the sample was introduced into the vessel can greatly impact the dissolution rate. As an example, if the tablet is off-center, the dissolution rate may be higher due to shear forces. If the tablet is in the center, coning may occur, and the dissolution rate will decrease. Film-coated tablets can be sticky and pose variability problems related to tablet position when introduced: therefore, USP Apparatus I (basket) may be used if there is no gelatinous or excipient build up. Another option to avoid the sticking of film-coated tablets would be using paddles with sinkers.

Suspensions can be introduced in a variety of ways: manual delivery using syringes or pipettes, pouring from a tared beaker, or automated delivery using a calibrated pipette. Each method has its own set of limitations; however, automated methods will have less variability. Mixing of the suspension sample will generate air bubbles and, thus, cause method variability. Therefore, the mixing time of suspension samples prior to introduction into the vessel must be strictly uniform to reduce erroneous or biased results [35].

7.6.2 Media Attributes

The dissolution media volume is a critical component of the dissolution test. Inaccurate results may be caused by a large volume of medium removed through multiple sampling without replacement, thereby adversely influencing sink conditions.

The level or lack of deaeration can influence the dissolution rate of particularly slow dissolving drugs. In USP <711> *Dissolution*, there is provided a preferred manual method for deaeration, though automated deaeration is useful if validated. Refer to Section 7.4.2 for further discussion of deaeration.

The use of surfactants in the dissolution media can present quite a cleaning problem, especially if the surfactant concentration is high (i.e. over 0.5%). In the sampling lines, surfactants such as sodium lauryl sulfate (SLS) may require several rinsing to assure no residues. The grade and age of SLS have dissolving effects that consequently change depending upon the surface-active impurities and electrolytes [36]. Because of foaming, it is hard to deaerate surfactants effectively, and some pumps used in automated equipment are not well suited for surfactant use.

Bubbles can be present on the fluid surface when surfactants are used. When lowering a basket into the surfactant medium, bubbles can attach to the bottom of the basket and prohibit the media from flowing into the basket, thus decreasing the dissolution rate. When high-performance liquid chromatography (HPLC) analysis is performed on dissolution samples containing surfactants, the autosampler will need additional needle washing steps to avoid carry-over. Other surfactants, especially Cetrimide, may be too viscous for accurate injector delivery. Surfactants can affect column packing to a high degree, giving extraneous peaks or poor chromatography [37]. Also, dissolution media above pH 8 may cause column degradation.

7.6.3 Observations

Making and documenting visual observations of each vessel during the dissolution analysis is important for identifying sources of error. A trained analyst can recognize analytical issues if he or she understands the cause and effect of certain observations. Accurate and meaningful dissolution rates occur when the product dissolves without disturbance from barriers to dissolution, or without disturbance of vessel hydrodynamics from any source. The particle disintegration pattern must show freely dispersed particles (i.e. no significant clumping). Anomalous dissolution usually involves some of the following observations: floating chunks of the tablet, spinning dosage units, coning or mounding of particulates under the paddle or basket, dosage forms gumming or swelling. Cross-linking of capsules will be evidenced by sacs, swollen/rubbery masses, or clear pellicles. The dosage form can exhibit capping, "clam-shell" erosion patterns. The sample can land in an off-center position or stick to vessel walls in random places, and particles may be seen adhering to apparatus or vessel walls. All the observations mentioned could lead to aberrant and variable dissolution results.

Along with good documentation, familiarity with the dissolution behavior of a product is crucial to identifying changes in stability studies or changes associated with a modification of the formulation quickly. One may notice a change in the size of the dissolving particles, excipients floating upward, or a slower erosion pattern. Changes in the formulation or an increase in strength may produce previously unobserved basket screen clogging. One might see the spindle assembly surge when the contents of the basket fall out and settle at the bottom of the vessel, or particles or air bubbles clinging onto vessel walls indicating the medium has not been properly deaerated.

Visual observations on all vessels and at useful timepoints are critical to dissolution testing, especially during formulation development; therefore, photography of the vessel contents may be useful. While the routine observation of dissolution tests may not be practical, at a minimum, observations should be made during method development, and whenever testing proceeds to Stage 2 or any aberrant data are present.

7.6.4 Sinkers

Sinkers are defined in the USP as "not more than a few turns of a wire helix." Recently, another sinker type was added to USP <711> *Dissolution*, in Figure 2A therein, named "alternative sinker." It is based on the Japanese Sinker found in the JP General Chapter on Dissolution. Other sinkers may be used, but the analyst should be aware of the effect that different types of sinkers may have on mixing [38], and the sinker type should always be specified in the method (e.g. catalog number or if handmade the procedure used).

7.6.5 Filters

Filters are used for almost all analyses; many types of different materials are used in automated and manual sampling. Validation of the pre-wetting or discard volume is critical for both the sample and standard solutions. Plugging of filters is a common problem, especially with automated devices. The pore size should be compared to the drug substance particle size. It is difficult to find a filter membrane with a pore size small enough to filter out undissolved particles for nanoparticle drug products.

7.6.6 Manual Sampling

Manual sampling techniques can introduce errors due to the strength and speed that an analyst uses to collect samples. The pulling velocity through the filter may vary considerably. Because the sampling zone prescribed in USP <711> *Dissolution*, is a large area, pulling a sample at the same place using manual sampling is difficult. If the mixing is not uniform throughout the vessel, especially in a low-speed basket procedure, the location where the sample can be pulled can be critical and create a source of error and variability.

7.6.7 Automation of Dissolution Sampling

While the automation of dissolution sampling is very convenient and labor-saving, errors often occur with these devices because the analysts tend to overlook problem areas [9, 39]. Sample lines are often a source of error for a variety of reasons: unequal lengths, crimping, wear beyond limits, disconnection, carryover, mix-ups or crossing, and inadequate cleaning.

The volume dispensed, purged, recycled, or discarded should be routinely checked. Pumping tubes can wear out through normal use or repeated organic solvent rinsings; therefore, they should be checked and replaced, if needed.

The use of flow cells may generate variability in absorbance readings. Air bubbles can become caught in the cell, either introduced via a water source containing bubbles or by inadvertently entering into poorly secured sample lines. The flow rate and dwell times should be evaluated so the absorbance reading can be determined to have reached a steady plateau. Cells need to be frequently cleaned to avoid a buildup of drug residue, excipient, surfactant, or buffer salts from the dissolution medium.

Auto sampling should always be validated against the manual sampling method. It is assumed in compendial methods that manual methods are used unless otherwise stated.

7.6.8 Cleaning of Dissolution Equipment

Cleaning the dissolution equipment (e.g. vessels, automated sample lines, and baskets) should be a routine part of the analytical method. Incomplete or improper cleaning of the apparatus is likely to be a source of error and contamination. Cleaning steps should be evaluated with method validation, especially in laboratories where different products are tested on the same equipment.

7.7 Method Development

As mentioned previously in this chapter, USP <1092> *The Dissolution Procedure: Development and Validation* [40] is a valuable guide for developing dissolution methods. Its scope addresses the development and validation of dissolution procedures, with a focus on solid oral dosage forms. There are many sources in the literature with ample guidance on method development [41–43].

The fundamental requirements for developing a good dissolution method are low variability, a good profile, and the ability of the test to show changes in the product. Low-to-moderate variability is critical; two dissolution curves are not distinguishable if the standard deviation is large. The test conditions should include general dissolution data that show significant changes in dissolution caused by variations in the formulation, manufacturing process, drug substance characteristics, and stability.

The hydrodynamic aspect of product mixing in the vessel is very important; this is where visual observations are necessary. Any artifacts such as tablet sticking, coning under the paddle of solid material, clogging of the basket screens, and floating chunks should be minimized, since these phenomena may affect the dissolution rates.

BCS classification is an important starting place when developing a dissolution testing method [1]. The four categories are described in Table 7.1.

Table 7.1 Biopharmaceutics classification system (BCS) classifications.

Class 1	Class 2	Class 3	Class 4
Highly soluble	Poorly soluble	Highly soluble	Poorly soluble
High permeability	High permeability	Low permeability	Low permeability

Source: Based on Food and Drug Administration [1].

7.7.1 Drug Properties

Method development starts with obtaining as much information as possible about the drug substance and applying FDA's quality-by-design (QbD) concepts [44]. The available information may be limited, but the key properties of the compound are the pK_a, particle size range, solubility as a function of pH and surfactants, and crystallinity.

7.7.2 Dosage Form Properties

Critical dosage form properties are the disintegration and erosion rate, the functionality of the coating (e.g. enteric-coated), modified release (e.g. extended, sustained, and delayed), the presence of solubility enhancers, and quality and ratios of excipients.

7.7.3 Dissolution Profile

Ideally, unless the drug is a BCS Class 1 drug that is 80% dissolved in 15 minutes using one of the three preferred media (0.1 N hydrochloric acid, acetate buffer pH 4.5, or phosphate buffer pH 6.8), it will be necessary to develop a method that yields a dissolution curve with a reasonable profile shape (Figure 7.9). In other words, the dissolution rate should be gradual so that results can be compared using several time points. There is a draft FDA guidance, Dissolution Testing and

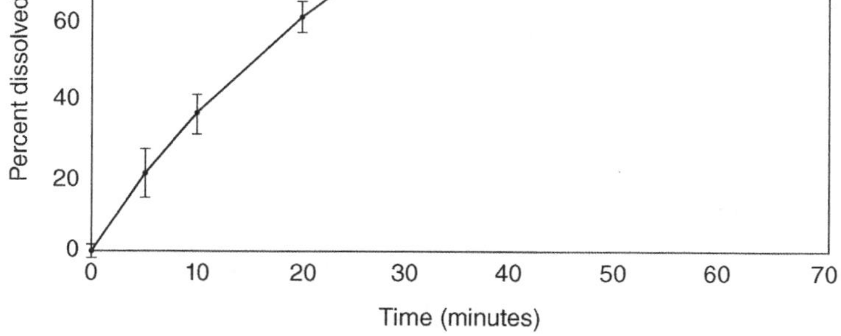

Figure 7.9 Typical dissolution curve for immediate release.

Specification Criteria for Immediate-Release Solid Oral Dosage Forms Containing Biopharmaceutics Classification System Class 1 and 3 Drugs, which gives concrete guidance on methods for BCS Class 1 and 3 [45].

There are many ways to slow down the dissolution rate, such as decreasing the apparatus speed or medium flow rate, changing the molarity of the buffers and acids used, altering the pH, or changing the apparatus. For example, 0.01 N hydrochloric acid medium with Apparatus 1 at 50 rpm can be used to slow down the dissolution rates of many dosage forms, but it only works if the product is compatible with pH 2 medium and does not cause clogs in the basket mesh.

The similarity factor, *f*2, discussed in many FDA guidance documents, use at least three points, with only one point allowed above 85% dissolution.

7.7.4 Dissolution Media

Choices of dissolution media include acids (hydrochloric acid 0.1–0.001 N); buffers (e.g. acetate [pH 4.1–5.5, 0.05 M] and phosphate [pH 5.8–8.0, 0.05 M]; use the USP preparation instructions); and simulated fluids with or without enzymes (gastric and intestinal). For Class 1 and 3 compounds, the preference would be 0.01 N HCl [45]; however, obtain profiles in the other media for future comparisons. As a dissolution test medium, water may not be appropriate as it provides no buffering capacity, and the pH cannot be measured accurately. The conductivity or pH may vary depending on the water source, as well. However, there are advantages to using water in that it is inexpensive, disposal is convenient, has, in some cases, shown discriminatory power.

For very poorly soluble compounds, aqueous solutions may be modified to contain a percentage of a surfactant to enhance drug solubility. The need for surfactants and the concentrations used must be justified by showing dissolution profiles at several different surfactant concentrations. The lowest surfactant concentration, which produces acceptable dissolution results, is generally preferred.

Molarity changes can change the dissolution rate.

Other media include mixtures of aqueous and organic components and buffers above pH 8. When looking for extensive bio-relevance in the dissolution media, fed and fasted, gastric and intestinal media are discussed in the literature [45–50].

7.7.5 Medium Volume

The medium volume is typically 900 mL, with 500 mL for low dosage strengths to increase sensitivity for detection. The volume may be increased to 1, 2, or 4 L. For the special needs of low dosage strengths, volumes of 200 mL or less may be necessary [51, 52].

7.7.6 Deaeration

Deaeration is a critical variable that needs to be performed if the presence of air bubbles interferes with dissolution testing. Deaeration of surfactants may not be practical due to foaming and may not be necessary [53].

There is a multitude of deaeration methods available, including, but not limited to, the USP method involving heat, filtration, and vacuum [10]; helium sparging; and automated methods.

7.7.7 Speed

The typical rotation speeds for the paddles are 50 rpm, 75 rpm to eliminate coning and variability, or 25 rpm or more for suspensions. A speed of 100 rpm or higher requires justification. However, 100 rpm is frequently used with extended-release products. For the basket apparatus, 50–100 rpm is preferred, but speeds of greater than 100 rpm are sometimes necessary.

7.7.8 Sinkers

Sinkers are needed with floating dosage forms, typically capsules. Sinkers can also help prevent a large film-coated tablet from sticking to the vessel wall or bottom. Since different sinkers may have significantly different hydrodynamic effects on the dosage form and can yield different dissolution results, the choice of sinker type may vary from case to case depending on the dosage form of the drug product [38]. The acceptable type(s) of sinkers should be specified in the method.

7.7.9 Filtration

Filter use is necessary for most products. Centrifugation is not preferred because the dissolution can continue in the tube, and centrifugation is time-consuming. Filters are made of many different materials (e.g. nylon, polyethylene, and glass fiber). When selecting a filter, compatibility with the media and formulation must be considered, and validation must occur before the filter is used routinely. There are several types and positions of filters: in-line; at the end of the probe or cannula; disk; or as was used in the past, a stainless-steel filter holder. The pore size of the filters typically ranges from 0.20 to 70 μm, with depth or full flow design.

It is important to filter samples as soon as possible to avoid post-sampling dissolution.

7.7.10 Time Points – Immediate Release

For immediate release dosage forms, a five-minute time point where disintegration occurs or is partially completed should be considered if the variability is not too high (≤20% RSD). This time point may give profile information, especially for suspensions, or be useful in accumulating the necessary three points for an *f*2 comparison. The other intermediate points are 10, 15, or 20 minutes; any of these points will be useful for a profile, and an *f*2 comparison, and in some cases, the specification will be at one of these earlier points. For example, a suspension may have a Q value at any of these points. The later points of 30, 45, and 60 minutes will be necessary for the specification of immediate release, and the test for a poorly soluble drug may go longer (up to three hours in some cases). If complete (100%) dissolution occurs at 30 minutes, the 60-minute time point will not be necessary. It is important, however, to keep one extra point past the 100% dissolved point in case there is a decrease in the dissolution rate of samples on stability testing.

7.7.11 Fast Stir or Infinity Point

After a sample has been drawn for the last time point, the rpm may be increased to 150–200 rpm for another 15–30 minutes. These conditions should completely dissolve the sample in the vessel. A sampling of each vessel will provide at least six sample readings. And thus, there is a data set that is appropriate for comparing with the content uniformity data for the product. A comparison of the fully dissolved samples versus label claim will give an early read on recovery and variability in the dissolution method. If the content uniformity data are different in either potency or variability, it provides additional information for assessing the method.

7.7.12 Time Points for Extended-Release Products

A minimum of three time points is required for extended-release products. There will be a time point in the first hour or two to measure the potential for dose dumping, a midway point at around 50% dissolved, and a Not-Less-Than endpoint where typically at least 80% is dissolved or an asymptote is reached. Other time points may be useful, especially if the test continues for longer than eight hours. With extended or modified-release dosage forms, it is sometimes difficult to achieve a 100% dissolved state. This event can be caused by the matrix holding on to the drug in such a way that not all of it is exposed to the media and readily dissolved. A fast stir is also not practical with a modified-release product unless a 100% dissolved state is achievable; then the information would be useful when compared to the content uniformity results.

7.8 Poorly Soluble Drugs

The BCS is the first step toward dissolution method development for poorly soluble drugs. BCS Class 2 is the most common type of drug and the most challenging when developing a discriminating dissolution test. Class 2 and 4 can provide *in vitro–in vivo* correlation (IVIVC) because the dissolution is the rate-limiting step for these drugs. To select media for poorly soluble drugs, any of the media listed for Class 1 will provide a good starting point.

As mentioned previously, surfactants are usually needed for poorly soluble drugs. Surfactants can be used either as wetting agents or, when the critical micelle concentration is reached, to solubilize the drug substance. Surfactants are cationic, anionic, or nonionic and are selected based on the chemical nature of the drug substance. A concentration of 1–2% is typically needed to achieve sink conditions. There are many surfactants available, including SLS, polysorbate 20–80, Cetrimide, lauryl dimethylamine oxide, bile salts, Brij™Triton X™, Solutol™, and Cremophor. SLS is one of the most widely used, and the technical grade of SLS is a mixture, thus the purest form is recommended to be used. SLS is not stable with a pH below 2.5. It also denatures the enzymes typically used in two-tier testing – pepsin and pancreatin – making it difficult to use when a capsule product shows failed dissolution results due to cross-linking. If using SLS in combination with pH 6.8 buffer, the phosphate sodium salt should be used instead of the potassium salt, because the potassium salt forms a precipitate at room temperature [54].

Combinations of surfactants and buffers/acids are also favored when the pH needs to be controlled, and the solubility is an issue.

7.8.1 Sink Conditions

Sink conditions are critical for dissolution testing for poorly soluble drugs. As mentioned before, USP <1092> *The Dissolution Procedure: Development and Validation* [9] states, "The quantity of medium used should be not less than three times that required to form a saturated solution of the drug substance."

Establishing and maintaining sink conditions during the dissolution test is an essential criterion for the dissolution method because the true dissolution rate should be measured and not be overlapping in the area of concentration equilibrium. As the concentration of the solution increases, the dissolution rate will decrease. To achieve sink conditions, the surfactant concentration can be changed, as discussed in previous sections, or the media volume can be increased by using 2- or 4-L vessels. The use of Apparatus 4 is also an option since the infinite sink is obtained with the constant flow of media over the dosage unit.

7.8.2 Apparatus Selection

USP Apparatus 1 or 2 should be the first choice for most dosage forms. Apparatus 3 is a research tool and may be useful for enteric-coated products and some other dosage forms like soft-gel capsules or chewable tablets. Apparatus 4, the flow-through cell, with the open system, can provide infinite sink conditions. In both

Apparatus 3 and 4, media can be changed during the test. Apparatus 7 has some utility for extended-release products (e.g. osmotic pump tablets), transdermal administration, and stents or surgical implants.

7.8.3 The Discriminatory Power of the Method

The regulatory agencies often ask sponsors to support the discriminatory ability of their selected methods. This discriminatory ability may be achieved by comparing different dissolution profiles of formulations that are developed with variations of the most relevant critical manufacturing parameters (e.g. ±10–20% change to the ranges of selected parameters). The design of experiments (DoE) may be useful to establish the critical factors that influence the dissolution rate. The DoE can evaluate the operational dissolution parameters [55, 56] or the product formulation release attributes [57, 58]. It may provide an understanding of the release mechanisms and determine if the dissolution method can show change to the critical quality attributes.

Developing a dissolution method is a team effort. Formulators provide information on the critical quality attributes and the actual dosage forms that reflect variables for the DoE. If an IVIVC is attempted, then the pharmacokinetics scientists will be involved in the *in vivo* data handling, and a clinical study director is involved in developing the clinical study design. Regulatory scientists are needed if bioequivalence studies or biowaiver requests are planned. The analytical scientists, those who develop the method, will interact with all these other scientists because dissolution testing must be a meaningful test rather than strictly a quality control test.

Methods in the past have been single point with a fast dissolution rate. Now, however, unless the product is a very rapidly dissolving BCS Class 1 or BCS Class 3, a slower profile test is typically needed to give pertinent information about any change in the product. Regulatory agencies prefer a clinically relevant method, that is those that show IVIVC or discriminatory power. Since clinically relevant specifications become a requirement for the characterization and regulation of drug products, laboratories are putting more resources into method development of dissolution testing and the establishment of an IVIVC relationship.

7.9 Setting Specifications

Manufacturing control is also a concern for the regulatory agencies; therefore, variability among batches is carefully examined. Specifications for dissolution testing should differentiate between the bioequivalent and bio-inequivalent batches; however, the specification should not be so tight that lots are unreasonably rejected (especially those on stability) which leads to unnecessary recalls. The FDA no longer accepts specification limits set at ±3 standard deviations and stipulates that accommodating the variability of the manufacturing process is no longer a consideration in setting dissolution specifications [59].

Variability with extended-release products can be troublesome as the practice of trying to include all the values within a specification window may lead to specifications greater than ±10%, as stated in the FDA guidance [5]. The challenge, then, is to develop methods with only moderate variability. By visually observing the dissolution of the product in the vessel, the cause of variability may often be identified as either the actual dissolution method, which often can be discerned by observation or variability from a formulation or manufacturing issue. Typical observable causes of variability are coning, sticking tablets, pellicles, erosion patterns, disintegration characteristics of both immediate- and extended-release products, and the dosage form position. In addition, observations should be made on with all six vessels to discern the uniformity of the dissolution pattern among all the dosage units tested.

For a product release, the specification may include a requirement at a single time point for a highly soluble, rapidly dissolving drug product, or requirements at multiple timepoints, for poorly soluble drugs or extended-release products. Extended-release products typically have a minimum of three-time points. The first time point is designed to show potential dose dumping, the second is the characteristic of the release rate, and the final time point shows the extent of dissolution.

The acceptance criteria generally follow the algorithms described in the Acceptance Tables in USP <711> *Dissolution* or USP <724> *Drug Release*. Tables in these General Chapters allow for multiple stages of testing; a drug product that meets the requirements of any of the three stages is acceptable. A drug product is expected to meet the specification throughout its shelf life.

Acceptance criteria for dissolution tests are often the subject of extensive discussions with regulatory authorities. Having a complete and well-documented database of dissolution results is often helpful for establishing appropriate acceptance criteria. Once the acceptance criteria are established, conformance to the compendial acceptance tables is expected, and any deviations should be thoroughly investigated and documented. It should be noted that occasional progression to second stage testing is not considered a deviation unless it represents a significant change from historical data.

The ICH document Q6A Decision Trees #7: Setting Acceptance Criteria for Drug Products Dissolution contains three decision trees. The first discusses the types of drug release acceptance criteria that are appropriate and discusses disintegration testing instead of dissolution testing. The second decision tree points to specific test conditions and acceptance criteria that are appropriate for immediate release; the topic of a dissolution test with or without discriminatory power is specifically addressed. The third decision tree deals with appropriate specifications for extended-release drug products. The subject of IVIVC relationships is covered.

7.10 Harmonization

ICH is a unique organization that, since 1990, has brought together the drug regulatory authorities of Europe, Japan, and the United States and experts from the pharmaceutical industry in those regions to discuss scientific and technical aspects of product registration.

The ICH recommended that the USP, European Pharmacopeia (Ph. Eur.), and the Japanese Pharmacopeia (JP) harmonize the general chapters on dissolution, disintegration, and drug release.

Additional information on the harmonization process can be found in USP General Chapter <1196> *Pharmacopeial Harmonization* [60]. Harmonization of the dissolution chapters of the pharmacopeias was completed in 2006. For the USP, harmonization consisted of adding elements of <724> *Drug Release* to USP <711> *Dissolution*. Now USP Apparatuses 1–4 are included in USP <711> *Dissolution*, along with the acceptance tables for delayed- and extended-release products from USP <724> *Drug Release*. USP <724> *Drug Release* now contains only the Transdermal Delivery System with Apparatus 5 (paddle over disk), 6 (cylinder), and 7 (reciprocating holder).

Some elements are still not harmonized since the JP does not recognize Apparatus 3 (reciprocating cylinder). The JP also follows a separate approach to delayed-release products, using serial versus concurrent testing and, therefore, the delayed-release methods and interpretation are different. Harmonizing the name for each release category was not accomplished.

The basket wire diameter dimensions were widened to 0.25–0.31 mm to accommodate all regions, which may cause method transfer issues due to the large range of basket dimensions. It was also discovered that the 0.25-mm bottom end of the range is the midpoint to which most baskets were manufactured; therefore, a revision to extend the range down to 0.22 mm is now in place.

Several other elements of the chapters were not harmonized. The use of PVT is not included in the Ph. Eur. and JP, although the Apparatus Suitability section is included in all Pharmacopeia.

The USP allows for 2- and 4-L vessels with the dimensions provided, but this allowance is not included in the Ph. Eur. or JP.

The USP also includes an extended discussion of adding an enzyme to the medium when there is a dissolution failure for gelatin capsules and gelatin-coated tablets. Where gelatin cross-linking has occurred, adding the enzyme to the medium allows many products to meet dissolution specifications [61, 62]. Since cross-linking does occur in products that are produced globally, the problem is handled on a case-by-case basis with filings in Japan and the European Union, so therefore this discussion is not presented similarly in the JP or Ph. Eur.

7.11 Method Validation

The level of validation depends on the phase of product development. For scouting, the linear range of standards may be sufficient, but as the need for "reportable" data approaches, other validation parameters must be evaluated. Full validation should be considered for products that are in late clinical Phase 2 or Phase 3 of product development. USP <1092> *The Dissolution Procedure: Development and Validation* [9] provides guidance for the general validation practices by industry dissolution experts.

Validation of dissolution testing is divided into two segments. The first part is the validation of the product performance test including robustness, ruggedness (intermediate precision), recovery (accuracy), selectivity (placebo interference), sample stability, sampling method, filtration, comparison dissolution results of manual versus automated, carryover in automation, and sinker validation [43]. This step is essentially the sample preparation step. The second part is the determinative-step validation, which is the validation of the analytical method that is used to analyze the sample aliquot from the product performance step. This determinative step validation is covered thoroughly in the literature [63] and will not be covered in any detail in this chapter. Critical parameters for the dissolution sample analysis are linearity, precision, and standard stability.

During the assessment of product performance with the dissolution method, some primary criteria must be achieved before proceeding to the validation of the sample analysis. There should be acceptable variability of the method, and the dissolution profile must be able to detect formulation and process changes. In other words, the method is meaningful, and results can be interpreted without being confounded by other factors. There should be no significant analytical solution to stability problems. In other words, the drug should not decompose in the media.

7.12 Validation of Product Performance Parameters

7.12.1 Accuracy/Recovery

Recovery experiments are typically conducted at 50, 100, and 125% concentration levels, and in addition, the lowest expected profile concentration. The placebo mixture should include all excipients, the capsule shells, coating blends, inks, and sinkers. The recovery experiment can be performed in a vessel or a flask on the benchtop with a preheated medium. During recovery experiments, the order of addition (drug versus excipient) may vary on a case-by-case basis depending on the physical characteristics of the excipients and drug substance. The drug is preferably added to the mixing flask directly as a powder; however, in the case that the amount of drug is small or weighing may be inaccurate (hydrostatic), the drug may be first dissolved in dissolution media or another suitable diluent and spiked into the vessel or flask. Poorly soluble drugs may require a more rigorous evaluation of the experimental steps. If the powder is used, longer mixing times and higher initial apparatus speed may be necessary if performed in a vessel. The acceptance criterion for recovery is typically 97–103% of the theoretical value.

7.12.2 Selectivity

Interference from placebo must be well understood. Interference from placebo is studied using the same placebo mixture as used in the recovery experiment. The placebo mixture should be stirred for at least one hour at high rpm. The wetting properties should be noted. For a product with multiple strengths, studies should be done comparing an equivalent amount of placebo mixture for the highest and lowest strength to those of the 100% standard. The acceptance criteria for the interference are typically not more than 2%.

7.12.3 Solution Stability

The stability of the sample should be studied on day one, and then at intervals ranging from 3 to 12 days. The length of solution stability depends on how many days a re-reading of the sample is allowed by approvals and/or mandated by Standard Operating Procedures (SOPs). The usual criterion is 98–102% of the fresh sample reading. If ultraviolet (UV) analysis is the analytical method of choice, an analysis of the UV samples by HPLC may be instructive in the event, there are hidden stability issues evidenced by separated impurities.

7.12.4 Filter

Filter validation is performed on both sample and standard solutions using a 100% solution, although a range is more comprehensive. Solutions of filtered and unfiltered standards are studied for standard solutions. For the sample solution filter validation, a 100% solution should be prepared. It is important to assure that 100% of the sample is dissolved, because earlier sampling may show ongoing dissolution during the centrifugation. The typical acceptance criterion is within 98–102% of the unfiltered standard and unfiltered/centrifuged sample solutions.

7.12.5 Robustness

The impact of small changes of parameters within the dissolution test constitutes the robustness validation parameter. The robustness provides very useful information wherein critical variables may be discovered. Dissolution tests can be very technique-dependent for some compounds, especially those of low solubility. The most critical aspects are typically deaeration, medium concentration (e.g. molarity, percentage of surfactant), and pH. For deaeration, a comparison of deaerated media versus non-deaerated media is studied because bubbles can affect the results. For media concentration, 80, 100, and 120% levels of the chosen media concentration may be used. For example, with a 1% SLS as the chosen medium, the variation of the media concentrations will include 0.8 and 1.2% SLS. Varying the medium pH by ±0.05 pH unit will adequately assess the effects of pH. Other optional variations include paddle height (±0.5 cm), water bath temperature (±1 °C), sample times (±2 minutes), and rpm (±4%). Assessing the relationship of the dosage unit position in a vessel (center versus off-center) to the dissolution results and variability is more challenging. Lastly, one should at least attempt to understand if the product is sensitive to vibration. The impact is usually discovered serendipitously, and rarely experiments are designed to assess this problem.

The usual criterion for robustness is within 3–5% of the values determined under the typical method conditions. It should also be pointed out that basket attachment design may affect the dissolution rate. It has been documented [18, 25] and deals with clipped (official USP design) versus O-ring attachment design. If both attachment methods are used or may be used in a transfer lab, it must be a part of validation.

7.12.6 Intermediate Precision

The ruggedness parameter is often referred to as intermediate precision. It is as close to a method transfer as one can get, and it is considered as an early indication of possible method transfer issues. Therefore, the test parameters should be varied as much as is feasible, that is a different analyst, dissolution tester, spectrophotometer, flow cell, media, standard and buffer preparations, and autosampler, on different days and in a second laboratory, if possible. The same sample batch should be tested using 12 units. All product strengths should be tested or a bracket is used when 3 or 5 strengths are present. The acceptance criterion stated in USP <1092> *The Dissolution Procedure: Development and Validation* is that the difference in the mean value for dissolution results between any two conditions, using the same strength, does not exceed an absolute 10% at time points with ≤85% dissolved and does not exceed 5% for time points > 85%. Acceptance criteria may be product-specific, and other statistical tests and limits may be used.

7.12.7 Automated Methodology

Validating a method that has an automated component requires special considerations. Automation can occur in many forms, from part of the equipment to fully automated systems. Automated systems can include fiber optics and in-residence probes, autosamplers, automated deaeration equipment, on-line UV testing, and robotics automation.

Regardless, the principles governing validation of an automated method involve doing a manual sampling method and comparing the dissolution results to those obtained using an automated method. Several sources of error that can come from automation; therefore, a comparison of automated sampling versus manual sampling is quite critical. The comparison experimental study for products with high variability would include simultaneous manual versus automated sampling at all time intervals. Calculations need to account for the duplicate volume lost. However, a strong caveat against this simultaneous manual versus automated sampling is that it will not assess sampling probe interferences. To better evaluate this critical parameter, concurrent testing is recommended. One to two runs of each dosage strength should be performed using manual and automated sampling. A typical acceptance criterion of the difference in the mean value for dissolution results between manual and automated dissolution results, using the same strength, does not exceed an absolute 10% at time points with ≤85% dissolved and does not exceed 5% for time points > 85%.

7.13 Validation of the Analytical Finish

For dissolution testing, the analytical finish (i.e. determinative step) includes linearity, range, and precision. Up to 5% organic solvents (2% organic component preferred) should be used to enhance the solubility of the drug in the final standard solution. The typical concentration range for validation is between 25 and 125% (three to five points) of the label claim concentration.

If the detector has flow cells, where the absorbance is read automatically, validation should be performed comparing standard absorbances using the flow cell versus those of standards read as samples that are placed in the flow cell manually. All solutions may be made from a common stock, and triplicate readings or duplicate injections are made for standards. The usual linearity acceptance criterion is a correlation coefficient of >0.997, with a Y-intercept of 2% or less of the 100% level of the standard concentration.

The determinative validation step for precision is acquired from linearity data. The usual acceptance criteria are 1–2% relative standard deviation (RSD) for UV analysis and 2% RSD for HPLC analysis. The stability of standard solutions is studied from Day 1 up to Day 12 following the sample preparation. The usual criterion for solution stability is 98–102% calculated from a fresh standard solution reading. A historical database of the typical absorptivity range of standard solution is useful for reference. For HPLC analysis, there are also usually retention time and precision criteria. Response factors are not too reliable but do afford some reassurance of a working system.

A robustness attribute for the UV analysis is studied by varying the wavelength (±2 nm). For the HPLC analysis, robustness is typically studied by varying the column manufacturers, the age of the column, different mobile phase compositions (±10%), and changes in pH.

7.14 Method Transfer Considerations

Problems that occur during the transfer of dissolution methods can often be traced to the difference in equipment or the set-up components such as baskets/shafts, sinkers, dispensing apparatus, or sampling methods or differences in SOPs between laboratories, sometimes related to lack of detail. A precise description of medium and standard preparation, including grade/purity of reagents, must be documented. For example, it is generally known that the use of alcohol can be helpful to avoid errors in preparing a standard, but if a standard was prepared without alcohol and the sonication step (if used) is longer than expected. These distinctions should be explicitly noted in the method transfer documentation to avoid introducing errors.

A common cause for the failure of a method transfer is that one site uses automation, and the other site uses manual sampling. Automation and manual testing are affected by different filters, sampling techniques, or deaeration levels. Addressing these in the transfer protocol is prudent.

7.14.1 Robustness

Findings from the robustness evaluation during validation can uncover flaws of the dissolution method and frequently leads to discovering sources of error. This knowledge can be instructive during and before method transfer.

7.14.2 Details of the Analytical Method

Analytical test methods that lack critical details will generally lead to transfer issues. For example, if the product is sensitive to dissolved gases, the deaeration technique is a critical step that should be described in detail in the procedure. Otherwise, there may be variable results from one lab to the next due to differences in deaeration practices. During method development, there should be a risk assessment to assist in the identification of those parameters requiring control, and the inclusion of a control strategy in the analytical procedure to avoid issues.

7.14.3 Other Considerations

The basket attachment type and mesh size should be thoroughly described in the procedure. The sinker type is vital for the reasons discussed in previous sections. If a sinker is handmade, the procedure used to create it should be included in the transfer documentation. In most cases, the sample introduction technique needs to be described, especially in the case of suspensions. With suspensions, it must be specified whether the paddle is running or not when the sample is introduced.

Rigorous method development and validation, proper calibration and operation of equipment, and thorough and frequent observations can assist in identifying and preventing sources of error associated with method transfer.

7.15 Good Manufacturing Practices (GMP) in the Dissolution Testing Laboratory

In the dissolution laboratory, there are many GMP issues given all the equipment, documentation, and validation involved in testing many products at different stages of development [64–66]. Multiple users of equipment, reagents, and solutions performing testing on the same and different products add complexities to the laboratory operations. Each lab could have 10–40 testing units with associated autosamplers; HPLCs including detectors, pumps, autoinjectors, and columns; UV spectrophotometers and auto shippers; deaeration equipment; and fully automated testing equipment. All these instruments have their calibration, maintenance, and operation procedures, including logbooks. The testing process requires extensive notebook documentation and witnessing as the profile test can have numerous data points with observations and pre- and post-run equipment checks. The variety of products requires constant validation and re-validation as formulations change and new test methods are written and revised. Constant monitoring of adherence to GMP is necessary to ensure compliance and successful audit results. Internal audits need to be a regular part of the laboratory operations. The training and documentation of training are becoming more critical in the modern lab where turnover can be high and the type of products varies.

7.15.1 Metrology

The metrology program is an essential function associated with the dissolution laboratory. The tracking of equipment identification, repairs, and calibration may be performed by qualified personnel outside the dissolution group. It involves frequent communication between the groups, especially in the realm of calibration timelines. The calibration of equipment at its due date is a good indicator of the efficiency of the laboratory operations. Missed or late calibration dates can accumulate and give the appearance of poor management of resources and priorities, even if the equipment is labeled appropriately from a calibration standpoint. The status of equipment, whether it is out of service for repairs, calibration, or under investigation, should be very clearly and boldly marked as to avoid any ambiguities as to the equipment condition and usability. Special circumstances, such as use for only one apparatus or new equipment waiting for validation, should be labeled accordingly.

Logbooks or any notebook associated with or assigned to equipment must be current and contain the most useful information – that is, observations of problems, how the problems were remedied, calibration results and failures, corrective action, and routine maintenance or performance checks. Each piece of equipment should have a custodian who is responsible for and enters information into the logbooks. Communication becomes critical, so the analyst knows when the equipment has had problems in the past as well as the nature of the problems. An accurate and current logbook can offer insight into the cause of aberrant data and support the repair, replacement, or upgrading of equipment.

The operational procedures need to have enough detail so an analyst can use the instrument to obtain accurate results without having to rely on additional information from more experienced users.

7.15.2 Notebook Documentation

A current SOP for documentation in notebooks is critical. The dissolution test does lend itself to inserts or template worksheets, and such practices are useful for several reasons. The analyst has many things to remember such as the rpm and temperature checks (before and after the run), the correct speed and apparatus, sinkers or no sinkers, deaeration or no deaeration, deaeration method, observations, sample and equipment IDs, and sample, standard, and reagent preparation. It is only a partial list of all the items that should be recorded. A template list where one fills in the blanks or makes a checkmark can serve to keep the information in an organized manner, which is very helpful to the witness. It also requires the analyst to double-check that all aspects of the test have been performed properly and documented. The treatment of inserts or template worksheets has to be spelled out in the SOP, and quality assurance personnel should have complete confidence that the documentation would meet all compliance concerns. Electronic notebooks are widely used and should have SOP's in place for guidance on their use.

The recording of sampling times is the subject of much discussion. The analyst can record in real-time every pull (using a traceable calibrated timepiece, of course), or refer to a test method and presume adherence to the prescribed sampling interval. With manual sampling, this can be a labor-intensive task. Autosampling alleviates this task as the instrument printout often documents when the sample was taken.

In the dissolution lab, where the testing could require multiple analysts to share the same standard solution and/or medium preparation, a use log is needed where analysts will initial and date when the solutions are prepared or used. The analyst then refers to the use log as part of the write-up of the experiment. The role of the witness should not be underestimated. In GMP, the witness role is to verify the accuracy, completeness, and compliance of the data generated. Therefore, the witness should be an analyst who has knowledge of the test and has performed the test previously so that the person can accurately pick up omissions, mistakes, and out-of-trend results. The witness, in addition to having in-depth familiarity with the method, should have some training in the witnessing process. A checklist of things to watch for would also be useful.

7.15.3 Equipment Qualification, Validation, and Method Critical Factors

One of the most frequently cited areas in 483 warning letters is a lack of validation or improper validation. With the frequent use of autosamplers and fully automated systems in the dissolution laboratory, the distinction between manual test method validation and automated test validation is paramount. The equipment also needs to be qualified, with a focus on the unique performance aspects of the specialized equipment. The instrument itself should go through performance checks that are part of the routine operation of the instrument, usually thought of as operation qualification (OQ). The installation qualification (IQ) is performed when the instrument is newly acquired and set up. When the OQ and IQ are satisfactorily completed, validation can be performed for the dissolution procedure of a drug product. Validation of the use of a simple autosampler may be a manual and automated run performed concurrently. The results will be compared against predetermined acceptance criteria based on the inherent variability of the product. A fully automated dissolution system is much more involved. All validation requires a validation report as part of the documentation. Any automated system validation should address contamination from previously tested compounds (cleaning validation) and buildup of surfactant. Pump dwell times, sample lines, and filter checks are often problem areas.

A thorough validation is very important for dissolution procedures. A "critical factors" section is a crucial part of the method. For example, in dissolution testing, the standard may be difficult to dissolve in an aqueous medium; therefore, additional information on the standard preparation can be emphasized. Instructions for the proper amount and addition order of a small amount of alcohol may be very critical to the proper dissolution of the drug substance. Examples of critical factors are listed in the Key Information section below. This information may already be in the method, but a highlighted critical factors section will alert the analyst to aspects of the test that are out of the ordinary.

Examples of critical factors

- The deaeration method
- Sinker type and, if handmade, instructions
- Standard preparation if alcohol is used, including sonication time
- Cleaning instructions for vessels and/or autosamplers
- Special precautions for cleaning autoinjectors when surfactants are used
- Septum replacement for autoinjector vials
- Filter type and discard volume
- Apparatus speed if not the typical speed
- Special instructions for the rotation of paddles before the test begins (this may be required for suspensions)
- Exact mixing procedures for dosage forms that need reconstitution
- Typical absorptivity values (UV) or response factors (HPLC)
- Precautions to protect from light

7.15.4 Good Manufacturing Practice Audits

Frequent internal audits are recommended to maintain good laboratory practices and to identify potential GMP issues. The checklist can be used as a training tool. The auditor should inform the group of his/her observations and make recommendations. For example, the source of vibration, especially external vibration, should be routinely checked as part of the audit. The analyst should verify the dissolution bath on the countertop is not near any shakers, hoods, or centrifuges. Local construction is also a source of vibration. Heavy foot traffic or slamming doors nearby can also impact the test. Routine audits can be performed by quality assurance or other analysts to maintain a high GMP compliance level as dissolution testing requires detailed knowledge and documentation.

7.15.5 Training

Training is critical in the dissolution lab because regulatory agencies typically scrutinize this area. The term training, in this case, refers to training a new analyst to perform dissolution testing properly on product-specific test methods.

7.15.5.1 Training on Dissolution Test

One option for training is to use the USP PVT calibration of the equipment as a training tool. However, this training can be done in tandem with an actual calibration performed by a well-trained analyst, so that a training failure will not impact the results generated using that equipment. An essential aspect of dissolution testing is making observations, which is useful for any investigation. Dissolution testing also has specific terminologies that an analyst should be familiar with when explaining aberrant data and exploring the correct method during method development.

The training of a new analyst should be documented when all the training elements are complete before that analyst can generate reportable results.

7.15.5.2 Training on Product Specific Dissolution Methods

In the quality control lab, an analyst should be qualified on product-specific dissolution methods. In this scenario, the analyst performs a training sample test, the results of which should agree with those obtained by an experienced analyst. The experience level of the second analyst and the difficulty or uniqueness of the test should be considered when planning the training. The level of training is heightened if the product requires detailed observations or complicated sample introduction (e.g. suspensions).

7.16 Summary

The performance of a drug product is usually characterized by the dissolution method, making it one of the critical tests in a product specification. Assuring the method is valid for its intended purpose throughout the lifecycle of the product requires an approach that starts with the development of the procedure and continues through each change that may occur, either to the drug product or the execution of the dissolution method. The validity can be challenged at each step, so it is necessary to do an evaluation which may include additional validation experiments to demonstrate that the method continues to function as intended and that the data are reliable.

Due to the importance of the test in characterizing the performance of the dosage form, and where compliance issues can be encountered, dissolution testing is often a target for regulatory or compliance scrutiny. Evaluating dissolution testing provides an inspector with insight into the procedures in a laboratory related to not only the dissolution results but also instrument qualification and calibration, method validation procedures, employee training, laboratory investigations, and change control procedures. Inspector scrutiny also provides a laboratory the opportunity to showcase its procedures, so long as laboratory personnel are well acquainted with regulatory expectations and have addressed those expectations in the generation of the dissolution data.

Acknowledgment

I appreciate the drawings for all figures that were supplied by the Hanson Research Corporation.

List of Abbreviations

AAPS	American Association of Pharmaceutical Scientists
BCS	Biopharmaceutics Classification System
CDER	Center for Drug Evaluation and Research
DOE	Design of Experiments
EMA	European Medicines Agency
FDA	Food and Drug Administration
FIP	International Pharmaceutical Federation
GMP	Good Manufacturing Practices
HPLC	High-Pressure Liquid Chromatography
ICH	International Conference on Harmonization
IQ	Installation Qualification
IVIVC	*In Vitro* and *In Vivo* Correlations
JP	Japanese Pharmacopeia
OQ	Operation Qualification
Ph.Eur	European Pharmacopeia
PVT	Performance Verification Testing
Q	Quantity
QbD	Quality-by-Design
RPM	Revolutions Per Minute
RSD	Relative Standard Deviation
SLS	Sodium Lauryl Sulfate
SOPs	Standard Operating Procedures
USP	United States Pharmacopeia
UV	Ultraviolet
WHO	World Health Organization

References

1 Food and Drug Administration (2016). *Guidance for Industry: Waiver of In Vivo Bioavailability and Bioequivalence Studies for Immediate-Release Solid Oral Dosage Forms Based on Biopharmaceutics Classification System*. Rockville, MD: Food and Drug Administration https://www.fda.gov/media/70963/download (accessed 5 November 2020).

2 Food and Drug Administration (1997). *Guidance for Industry: Dissolution Testing of Immediate-Release Solid Oral Dosage Forms*. Rockville, MD: Food and Drug Administration https://www.fda.gov/media/70936/download (accessed 5 November 2020).

3 Food and Drug Administration (2002). *Guidance for Industry: Bioavailability and Bioequivalence Studies for Orally Administered Drug Products-General Considerations*. Rockville, MD: Food and Drug Administration https://www.fda.gov/files/drugs/published/Guidance-for-Industry-Bioavailability-and-Bioequivalence-Studies-for-Orally-Administered-Drug-Products---General-Considerations.PDF (accessed 5 November 2020).

4 Food and Drug Administration (1995). *Guidance for Industry: Immediate-Release Solid Oral Dosage Forms, Scale-Up, and Post-Approval Changes: Chemistry, Manufacturing, and Controls; in vitro Dissolution Testing and in vivo Bioequivalence Documentation*. Rockville, MD: Food and Drug Administration https://www.fda.gov/media/70949/download (accessed 5 November 2020).

5 Food and Drug Administration (1997). *Guidance for Industry: Extended-Release Oral Dosage Forms: Development, Evaluation, and Application of in vitro/in vivo Correlations*. Rockville, MD: Food and Drug Administration https://www.fda.gov/media/70939/download (accessed 5 November 2020).

6 USP <1088> (2020). *In Vitro and In Vivo* evaluation of dosage forms. *USP 43–NF 38*. Rockville, MD: United States Pharmacopeial Convention.

7 USP <701> (2020). Disintegration. *43–NF 38*. Rockville, MD: United States Pharmacopeial Convention.

8 Noyes, A. and Whitney, W. (1897). The rate of solution of solid substances in their own solutions. *Journal of the American Chemical Society* 19: 930.

9 USP <1092> (2020). The dissolution procedure: development and validation. *USP 43-NF 38*. Rockville, MD: United States Pharmacopeial Convention.

10 USP <711> (2020). Dissolution. *USP 40–NF 35*. Rockville, MD: United States Pharmacopeial Convention.

11 Gray, V. (2002). Identifying sources of error in calibration and sample testing. *American Pharmaceutical Review* 5: 8–12.

12 Nithyanandan, P., Deng, G., Brown, W. et al. (2006). Evaluation of the sensitivity of the USP prednisone tablets to dissolved gas in the dissolution medium using USP Apparatus 2. *Dissolution Technologies* 13 (3): 15–18.

13 Deng, G., Hauck, W.W., Brown, W. et al. (2007). Perturbation study of the dissolution apparatus variable – a design of experiment approach. *Dissolution Technologies* 14 (1): 20–26.

14 Moore, T.W. (1998). Dissolution testing: a fast, efficient procedure for degassing dissolution medium. *Dissolution Technologies* 3 (2): 3–5.

15 Queshi, S.A. and McGilveray, I.J. (1994). Impact of different deaeration methods on the USP dissolution apparatus suitability test criteria. *Pharmacopeial Forum* 20 (6): 8565–8566.

16 Degenhardt, O.S., Waters, B., Rebelo-Cameirao, A. et al. (2004). Comparison of the effectiveness of various deaeration techniques. *Dissolution Technologies* 11 (1): 6–11.

17 Collins, C.C. (1998). Vibration: what is it, and how might it affect dissolution testing. *Dissolution Technologies* 5 (4): 16–18.

18 Crist, B. and Spisak, D. (2005). Evaluation of induced variance of physical parameters on the calibrated USP dissolution apparatus 1 and 2. *Dissolution Technologies* 12 (1): 28–34.

19 Vangani, S., Flick, T., Tamayo, G. et al. (2007). Vibration measurements on dissolution systems and effects on dissolution prednisone tablets RS. *Dissolution Technologies* 14 (1): 6–14.

20 Gao, Z., Moore, T., Doub, W. et al. (2008). Vibration effects on dissolution tests with USP Apparatus 1 and 2. *Journal of Pharmaceutical Sciences* 97 (8): 3335–3343.

21 Liddell, M.R., Deng, G., Hauck, W.W. et al. (2007). Evaluation of glass dissolution vessel dimensions. *Dissolution Technologies* 14 (1): 28–33.

22 Scott, P. (2005). Geometric irregularities common to the dissolution vessel. *Dissolution Technologies* 12 (1): 18–21.

23 Cox, D.C., Wells, C.E., Furman, W.B. et al. (1982). Systematic error associated with apparatus 2 of the USP dissolution test II: effects of deviations in vessel curvature from that of a sphere. *Journal of Pharmaceutical Sciences* 71: 395–399.

24 Tanaka, M., Fujiwara, H., and Fujiwara, M. (2005). Effect of the irregular inner shape of a glass vessel on prednisone dissolution results. *Dissolution Technologies* 12 (4): 15–19.

25 Gray, V., Beggy, M., Brockson, R. et al. (2001). A comparison of dissolution results using O-ring versus clipped basket shafts. *Dissolution Technologies* 8 (4): 8–11.

26 Kraemer, J. and Schwan, R. (2011). Practical aspects of dissolution instrument qualification – a European perspective. *Dissolution Technologies* 18 (2): 11–15.

27 Borst, I., Ugwu, S., and Beckett, A.H. (1995). New and extended application for USP drug release apparatus 3. *Dissolution Technologies* 2 (2): 1–8.

28 Takiar, N.B. and Hollenbeck, R.G. (1997). *in vitro* evaluation of drug release from modified-release delivery systems: initial experiences with calibrators for the USP dissolution apparatus 3. *Dissolution Technologies* 4 (3): 5–8.

29 Rohrs, B.R. (1997). Calibration of the USP 3 (reciprocating cylinder) dissolution apparatus. *Dissolution Technologies* 4 (2): 11–18.

30 USP liothyronine sodium tablets. *USP 40–NF 35*. Rockville, MD: United States Pharmacopeial Convention.

31 Kraemer, J. and Stippler, E. (2005). Experiences with USP Apparatus 4 calibration. *Dissolution Technologies* 12 (3): 33–39.

32 Brown, W. (2005). Apparatus 4 flow-through cell: some thoughts on operational characteristics. *Dissolution Technologies* 12 (3): 28–30.

33 Fotaki, N. (2011). Flow-through cell apparatus (USP Apparatus 4): operation and features. *Dissolution Technologies* 20 (4): 46–49.

34 USP <724> (2020). Drug release. *USP 40–NF 35*. Rockville, MD: United States Pharmacopeial Convention.

35 Palmieri, A. and Gray, V. (2007). *Dissolution of Heterogeneous Dosage Forms. Dissolution Theory, Methodology, and Testing* (ed. A. Palmieri), 232–240. Hockessin, DE: Dissolution Technologies.

36 Crison, J.R., Weiner, N.D., and Amidon, G.L. (1997). Dissolution media for *in vitro* testing of water-insoluble drugs, effect of surfactant purity, and electrolyte on *in vitro* dissolution of carbamazepine in aqueous solutions of sodium lauryl sulfate. *Journal of Pharmaceutical Sciences* 86 (3): 384–388.

37 Wueling, P., Kosuda, K., Templeton, A. et al. (2006). Polysorbate 80 UV/Vis spectral and chromatographic characteristics-defining boundary conditions for use of the surfactant in dissolution analysis. *Journal of Pharmaceutical and Biomedical Analysis* 41: 774–782.

38 Soltero, R.A., Hoover, J.M., Jones, T.F., and Standish, M. (1989). Effects of sinker shapes on dissolution profiles. *Journal of Pharmaceutical Sciences* 78 (1): 35–39.

39 Crist, B.G. (2013). Considerations for automating the dissolution test. *Dissolution Technologies* 20 (2): 44–47.

40 Skwierczynski, R., Curry, P., Gray, V. et al. (2014). Revision of the dissolution procedure: development and validation <1092>. *Dissolution Technologies* 21 (1): 6–8.

41 Brown, C.K., Chokshi, H.P., Nickerson, B. et al. (2004). Dissolution testing of poorly soluble compounds. *Pharmaceutical Technology* 28: 56–43.

42 Wang, Q., Fotaki, N., and Mao, Y. (2009). Biorelevant dissolution: methodology and application in drug development. *Dissolution Technologies* 16 (3): 6–12.

43 Hanson, R. and Gray, V. (2004). Solving practical problems, method development, and method validation. In: *Handbook of Dissolution Testing*, 3e, 136–139. Hockessin, DE: Dissolution Technologies.

44 Woodcock, J. (2004). The concept of pharmaceutical quality. *American Pharmaceutical Review* 7 (6): 10–15.

45 Food and Drug Administration (2018). *Guidance for Industry: Dissolution Testing and Specification Criteria for Immediate-Release Solid Oral Dosage Forms*

Containing Biopharmaceutics Classification System Class 1 and 3 Drugs. Rockville, MD: Food and Drug Administration https://www.fda.gov/media/92988/download (accessed 5 November 2020).

46 Dressman, J. (2014). Evolution of dissolution media over the last twenty years. *Dissolution Technologies* 3: 6–10.

47 Marques, M. (2004). Dissolution media simulating fasted and fed states. *Dissolution Technologies* 11 (2): 16–17.

48 Nicolaides, E., Symillides, M., Dressman, J.B., and Reppas, C. (2001). Biorelevant dissolution testing to predict the plasma profile of lipophilic drugs after oral administration. *Pharmaceutical Research* 18 (3): 380–388.

49 Horter, D. and Dressman, J.B. (2001). Influence of physicochemical properties on dissolution of drugs in the gastrointestinal tract. *Advanced Drug Delivery Reviews* 46: 75–87.

50 Kostwicz, E.S., Brauns, U., Becker, R., and Dressman, J.B. (2002). Forecasting the oral absorption behavior of poorly soluble weak bases using solubility and dissolution studies in biorelevant media. *Pharmaceutical Research* 19 (3): 345–349.

51 Lobenberg, R., Kramer, J., Shah, V.P. et al. (2000). Dissolution testing as a prognostic tool for oral drug absorption: dissolution behavior of glibenclamide. *Pharmaceutical Research* 17: 439–444.

52 Craig, D.J., Tuis, A., and Dansereau, R. (2004). Is the use of a 200 mL vessel suitable for dissolution of low dose drug procedures? *International Journal of Pharmaceutics* 269 (1): 203–209.

53 Klein, S. (2006). The mini paddle apparatus – a useful tool in the early developmental stage? Experiences with immediate release dosage forms. *Dissolution Technologies* 13 (4): 6–11.

54 Flisar, K.A., Forsyth, R.J., Li, Z., and Martin, G.P. (2005). Effects of dissolved gases in surfactant dissolution media. *Dissolution Technologies* 12 (3): 6–10.

55 Stippler, E., Kopp, S., and Dressman, J.B. (2004). Comparison of US pharmacopeia simulated intestinal fluid TS (without pancreatin) and phosphate standard buffer pH 6.8, TS of the international pharmacopoeia with respect to their use in the *in vivo* dissolution testing. *Dissolution Technologies* 11 (2): 6–10.

56 Eaton, J., Deng, G., Hauck, W. et al. (2007). Perturbation study of dissolution apparatus variables – a design of experiment approach. *Dissolution Technologies* 14 (1): 20–26.

57 Kikawai, L., Tran, D., Hauck, W. et al. (2012). Effect of various operational parameters on drug release from a 1% hydrocortisone semisolid dosage form using the vertical diffusion cell apparatus. *Dissolution Technologies* 19 (3): 6–12.

58 Villalobos, R., Viquez, H., Hernandez, B. et al. (2012). Parameters affection drug release from inert matrices. Monte Carlo simulation. *Pharmaceutical Development and Technology* 17 (3): 344–352.

59 Chowdary, K.P., Rao, R., and Surya Prakasa, K. (2011). A factorial study on the effects of β-cyclodextrin and poloxamer 407 on the dissolution rate of valsartan from CD complexes and their tablets. *Journal of Pharmaceutical, Biological and Chemical Sciences* 2 (3): 438–444.

60 Marroum, P. (2012). Clinically relevant dissolution methods and specifications. *American Pharmaceutical Review* 15 (1): 36–41.

61 USP <1196> (2020). Pharmacopeial harmonization. *USP 43–NF 38*. Rockville, MD: United States Pharmacopeial Convention.

62 Gelatin Capsule Working Group (1998). Collaborative development of the two-tier dissolution testing for gelatin capsules and gelatin-coated tablets using enzyme-containing media. *Pharmacopeial Forum* 24 (5): 7045–7050.

63 Gray, V., Cole, E., Toma, J. et al. (2014). Use of enzymes in the dissolution testing of gelatin capsules and gelatin-coated tablets-revisions to dissolution <711> and disintegration and dissolution of dietary supplements <2040>. *Dissolution Technologies* 21 (4): 6–19.

64 USP <1225> (2020). Validation of compendial methods. *USP 43–NF 38*. Rockville, MD: United States Pharmacopeial Convention.

65 Gray, V. and Miller, B. (2002). Current good manufacturing practices in the dissolution laboratory. *Pharmaceutical Canada* 3 (3): 19–21.

66 Martin, G. and Gray, V. (2011). Addressing compliance concerns during dissolution method development, validation, and transfer. *Journal of Good X Practices. The X could be laboratory, clinical, or manufacturing practices* 15 (3): 11–16.

Bibliography of Important Books Related to Dissolution Testing

Dressman, J.B. and Krämer, J. (2005). *Pharmaceutical Dissolution Testing*. Boca Raton, FL: Taylor & Francis.

Dressman, J.B. and Lennernäs, H. (2000). *Oral Drug Absorption: Prediction and Assessment*. New York: Dekker.

Hanson, R. and Gray, V. (2004). *Handbook of Dissolution Testing*. Hockessin, DE: Dissolution Technologies, Inc.

Marques, M. and Brown, W. (2010). *Dissolution Technologies Questions and Answers*. Hockessin, DE: Dissolution Technologies, Inc. http://dissolutiontech.com/DTresour/201002Articles/DT201002_QA.html (accessed 5 November 2020).

Palmieri, A. III (2007). *Dissolution Theory, Methodology, and Testing*. Hockessin, DE: Dissolution Technologies.

Webster, G.K., Jackson, J.D., and Bell, R.G. (2017). *Poorly Soluble Drugs: Dissolution and Drug Release*. Singapore: Pan Stanford Publishing.

8

Analytical Data and the Documentation System

Kim Huynh-Ba

Pharmalytik LLC, Newark, DE, USA

TABLE OF CONTENTS

Analytical Testing for the Pharmaceutical GMP Laboratory, First Edition. Edited by Kim Huynh-Ba.

8.1 Introduction

A documentation system is the key to ensuring compliance with Good Manufacturing Practices (GMPs) and traceability of all manufacturing, production, and testing activities. A variety of functions that require documentation are

Table 8.1 Typical purposes of documentation.

Type	Purpose
Recording information or data	• Record how a sample is tested. • Record data for a certain batch or lot. • Report results of a sample. • Record how a sample is tested. • Record data for a certain batch or lot.
Providing instructions	• Provide instruction to how a task is performed. • Provide a plan on how a study to be done. • Provide product specifications. • Provide details of policies or practices.
Providing proof	• Provide traceability of events or activities. • Provide evidence how a product was made. • Record how an investigation was done. • Record an action to be taken. • Record a qualification of an instrument or an analyst. • Report evaluation of a study.

illustrated in Table 8.1. These documents provide a route to assess the overall quality of the drug product manufacturing process as well as operations within the company. This chapter discusses the fundamental requirements of recording analytical data and maintaining a documentation system for the laboratory in a GMP environment. Analysts are trained in these functions and follow these practices when working with companies that manufacture foods and drugs regulated and inspected by the US Food and Drug Administration (FDA) and other regulatory agencies [1, 2].

8.1.1 Types of Documents

There are various types of procedures that a GMP facility follows. Below is a list of the most common types of documents.

Quality Manual – The quality manual describes the overall regulations or parts of the regulations that the company is required to follow.

Policies – Policy documents describe the specific GMP aspects such as security, documentation, health, and responsibilities, implemented at the company. These documents are often written in general terms and not with step-by-step instructions.

Standard Operating Procedures (SOPs) – Detailed instructions describe how a task or activity is performed. There should be an SOP for any activity that is deemed critical to the organization.

Batch Records – Batch records describe production-related tasks and activities about a specific batch or lot manufactured at the company or site. This document also includes the analytical data generated.

Certificates of Analysis (CoA) – CoA documents record analytical results for a batch of material such as raw materials, components, reagents, drug substance, and drug product. These documents typically include the specifications of the materials and the test methods used for the analysis.

Test Procedures – Detailed instructions describing the testing of supplies, materials, products. Once the product is approved, most companies write test procedures into an SOP. An analyst documents the testing of the materials and results either in an assigned notebook or controlled blank forms.

Specifications – Specification documents list the requirements that a supply, material, or product must meet before being released for use or sale. This requirement is proposed by the manufacturer and agreed upon by the regulatory agencies, where the product will be marketed. The quality control (QC) department will compare their analysis test results to the specifications to determine if the testing product meets the requirements that have been established. The specifications also link with the test procedures used to generate such data. This test procedure must be validated and approved by the regulatory agencies.

Logbooks – A bound collection of forms used to document activities such as the operation, maintenance, and calibration of a piece of equipment. Logbooks are also used to record critical activities (e.g. monitoring cleanrooms, solution preparation, recording of deviations, change controls, and corrective action assignment) where needed. Logbooks typically document activities chronologically to track certain activity performance.

Protocols – Protocol documents list specific instructions to be performed and predetermined acceptance criteria for a scientific study. Typically, protocols are developed and executed for method validation, method transfer, and investigation. They contain more specific details than what is listed in an SOP, and it must be approved by a quality assurance (QA) department.

8.2 GMP for Records and Reports–Subpart J

Information regarding records and reports used in GMP operations are provided in Subpart J of Title 21 CFR 211. These requirements typically apply to the production area and the QC area. There is a distinct difference between records and reports. A "record" is a document stating results achieved or providing evidence of activities performed. Therefore, a record is a matter of fact; records cannot be modified or changed without justification. A "report" is a document evaluating the results or providing directions for the tasks to be completed. In this sense, a report would depend greatly on the author, the process of interpreting the data, and the evaluation of the data. Table 8.2 gives an example of records and reports to illustrate the differences [1].

Regulatory agencies focus on the integrity and accuracy of the data through the documentation system. The generation of records and reports is impacted by the time the data are generated, the people performing the activities, the location or facility where this activity happens, and the process or procedure that is used to generate such data [1, 2].

The following issues are typically observed during laboratory inspections or regulatory audits:

- Lack of assignment of a delegate of QA manager
- Lack of QA-related SOP authorization by QA manager
- Lack of documentation for sample sequence tables
- Lack of out-of-spec investigations, such as correction of errors without proper reasoning, signature/initial, and/or date
- Use of incorrect corrections such as overwrites or using whiteouts.
- Back-entering data without proper traceability (data entry without initials/dating)
- Backdating the data

Table 8.2 Examples of records vs reports.

Type	Document
Records	• Analytical data • Batch records • Validation data • Instrument logbook
Reports	• Validation protocols • Validation reports • Stability study • Process validation plan • Master validation plan • Investigation reports

Source: Based on Nally [1].

- Creating, altering, or deleting information to meet acceptable criteria
- Signing on behalf of another person
- Hiding/discarding undesired data

8.2.1 General Requirements

The general requirements for records and reports used in GMP operations are listed in Table 8.3 Title 21 CFR 211.180 of the General Requirements. These sections require the company to have records and reports available at any time to review the study during an audit or inspection. These records include any data

Table 8.3 21 CFR 211 Subpart J – general requirements.

Section	Regulation
Section 211.180(a)	Any production, control, or distribution record that is required to be maintained in compliance with this part and is specifically associated with a batch of a drug product shall be retained for at least 1 year after the expiration date of the batch or, in the case of certain OTC drug products lacking expiration dating because they meet the criteria for exemption under 211.137, 3 years after distribution of the batch.
Section 211.180(b)	Records shall be maintained for all components, drug product containers, closures, and labeling for at least 1 year after the expiration date or, in the case of certain OTC drug products lacking expiration dating because they meet the criteria for exemption under 211.137, 3 years after distribution of the last lot of drug product incorporating the component or using the container, closure, or labeling.
Section 211.180(c)	All records required under this part, or copies of such records, shall be readily available for authorized inspection during the retention period at the establishment where the activities described in such records occurred. These records or copies thereof shall be subject to photocopying or other means of reproduction as part of such inspection. Records that can be immediately retrieved from another location by computer or other electronic means shall be considered as meeting the requirements of this paragraph.
Section 211.180(d)	Records required under this part may be retained either as original records or as true copies such as photocopies, microfilm, microfiche, or other accurate reproductions of the original records. Where reduction techniques, such as microfilming, are used, suitable reader and photocopying equipment shall be readily available.

about production, control, or distribution of products, components, containers, closures, and labels. It is required that documents be retained for at least a year after batch expiration; however, some companies may keep them for a longer period of time, depending on their liability to the customer.

Sections(c) and (d) listed in Table 8.3 indicate that all records must be available at any time at the location where they were generated, and readily accessed on demand. Records can be in paper or electronic format. Electronic copies are readily acceptable, or records can be made available after electronic transmission. These sections allow the FDA to have a copy of the records. These copies are released under the Freedom of Information Act unless they contain information related to the disclosure of trade-secret information, where a court order may be needed to reveal such information [1, 2].

Test data and records are critical because they provide the basis to evaluate the study against established quality standards. Therefore, Section(e) in Table 8.4 requires the annual evaluation of these records to ensure that the product is maintained within these standards. Procedures should be available to show how the evaluations, including deviations, rework, rejects, recalls, complaints, and

Table 8.4 21 CFR 211 Subpart J – 211.180(e) and (f) general requirements.

Section	Regulation
Section 211.180(e)	Written records required by this part shall be maintained so that data therein can be used for evaluating, at least annually, the quality standards of each drug product to determine the need for changes in drug product specifications or manufacturing or control procedures. Written procedures shall be established and followed for such evaluations and shall include provisions for: 1) A review of a representative number of batches, whether approved or rejected, and, where applicable, records associated with the batch. 2) A review of complaints, recalls, returned or salvaged drug products, and investigations conducted under 211.192 for each drug product.
Section 211.180(f)	Procedures shall be established to assure that the responsible officials of the firm, if they are not personally involved in or immediately aware of such actions, are notified in writing of any investigations conducted under 211.198, 211.204, or 211.208 of these regulations, any recalls, reports of inspectional observations issued by the Food and Drug Administration, or any regulatory actions relating to Good Manufacturing Practices brought by the Food and Drug Administration.

investigations, are done against currently established quality standards and how the changes are implemented when needed. Additional information for annual reporting requirements is also mentioned in Section 21 CFR 314.70(d) and 314.81(b). It is an activity for management to have a complete review of the lifecycle of the product quality, identify any trends, and drive improvement [2].

Review of batch records can be done continuously with an electronic laboratory information management system (LIMS) system. The FDA also issues a series of guidance documents to assist applicants in making regulatory submissions for NDAs or ANDAs by electronic format. Chapter 10 will discuss in more detail electronic data and the LIMS system.

8.2.2 Equipment Cleaning and Use Log

Section 211.182 requires that records of major equipment use, cleaning, sanitization and/or sterilization, and maintenance are kept (Table 8.5). Entries in logbooks should be kept in chronological order with date, time, products, and batch number processed in the equipment. Cross-contamination must be avoided by implementing appropriate technical and organizational measures such as cleaning, and decontamination procedures performed by qualified individuals. Because dedicated equipment does not typically require a logbook for usage, it is easier to retrieve these records. Qualification, calibration, and routine maintenance of analytical equipment are discussed in Chapter 2. Electronic systems can be used to

Table 8.5 21 CFR 211 Subpart J – equipment cleaning and use log.

Section	Regulation
Section 211.182	A written record of major equipment cleaning, maintenance (except routine maintenance such as lubrication and adjustments), and use shall be included in individual equipment logs that show the date, time, product, and lot number of each batch processed. If equipment is dedicated to manufacture of one product, then individual equipment logs are not required, provided that lots or batches of such product follow in numerical order and are manufactured in numerical sequence. In cases where dedicated equipment is employed, the records of cleaning, maintenance, and use shall be part of the batch record. The persons performing and double-checking the cleaning and maintenance (or, if the cleaning and maintenance is performed using automated equipment under 211.68, just the person verifying the cleaning and maintenance done by the automated equipment) shall date and sign or initial the log indicating that the work was performed. Entries in the log shall be in chronological order.

Table 8.6 21 CFR 211 Subpart J – component, drug product container, closure, and labeling records.

Section	Regulations
Section 211.184(a)	These records shall include the following: a) The identity and quantity of each shipment of each lot of components, drug product containers, closures, and labeling; the name of the supplier; the supplier's lot number(s) if known; the receiving code as specified in 211.80; and the date of receipt. The name and location of the prime manufacturer, if different from the supplier, shall be listed if known.
Section 211.184(b)	b) The results of any test or examination performed (including those performed as required by 211.82(a), 211.84(d), or 211.122(a)) and the conclusions derived therefrom.
Section 211.184(c)	c) An individual inventory record of each component, drug product container, and closure and, for each component, a reconciliation of the use of each lot of such component. The inventory record shall contain sufficient information to allow determination of any batch or lot of drug product associated with the use of each component, drug product container, and closure.

track equipment records and evaluate equipment performance. More information on LIMS and data integrity can be found in Chapter 10.

8.2.3 Component, Drug Product Container, Closure, and Labeling Records

Section 211.184 describes the requirement set forth for the container, closure, and labels of the finished products (see Table 8.6). This section requires the origin of components, drug product containers, closures, and labeling identified; thus, the names of the manufacturers should be listed. The supplier name must be available, especially if the manufacturer is not known. Testing and disposition of these materials are required. Section(c) requires that individual inventories of product containers and closures be maintained, including a record and reconciliation of usage. Investigations are required if any usage falls outside of the specified acceptance criteria to evaluate potential production problems.

8.2.4 Master Production and Control Records

Master production and control records for each drug product describe all aspects of product manufacturing, packaging, and control. Table 8.7 shows the requirement of Section 211.186. As required by GMP, to ensure compliance with expectations

Table 8.7 21 CFR 211 Subpart J – master production and control records.

Section	Regulation
Section 211.186(a)	To assure uniformity from batch to batch, master production and control records for each drug product, including each batch size thereof, shall be prepared, dated, and signed (full signature, handwritten) by one person and independently checked, dated, and signed by a second person. The preparation of master production and control records shall be described in a written procedure and such written procedure shall be followed.
	Master production and control records shall include:
	1) The name and strength of the product and a description of the dosage form.
Section 211.186(b)	2) The name and weight or measure of each active ingredient per dosage unit or per unit of weight or measure of the drug product, and a statement of the total weight or measure of any dosage unit.
	3) A complete list of components designated by names or codes sufficiently specific to indicate any special quality characteristic.
	4) An accurate statement of the weight or measure of each component, using the same weight system (metric, avoirdupois, or apothecary) for each component. Reasonable variations may be permitted, however, in the amount of components necessary for the preparation in the dosage form, provided they are justified in the master production and control records.
	5) A statement concerning any calculated excess of component.
	6) A statement of theoretical weight or measure at appropriate phases of processing.
	7) A statement of theoretical yield, including the maximum and minimum percentages of theoretical yield beyond which investigation according to 211.192 is required.
	8) A description of the drug product containers, closures, and packaging materials, including a specimen or copy of each label and all other labeling signed and dated by the person or persons responsible for approval of such labeling and
	9) Complete manufacturing and control instructions, sampling and testing procedures, specifications, special notations, and precautions to be followed.

and uniformity from batch to batch, the record must be prepared by a single person and verified by a second person. These activities are completed with the proper attribute of signature and date. Any amendment or deviation in these records must be documented with justification to ensure the changes do not impact the quality of the materials. Although outdated master production records may be retained for reference, copies of these records should be prepared in a manner that eliminates any possibility of transcription error.

Section 211.186(b) lists the minimum information that is required for master production and control records. Information from these sections is included in the company's SOP to provide directions on how data are collected. Tracking lists should be available to provide document controls of the data recorded and avoid unintended changes to the records [1, 2].

8.2.5 Batch Production and Control Records

Batch production records are prepared for all materials produced (e.g. intermediates, active pharmaceutical ingredient(s) (API), and drug products). This record follows every production batch and provides detailed descriptions for all processing operations and controls, according to the instructions of the master production used. Before any production, a thorough check should be performed and recorded to ensure that the related area and equipment are free from cross-contamination and mix-ups. Table 8.8 lists Section 211.199 with the documents required for batch production and control records. All measurements should include unit measurement systems to prevent errors. Ideally, one measurement system should be used in an organization-wide policy of handling significant values and decimal points [1, 2].

8.2.6 Production Record Review

Table 8.9 lists requirements of Section 211.192 for the production record. Written procedures should be established for the review and approval of batch production and laboratory control records, including packaging, and labeling before the batch is released for distribution. The review and approval are done by the production area and followed by the QA unit.

Any discrepancy must be investigated and documented. These investigations should extend to other batches of the same product or similar products made in the same period. All deviations, investigation, and discrepancy reports should be reviewed and included as a part of the batch record review [1, 2].

Table 8.8 21 CFR 211 Subpart J – batch production and control records.

Section	Regulations
	Batch production and control records shall be prepared for each batch of drug product produced and shall include complete information relating to the production and control of each batch. These records shall include:
	a) An accurate reproduction of the appropriate master production or control record, checked for accuracy, dated, and signed.
	b) Documentation that each significant step in the manufacture, processing, packing, or holding of the batch was accomplished, including:
Section 211.188	1) Dates
	2) Identity of individual major equipment and lines used
	3) Specific identification of each batch of component or in-process material used
	4) Weights and measures of components used in the course of processing
	5) In-process and laboratory control results
	6) Inspection of the packaging and labeling area before and after use
	7) A statement of the actual yield and a statement of the percentage of theoretical yield at appropriate phases of processing
	8) Complete labeling control records, including specimens or copies of all labeling used
	9) Description of drug product containers and closures
	10) Any sampling performed
	11) Identification of the persons performing and directly supervising or checking each significant step in the operation, or if a significant step in the operation is performed by automated equipment under 211.68, the identification of the person checking the significant step performed by the automated equipment.
	12) Any investigation made according to 211.192 and
	13) Results of examinations made in accordance with 211.134.

8.2.7 Laboratory Records

Laboratory records include complete data generated from all tests done on the sample(s) according to established specifications and standards. Section 211.194 is important to the pharmaceutical analysis; thus, the regulations are very specific. The analysts working in this area typically documents information and all modifications necessary to reconstruct and evaluate the study. The laboratory must

Table 8.9 21 CFR 211 Subpart J – 211.192 production record review.

Section	Regulation
Section 211.192	All drug product production and control records, including those for packaging and labeling, shall be reviewed and approved by the quality control unit to determine compliance with all established, approved written procedures before a batch is released or distributed. Any unexplained discrepancy (including a percentage of theoretical yield exceeding the maximum or minimum percentages established in master production and control records) or the failure of a batch or any of its components to meet any of its specifications shall be thoroughly investigated, whether the batch has already been distributed. The investigation shall extend to other batches of the same drug product and other drug products that may have been associated with the specific failure or discrepancy. A written record of the investigation shall be made and shall include the conclusions and follow-up.

include justification and verification data to support any modifications made. Records for any testing and standardization, equipment-related records, stability data, and investigations are maintained according to company and department policies. Table 8.10 shows the regulations for laboratory records.

A description of the sample is important for identifying samples and monitoring quality and integrity before and during the testing. This information is especially critical if the investigation is necessary.

Test procedures are used to monitor the product quality; therefore, test methods, acceptance criteria, the version of the procedures, and information on how the test is performed are essential for the GMP operations.

Analytical data are documented in graphs, charts, spectra, documents, forms, or notebooks. Recently, electronic data has become the focal point as technology in this area has advanced significantly in current laboratories. A record of calculations must be included with appropriate units, and the metric system is used in GMP laboratories.

Regardless of paper or electronic data, signatures of the analyst and the witness are required for all laboratory operations. The analyst signature signifies the authorship of the work, while the witness signature will verify the authenticity and accuracy of the work. The witness must also verify the adherence of the work to current applicable standards about the sample analysis.

Validation is required for analytical test methods used in GMP laboratories; therefore, testing must be conducted as written in the analytical procedure. All information related to the testing, standards, materials, equipment must be recorded in a notebook or worksheet. Raw data must be maintained to assure data

Table 8.10 21 CFR 211 Subpart J – 211.194 laboratory records.

Section	Regulation
Section 211.194(a)	a) Laboratory records shall include complete data derived from all tests necessary to assure compliance with established specifications and standards, including examinations and assays, as follows: 1) A description of the sample received for testing with identification of source (that is, location from where sample was obtained), quantity, lot number or other distinctive code, date sample was taken, and date sample was received for testing. 2) A statement of each method used in the testing of the sample. The statement shall indicate the location of data that establish that the methods used in the testing of the sample meet proper standards of accuracy and reliability as applied to the product tested. (If the method employed is in the current revision of the United States Pharmacopeia, National Formulary, AOAC INTERNATIONAL, Book of Methods, or in other recognized standard references, or is detailed in an approved new drug application and the referenced method is not modified, a statement indicating the method and reference will suffice). The suitability of all testing methods used shall be verified under actual conditions of use. 3) A statement of the weight or measure of sample used for each test, where appropriate. 4) A complete record of all data secured in the course of each test, including all graphs, charts, and spectra from laboratory instrumentation, properly identified to show the specific component, drug product container, closure, in-process material, or drug product, and lot tested. 5) A record of all calculations performed in connection with the test, including units of measure, conversion factors, and equivalency factors. 6) A statement of the results of tests and how the results compare with established standards of identity, strength, quality, and purity for the component, drug product container, closure, in-process material, or drug product tested. 7) The initials or signature of the person who performs each test and the date(s) the tests were performed. 8) The initials or signature of a second person showing that the original records have been reviewed for accuracy, completeness, and compliance with established standards.

Table 8.10 (Continued)

Section	Regulation
Section 211.194(b)	b) Complete records shall be maintained of any modification of an established method employed in testing. Such records shall include the reason for the modification and data to verify that the modification produced results that are at least as accurate and reliable for the material being tested as the established method.
Section 211.194(c)	c) Complete records shall be maintained of any testing and standardization of laboratory reference standards, reagents, and standard solutions.
Section 211.194(d)	d) Complete records shall be maintained of the periodic calibration of laboratory instruments, apparatus, gauges, and recording devices required by 211.160(b) (4).
Section 211.194(e)	e) Complete records shall be maintained of all stability testing performed in accordance with 211.166.

accuracy and integrity. Modification can only be done if they are included as part of the validation. Then when it is needed for the method to be suitable for the analysis, these modifications are documented in the analytical form or the notebook with appropriate justifications.

Information on analytical equipment used in the testing must also be recorded. A metrology program is required for all pieces of GMP equipment, including qualification and calibration. More information is available in Chapter 2, where different quality systems are discussed in more detail.

All critical areas of the stability program are discussed in Chapter 9, with complete information on different stability studies, sample management, stability chambers, study protocols, and stability reports. Additional information can also be found in the *Handbook of Stability Testing in Pharmaceutical Development: Regulations, Methodologies, and Best Practices* [1–4].

8.3 Keeping Good Records

Good documentation practices (GDocP) is the foundation of recording data for any industry. In the analytical laboratory, GDocP requires written procedures to define how data will be recorded. It also recommends using preprinted or prenumbered worksheets or forms to minimize transcription errors or templates for electronic documents.

In a regulated laboratory, data must be recorded accurately and thoroughly. Even for an R&D laboratory, it is important to record what works and what does not work. This information can be critical to developing procedures or setting control strategies later. Therefore, GMP requires good records to enable a reviewer to reconstruct and evaluate data effectively. SOPs are used to describe a series of activities to ensure controlled and consistent performances; any deviation from a written procedure would need to be documented and investigated. The time allotted for reviewing data should also be built into the documentation process. The reviewer can catch errors, check the transcription, verify the calculations, and validate the results. The reviewer must understand the activities that he/she verifies. A GMP laboratory must standardize the documentation process to improve efficiency and increase compliance [3, 4].

8.3.1 Writing Good Procedures

Any critical activity must be described in a SOP; thus, the SOP should include important directions for an analyst to follow successfully. A flow chart is a useful visual tool to describe the sequence of a process. Checklists can be used to ensure that essential steps are not missed inadvertently. Language should be straight-forward and precise to increase readability. Too much information can cause confusion, and too little information can affect traceability and cause difficulty in recreating the event.

8.3.2 Following Procedures

To keep current with industry practices and changes in regulations, most companies have a two-year review cycle for their SOPs; however, this frequency can be set depending on the process. Organizations have different ways to structure their SOPs to ensure all procedures are followed as written, therefore ensuring a controlled and consistent performance. Once the procedure is approved by the QA group, it should not be deviated from without the approval of a supervisor or the quality department. Any deviation from an SOP should be properly documented with appropriate justification or investigation [2, 4].

8.4 Raw Data Documentation

Data documentation is critical. Each company should define the responsibilities of documenting data and communicating data through records and reports within their unit. Document owners are required to follow SOPs and ensure that all GMP activities are performed according to the relevant SOPs. The QA unit should ensure these SOPs comply with GMP requirements and current industry practices. Any deviation in the procedure must be reported and adequately documented. Analysts must be trained on these procedures before making data entries.

Raw data are defined as entries in laboratory notebooks, logbooks, forms, worksheets, computer-collected data, printouts/charts from instruments, photographs, or any original observations and activities, making it complicated depending on whether the company has a paper or electronic documentation system. The first observation of an experiment is known as "raw data." Original raw data must be securely stored in a validated electronic database or must be recorded in a notebook or a binder by the analyst generating the data. Notebooks are assigned and managed by the QA unit or equivalent [1, 2, 4].

The analyst should only have one active notebook written in chronological order. No other analyst may enter any data in another person's notebook. The direct supervisor, or line of management, of the analyst can be authorized to make entries in the analyst's notebook to complete unfinished work or make corrections when the analyst is unavailable.

Complete notebooks should be stored and archived in a secured place, with access limited to authorized personnel only. However, they must be stored and archived in a matter that permits ready referencing or retrieval. The storage location must ensure adequate protection from loss, destruction, or falsification, and any natural damage such as fire or flood. The retention time for archiving notebooks depends on the work and applicable regulations. Worksheets also follow a similar process to ensure the integrity of the data. All sheets must be prenumbered and tracked. Loose sheets are not acceptable for data documentation. An electronic system to print out worksheets must assure that duplicate copies cannot be generated.

The date may be recorded by electromagnetic or photographic means, but detailed procedures relating to the system adopted must be available. Accuracy of the record should be checked as per the defined procedure and a vigorous audit trail must be maintained.

If documentation is handled by electronic data processing methods, it is critical to establish strict procedures to assure the electronic information is as trustworthy as paper records and signatures. Electronic records should be linked to electronic closed or open systems. Electronic signatures must be unique and attributable to one person as it is legally binding. Components of electronic signature must also be available such as printed name, time/date, the purpose that the signature was made for. Only authorized persons should be able to enter or modify data in the computer. Access must be restricted by passwords or other means, and the entry of critical data must be independently verified. Overall, it must follow Title 21 CFR Part 11 compliance [5, 6]. More information is available in Chapter 10.

8.4.1 Data Recording Practices

Data integrity is critical, and procedures must be in place to control all phases of the data lifecycle. Figure 8.1 shows the life cycle of the data recording process. The system must be designed, and controls are needed to enable easy detection of errors, omissions, and aberrant results through the lifecycle for analytical data [7, 8].

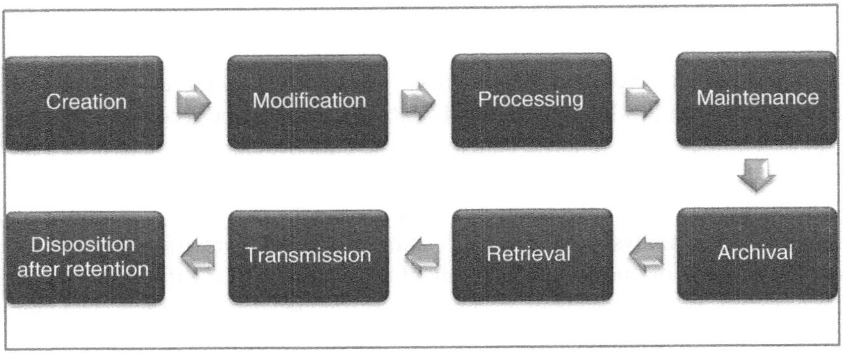

Figure 8.1 Lifecycle of analytical data.

Data are typically recorded in a *laboratory notebook* or *worksheet*. All handwritten entries must be legibly written with black or blue ink pens; however, this should be standardized, as a notebook should not have multiple colors of data entries. All blank spaces in a notebook page must be crossed out to avoid additional entry. If a form is used, all blank spaces must be filled out.

The bound notebook must be preserved intact. No page or part of a page can be removed from a bound notebook. There should be a written procedure for assigning and maintaining notebooks.

Worksheets – are used to record analytical data such as test procedures, instrument logbooks, or calibration forms. These worksheets must contain a pre-assigned unique identification number to track for integrity and traceability of the raw data. There should be a written procedure for working with worksheets.

Calculations – must be complete and accurate. Example calculations should be provided unless they are performed by a validated electronic system such as a LIMS.

Stand-alone Paper Printouts – During the data collection, instruments such as a pH meter, balance, or UV spectrometer can create a paper printout or status record as an original record. This paper printout or record must be retained. If it is printed on thermal paper, then a true copy must be kept. These records are subjected to a second person's verification to ensure all test results and associated information are properly reported. This also applies to a microbiological laboratory; thus, a contemporaneous written record of the colony counts in a petri dish must be maintained and verified with a second person's review.

GMPs require the integrity and traceability of all data; therefore, entries in the lab notebook or worksheet must be accompanied by the *signature* of the person who performs the task and the date that the entry was made. Most signatures are illegible; therefore, it is recommended that names are also printed. The analyst should document his/her work. All steps should be recorded contemporaneously before moving to the next step. A task should not commence, if documentation or a form is not available.

Each signature and initial must be unique, so the person who performed the activity can be identified. Some companies maintain a record of signatures and initials of their employees.

Each notebook should have a *table of contents,* which is kept up to date for ease of data retrieval. This table of contents should start with a reference of the most recently completed notebook, and end with the next notebook to be used to ensure continuity of documentation and audit trail.

Inserts are often used in the recording of data in notebooks. An insert should have writing only on one side and be attached to the page in such a way that the entire insert, when opened or unfolded, is legible within the boundaries of the open notebook. They must not be placed over any writing in the notebook and must not obscure written entries on the page to which they are attached. Procedures must be developed to ensure that inserts cannot be removed and replaced. It is recommended that the witness also initials the insert [7, 8].

Overall, Figure 8.2 illustrates the requirements of GDocP, covering four critical elements – concise, legible, accurate, and traceable. These are applicable for both paper and electronic data; thus, the documentation system must be built to ensure the integrity of the data and allow for data evaluation.

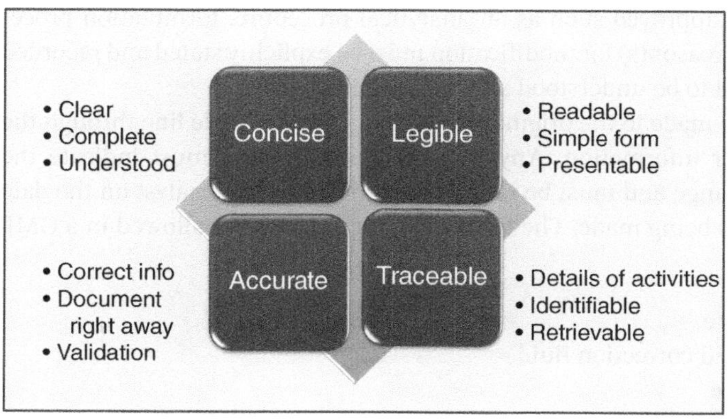

Figure 8.2 Documentation practices.

8.4.2 Witness Responsibilities

The witness must be an individual who understands the nature of the project and the work performed but not necessarily one who generates the data. It is typically the responsibility of the supervisor; however, it can be delegated to another qualified individual who has the necessary technical experience and training to perform the witnessing activity.

The witness must verify:

- All calculations are correct and accurate unless it is done by a validated electronic system.
- Transcriptions of all numbers such as weights, concentrations, sample factors, as listed in the procedures.
- Continuity of the study.
- All entries are done in compliance with applicable SOPs.
- All attachments such as spectra, chromatograms are included.
- Results of the analysis according to the procedure.

A second signature is required to verify that the original work is accurate, complete, and in compliance within a reasonable timeline before working on the next step. This second signature should be accompanied by the date that the witness verified this activity.

8.4.3 Changes in Notebook

No erasures or obliterations using whiteout or markers are to be made in laboratory notebooks. Any modification of a referenced procedure that has been previously documented and approved such as an analytical procedure, formulation procedure, etc. and the reason(s) for modification must be explicitly stated and recorded in sufficient detail to be understood and/or reproduced later.

Changes can be made to the original record by placing a single line through the incorrect entry or information. Any corrections or changes must indicate the reason for the change and must be initialed and dated by the analyst on the date that the change is being made. The following practices are not allowed in a GMP environment:

- Do not overwrite.
- Do not use liquid correction fluid.
- Do not backdate.
- Do not record data before the action or event has occurred.
- Do not leave any blank space.

No handwritten changes or corrections should be made to the printed text of an approved GMP document. The original copy should be evaluated if an error is discovered. Changes can be made through a document change control system.

The notebook should be kept in chronological order. If information is not entered at the time of completing the task, information may be added later with proper documentation explaining the reason that the information is missing and the date the event occurred. If a form is used, the missing information can be added and an explanation should be initialed and dated at the time of recording. Traceability of data must be maintained in any operation.

8.4.4 Recording Date and Time

Each company should standardize how dates and times are recorded, taking into consideration their employees and customers. As many companies work globally, this activity is critical. Also, this would be necessary when the LIMS system is developed.

There are two ways that time can be recorded. Either way is acceptable for a GMP record.

- *Military time:* Two (2) digits to indicate the hour (00–23) followed by two (2) digits to indicate the minutes (00–59). For example: 1030 or 1545.
- *Meridian time:* One or two digits to indicate the hour (1 or 12) followed by two digits to indicate the minutes (00–59) and then the morning (AM) or afternoon (PM) designation. For example: 10:30 a.m. and 03:45 p.m.

Dates can be recorded in several ways:

- MM/DD/YYYY as example 01/25/2013
- DD/MM/YYYY as example 25/01/2013
- DD-MON-YYYY as example 25-JAN-2013 (suggestion for global business.)
- YYYY-MM-DD as example 2013-01-25 (suggestion for computer files for easy sorting.)

8.4.5 Voiding and Restoring GMP Records

If errors are discovered after completing all entries on a record, a new form can be obtained; however, the original form should also be retained, crossed out, and marked invalid, stating the reason for it being invalid. This form should be attached to the new document for tracking purpose.

On rare occasions, it may be necessary to recreate or restore GMP records. The new document must be marked and supported with reference sources of information. If the original record is available, it must be attached to the new document. There are several reasons that recreating a record may be necessary, such as:

- the original record is not legible
- the original data is printed on thermal paper, which becomes faded over time
- the incorrect form or test method was used
- the original record is damaged

An evaluation should be performed to see if corrective action or preventive action is needed to prevent a reoccurrence of original record to be missing and needed to be recreated [2, 4, 7].

8.4.6 Computer-Collected Data

Electronic data are considered raw data for all computer systems that generate, process, or maintain information to support regulatory submissions. The electronic raw data should be available in both electronic and human readable form. The system must be validated so that data cannot be deleted or overwritten. Procedures should be developed to ensure that raw data are protected from loss or corruption with limited access. Raw data should be available throughout the entire required retention period.

CGMP workflow on a computer system must be validated for the intended use and based on risks. Access to the GMP computer systems must be restricted, and shared login accounts for computer systems is prohibited. Computer systems are not restricted only to an electronic system for data acquisition or calculation, it can also include EXCEL tracking lists, data reporting forms, electronic batch records, etc.

Electronic signatures must have appropriate controls to be used in lieu of handwritten signatures in any GMP record or document. Title 21 CFR 211 Part 11 establishes criteria for when electronic signatures are considered the legally binding equivalent to handwritten signatures and manufacturers must document the controls used to ensure that they are able to identify the specific person who signed the records electronically [1, 2, 5].

An electronic audit trail should be available to track changes to, and the processing of, the raw data. Information including the person making changes, timestamps of when changes occur, action performed and value before changes made must be recorded. This system must be secured so that existing entries cannot be modified or deleted. The audit trail should be attached to the data throughout the required retention period and the procedure for data backup must be available. Additional discussion of electronic data can be found in Chapter 10 [5, 6].

8.4.7 Reporting Analytical Results

Consistency of reporting analytical data is important in a GMP environment; therefore, a procedure should be developed to determine reportability and the format in which the date can be reported in an analytical laboratory, except microbiological assays.

8.4.7.1 Decimal Places and Significant Figures

For data evaluation, the significant level of reportable results should agree with those listed in the specification documents; therefore, the number of significant figures or decimal places of reportable data should agree with the number of significant figures or decimal places listed in the specification documents.

Table 8.11 lists the typical decimal places or significant figures for common analytical methods. The sensitivity of the equipment used is also a major factor in determining the reporting format.

8.4.7.2 Mathematical Operations

Addition and Subtraction: When adding or subtracting analytical results, the number of figures to the right of the decimal in the result should be kept the same, as its term having the fewest number of figures to the right of the decimal.

Example:

$$\text{The sum of } 314.2 + 62.63 + 1.749 = 378.6$$

$$\text{The difference of } 45.7 - 24.57 = 21.1$$

Multiplication and Division: When multiplying or dividing analytical results, the number of significant figures should be kept the same, as it is the factor with the fewest number of significant figures.

Example:

$$\text{The result of } 27.3 \times 0.0012467 = 0.34035, \text{which is rounded to } 0.340$$

8.4.7.3 Rounding Rules

When required, rounding typically follows USP General Notices Section 7.20, indicating that the observed or calculated values shall be rounded off to the number of decimal places that agree with the decimal places listed on the specifications. The number should not be rounded off until the final calculation is

Table 8.11 Typical decimal places and significant figures in specification documents.

Description	Decimal place	Examples
Drug substance assay	Use one decimal place	98.0–102.0%
Drug product assay	Use one decimal place	90.0–110.0% of label
Average capsule content	Use three significant figures	375 mg or 96.0 mg
pH	Use one decimal place	3.5–5.0
Residue on ignition (ROI), Loss on Drying (LOD), Water content	Use one decimal place	NMT 0.1% NMT 2.0%
Heavy metals	Use one significant figure	NMT 0.002%
Melting point	Use one decimal place	137.0–141.0 °C 79.5–83.5 °C
DSC	Use one decimal place for temperature and purity, three significant figures for area.	137.0 °C 99.70% 45.4, 154 J/g
TGA	Use two decimal places for percent weights, no decimal for temperature, and one decimal place for moisture	12.42, 0.32% 80–110 °C Ambient to 80 °C 3.4% moisture
ICP	Use one decimal place	15.9 ppm
Particle size	Use whole number	15 μm
Density	Use two decimal places	0.92 g mL^{-1}
Surface area	Use one decimal places	1.1 m g^{-2}
Impurities Individual Total quantitated	Use two decimal places	NMT 0.50% NMT 2.00%
Residual solvents	Use two decimal places	NMT 0.50% NMT 1.00%
Dissolution Mean Standard deviation	Record to the nearest whole number	85 ± 2%

completed. When rounding is necessary, one digit in the decimal place to the right of the last place is required. If the digit to be removed is less than 5, the preceding digit stays the same. If the digit to be removed is greater than or equals to 5, the preceding digit is increased by 1.

Example:

1.84 is rounded to 1.8
4.75 is rounded to 4.8
2.78 is rounded to 2.8

In a series of calculations, the extra digits are carried through to the final calculation and, the result will be rounded off.

8.4.7.4 Reporting Impurity Results

Impurity results are usually collected at a low level of validation; therefore, each company should develop a procedure to report impurities consistently. It is recommended that a practical quantitation limit (QL) be established for every component tested and set as the reporting limit (RL) for the method. The value of the reporting limit should always be listed on the report form. Values of qualification threshold (QT) and RL depend on the daily drug intake and should be set according to ICH Q3A and Q3B. The summary is listed in Table 8.12 [9, 10].

If the method requires impurity testing in duplicate, the mean of two impurity replicates can be calculated and reported based on the values received. Table 8.13 shows examples of reporting the impurities of the drug product for which the reporting limit for impurities is 0.10%.

As we can see from the above examples, it is critical to have the calculating and reporting procedure written, and the method shows the specific calculation examples to ensure the consistency of reporting analytical results of impurities. This is critical for data evaluation and product monitoring. A result should be reported for every component tested as per the specification even when its result is less than the reporting limit (<RL). When reporting total impurities, only those that are greater than or equal to the reporting limit (≥RL) will be included in the calculation [1, 2, 4, 7, 8].

8.5 Samples, Reagents, Standards, Reference Standards

8.5.1 Samples

A procedure should be established for receiving the materials delivered to the receiving laboratory for testing. Materials should be logged in before any testing can be performed. Information such as identity, batch number, total quantity, and date of receipt should be recorded. Other information such as storage condition, material handling instructions, and names of the test(s) that are to be performed

Table 8.12 Limit of impurity results.

	Individual impurity	Total impurities
Greater than or equal the QT	Report value	Include in total
Between QT and RL	Report value	Include in total
Equal with the RT	Report value	Include in total
Between RT and QL	Record value Report as <RL	Do not include in total
Below QL	Record value Report as <RL	Do not include in total

Source: International Council for Harmonisation of Technical Requirements for Pharmaceuticals for Human Use [9, 10].

should be provided. Samples should be examined, and results should be recorded for the container, package, and/or seal integrity, or any visible damages.

There should be designated areas where samples can be tested, approved, and rejected. Containers of these samples should be labeled with a unique tracking number.

8.5.2 Reagents

A written procedure should be established to manage the reagents that are used in the laboratory. All reagents must be labeled with proper identification, storage condition, and expiration date.

Working reagents that are prepared for analysis must be labeled with name, concentration, date prepared, date expired, and the method that was used to prepare the material.

8.5.3 Analytical Standards

Each company should have a standard program to manage reference materials used in the analytical laboratory. The analytical standard can be used for qualitative or quantitative analysis. Standards can also include materials used for calibration or maintenance of laboratory equipment or facility. Procedures should include purchasing, qualification, handling, labeling, and storing reference materials. A list of reference standards should be available that includes the standard name, lot number, amount available, recertification date, use-as value, and classification (e.g. bulk reference standard). All calibration and standardization

Table 8.13 Examples of different options to report impurity results.

Option	Type of reporting	Example of reporting
Option A	Individual values are rounded to the reporting threshold and then calculations are performed. All specified components were detected at <0.10%.	• Impurity A: Values <0.10 and <0.10%. Mean is <0.10% • Impurity B: Values <0.10 and <0.10%. Mean is <0.10% • Other impurity: <0.10 and <0.10%. Mean is <0.10% • Total impurities: <0.10%
Option B	Some individual values are less than the reporting threshold. The mean was calculated with unrounded results.	• Impurity A: Values 0.12 and 0.14%. Mean is 0.13% • Impurity B: Values 0.11 and <0.10%. Mean is <0.10% • Other impurity: <0.10 and <0.10%. Mean is <0.10% • Total impurities: 0.13%
Option C	Only one impurity has reportable results. Other specified and unspecified impurities were detected are at >0.10% level.	• Impurity A: Values 0.12 and 0.14%. Mean is 0.13% • Impurity B: Values <0.10 and <0.10%. Mean is <0.10% • Other impurity: 0.11 and <0.10%. Mean is <0.10% • Total impurities: 0.13%
Option D	Specified impurities are above reporting levels, and unspecified impurities are below reporting level.	• Impurity A: Values 0.12 and 0.14%. Mean is 0.13% • Impurity B: Values 0.10 and 0.14%. Mean is 0.12% • Other impurity: <0.10 and <0.10%. Mean is <0.10% • Total impurities: 0.13 + 0.12 = 0.25%

materials should be traceable to an acceptable national or international standard. Suppliers of analytical standards should be qualified.

If the product is developed in-house, a lot of the active ingredient in sufficient quantity is used as reference material. This lot is thoroughly tested and characterized according to the established specifications. Otherwise, the Standard Coordinator can purchase a bulk bottle, qualify it, and distribute it in smaller portions as a working bottle for testing to avoid cross-contamination. This bottle is typically a small amber glass bottle to minimize contamination and degradation. All standards are typically stored in a room temperature desiccator unless otherwise specified. The bulk and working bottles are properly labeled and stored under appropriate conditions according to their label. When the recertification date is reached, these bulk and working bottles must be discarded. The Standard Coordinator maintains a computerized inventory of all analytical standards [1, 4].

8.5.4 Reference Standards

Reference standards are materials of known purity that are used to quantitate active ingredients during testing. Reference standards are used for release and stability testing of drug substance and dosage form. A comparator compound can be obtained from an outside source to be used as a Reference Standard.

The proposed reference standard is analyzed to determine that it meets identity (e.g. infrared [IR], nuclear magnetic resonance [NMR], mass spectrometry [MS], atomic absorption [AA] spectroscopy, elemental analysis, functional group tests), or contains impurities (e.g. thin-layer chromatography [TLC], high-performance liquid chromatography [HPLC], gas chromatography [GC], Karl Fischer moisture analysis, thermal gravimetric analysis [TGA]), and is tested with absolute assays for purity (such as titration, phase solubility, and differential scanning calorimetry [DSC]).

It is also acceptable to use a certificate of analysis (CoA) if comparator compounds are used as reference standards. This material must be used before the manufacturer's expiration date. Other standards, such as USP, ASTM, etc. can also be used as reference standards. A certificate of analysis typically accompanies them [1, 4].

8.6 Drug Substance Analysis

In general, registration specifications are established for a drug substance to control the material quality. An evaluation will be performed and data will be recorded for each API lot. "Record results" are no longer acceptable specifications for any test. The analytical procedures cover physical properties, chemical properties, and microbiological properties. All test methods must be validated for the intended purpose. Table 8.14 lists typical testing to be done for an API and drug products at release.

Table 8.14 Typical release tests for new drug substance and drug product.

Type of sample	Testing to be performed
Drug substances	• Description/appearance (physical observation) • Color • Identity, infrared spectroscopy (IR) • Identity, chiral high-performance liquid chromatography (chiral HPLC), if applicable • Water content • Chemical tests, i.e. potency assay, impurity profile, chiral purity • Physical tests, i.e. particle size, bulk and tap density, surface area, polymorphic forms, specific rotation • Impurities • Particle sizes • Residual solvents (final solvents only), organic volatile impurities (OVIs). • Inorganic impurities (heavy metals and sulfated ash) or residue on ignition (ROI) • Microbiological tests, particularly for APIs to be used in parenteral formulations • Polymorphs, if applicable • Microbial limit, if applicable • Inductively coupled plasma (ICP) test, if applicable.
Drug products	• Description/appearance • Chemical tests, i.e. potency assay, Degradation product profile, chiral purity • Identification test(s) • Dissolution or drug release (solid dosage form) • Uniformity of dosage units • pH (liquid dosage form) • Moisture determination • Clarity/particulate matter • Preservative or antioxidant • Physical tests, i.e. hardness, disintegration • Micro tests, i.e. bacterial endotoxins, sterility, antimicrobial preservative effectiveness • Functional tests

8.7 Drug Product Analysis

The typical tests that are done to release a finished product are listed in Table 8.14. All test methods must be qualified for their intended purpose.

Most of the QC activities are testing stability samples. The stability program consumes a lot of resources. More information is available in Chapter 9.

8.8 Batch Release

Each company must have a written procedure to ensure that a system is in place to control and monitor incoming material during and between processing stages. This procedure defines the status of material and requirements for the final material disposition. Typical categories for dispositions are listed in Table 8.15.

Production and quality control records are part of the batch release. Data must meet all established quality standards before the batch can be released for distribution.

Any failure or divergence must be thoroughly investigated. The batch must be placed in quarantine until the investigation is completed. This investigation should, if necessary, extend to other batches of the same product and other products that may be associated with this discrepancy or failure. A record of this investigation should be available with the conclusion of the batch disposition along with any follow-up action taken [1, 4, 7, 11, 12].

The following records are part of the batch record.

Table 8.15 Typical disposition categories for materials.

Category	Disposition of materials
Approved (A)	Materials are accepted to be released for distribution or additional processing
Conditional (C)	Materials to be used at risk
Quarantine (Q)	Materials that are isolated physically pending a decision on subsequent approval, rejection, or further processing
Rejected (R)	Materials that are not acceptable for distribution or additional processing
Restricted (N)	Materials for specific use

8.8.1 Batch Packaging Record

There should be packaging instructions for each product including all aspects of the activity. All packaging materials must meet their established quality standards. Records of processing and reprocessing of packaging materials must be available. Details of in-process controls should also be available. Upon completion of packaging and labeling, reconciliation must be made between the number of labeling and packaging units issued and the number used to ensure there is no contamination or mix-up. Any discrepancy should be investigated before the batch can be released.

8.8.2 Batch Processing Record

A batch processing record should be available for each product, based on the relevant parts of the current master batch record and checked to avoid transcription errors.

8.8.3 Distribution Record

Before distribution, all batches must be tested; data are approved and released by the quality control group. Records for distribution are maintained for traceability in the event of recall, if needed.

8.9 Establishment of Specifications

Specifications are part of a total control strategy designed to assure the identity, strength, quality, and purity of the API and drug product. These values are set based on the characterization of the API and drug product during development, and adherence to GMPs for suitable facilities, validated manufacturing processes, validated analytical methods, quality of raw materials in-house, and stability testing. The values are established and approved by the appropriate analytical organization and the QA unit. These specifications can be numerical values, ranges, or other criteria such as description and are set based on available data, reasonable analytical and manufacturing variability, suitable facilities and equipment, control materials, and processes [1, 4, 9, 13, 14].

8.9.1 Content of Specifications

The documents for specifications include a list of tests to be performed, specific step-by-step procedures to be carried out for each test, and acceptance criteria for each product to meet. The acceptance criteria are defined for all materials such as raw materials, in-process samples, intermediates, drug substances, drug products, and labeling and packaging materials supported with the procedure for manufacture and control steps to serve as a basis for quality evaluation. There is a distinction between clinical release specifications in support of clinical studies and registration specifications for registration studies and marketed products. The specifications for marketed products are quality attributes recommended by the manufacturer and approved by the regulatory agency where the materials will be marketed. Many companies also have internal limits to provide tighter control limits in-house.

8.9.2 Setting Specifications

Specification development is a critical multi-functional activity that must be integrated across the development of a drug substance, drug substance manufacturing process, drug product formulation, drug product process, packaging configuration, and analytical methods. This evaluation must take into consideration the contributions from other aspects beyond data, from the test methods to justifying the registration specification for the commercial product. Table 8.16 lists the inputs used in the evaluation of setting specifications for the drug product. Control strategies should be developed to establish product specifications [11, 12]. Chapter 4 will discuss this topic in more detail.

For the products that have an existing pharmacopeia monograph, specifications are typically listed in the monograph. Extrapolation of stability data at accelerated

Table 8.16 Data used in the evaluation of specifications.

Type of study	Data to be evaluated
Clinical	• Release data of clinical materials • Stability data supporting clinical studies • Toxicology qualification
Product development	• Dosage form and dosing application • Drug substance process capability
Analytical development	• Drug product process capability • Drug substance stability • Drug product stability • Analytical method capability • Drug substance physical characterization • Stability condition (storage and expiry)

conditions can also be used to justify product specifications and predict the shelf life of the product for a long-term storage condition.

8.9.3 Periodic Revisions

Specifications must be periodically reviewed whenever changes are done to the product, to comply with new official compendia or new editions of national pharmacopeia. Data should be evaluated annually versus the quality standards of each drug product to determine the need for changes in drug product specifications, manufacturing, or control procedures. This evaluation should include a representative number of approved and rejected batches, and a review of complaints, recalls, returned, reworked, or salvaged drug products. Any deviations must be recorded and justified to assure that there is no impact on the quality of the drug product.

8.9.4 Test Procedures

The analytical laboratory receives samples to be tested for several purposes. Figure 8.3 shows an example of samples received by the laboratory for testing and tracking the chain of custody of the sample to be tested. Test procedures are required to release the product, and for testing stability samples, and these procedures are included in the specifications of drug products. Most companies have the same specifications for release and stability; however, these specifications can be set differently for parameters that may change over time, such as assay and impurities/degradation products. In such cases, release specifications are usually more stringent than shelf-life specifications. Testing procedures must be validated for their intended purposes. More information on validation is included in Chapter 5.

Analysis of the API and impurities are typically done with high performance liquid chromatography (HPLC) or ultra-high-performance liquid chromatography (UHPLC) methods. Many companies strive to develop one method to analyze both the main active ingredient(s) and impurities. Figure 8.4 shows the flow of samples once they arrive in the laboratory.

Typically, the test procedure for API can be adopted for analyzing drug product unless the excipients cause interference. Table 8.17 shows an example of the release specifications of a tablet formulation [1, 4, 13, 14].

Procedures must be available to monitor impurities in APIs and drug products at levels greater than the limit of quantitation (LOQ). Newly discovered impurities must be monitored and identified if they occur above a certain level (frequently 0.10%). Genotoxic impurities must also be monitored appropriately

Analytical Testing Request and Report Form		
Reference SOP no.1234 v 2		
Originated by:	Department:	Tracking Number:
Product Name:	Dose:	Manufacturer:
Batch/Lot no.	no. of units:	Manufacturing Date:
Protocol no.	Receiving no.	Source:
LIMS no.	Received Location:	Target Completion Date
Sampled by:	Sampling Date:	Submitted Dates:
Received by:	Received Date:	Label and Sample Check:
Accepted by:	Accepted Date:	Label and Sample Check:
Comments/Special instructions:		
Specifications no.		
Samples types (check one):	Method Development	Product Development
	Method Validation	Investigation Testing
	Test and Report	In-Process Testing
	Release Testing	QC Testing
	Raw Materials	Other: __________
Tests requested:		
Data References:	LIMS no.	NB no./page:
	LIMS no.	NB no./page:
	LIMS no.	NB no./page:
	LIMS no.	NB no./page:
	LIMS no.	NB no./page:
	LIMS no.	NB no./page:
	LIMS no.	NB no./page:
	LIMS no.	NB no./page:
Analyst Name:	Analyst Signature:	Date:
Witness Name:	Witness Signature:	Date:
QA Name (if applicable):	QA Signature:	Date:

Figure 8.3 Example of a form to request for analytical testing.

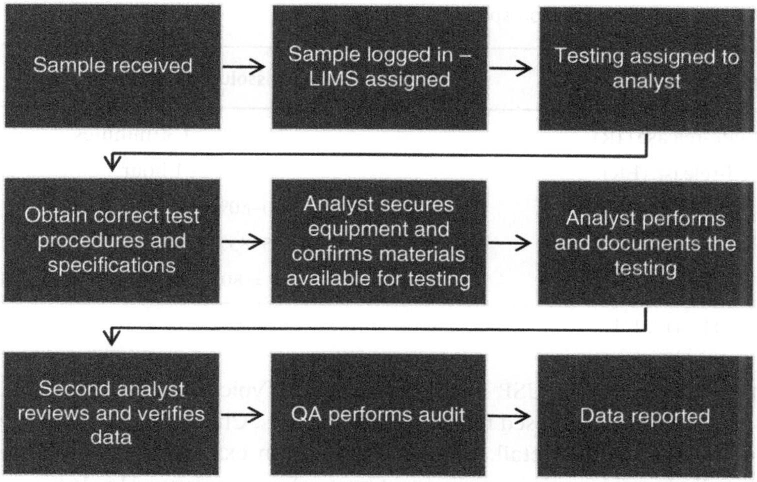

Figure 8.4 Flow of laboratory testing.

Table 8.17 Example of a tablet specification.

Test	Specification
Appearance/description	White to off-white tablets
Assay of active A	90.0–110.0% of label
Degradation products	Individual <0.10%
	Total degradation products <1.0%
Dissolution	$Q = 70\%$ at 30 minutes
Moisture determination	Less than 0.20%
Content uniformity	80.0–110.0% of label
Disintegration	Less than 3 minutes

Source: Nally [1]; International Council for Harmonisation of Technical Requirements for Pharmaceuticals for Human Use [13, 14].

according to the corresponding guidelines. ICH Q3 requires specifications to be set for individual and total degradation products for drug products. Specified degradation products can be tracked using their name, if known, or relative retention time, if unknown. Quantitation of degradation products can be done using response factor as compared with a known compound or calculated against an authentic substance (if one is available), or by area percent of the analysis [9, 10].

Table 8.18 Example of dissolution specifications.

Dosage forms	Dissolution specifications
For immediate-release (IR)	Q = 80% at 30 minutes
For extended-release (ER)	20–40% at 1 hour
	40–60% at 2 hours
	65–85% at 4 hours
	Q = 80% at 8 hours

Source: USP <711> [15, 16].

Dissolution testing follows USP <711> and <724>. Typically, USP Apparatus, Basket (1) and Paddle (2) are used for solid dosage forms. Chapter 7 discusses the dissolution testing in more detail. Table 8.18 shows an example of dissolution specifications that could result from different types of dosage forms [15, 16].

8.10 Out-of-Specification (OOS) Results

The safety and efficacy of the drug product are set during clinical studies. Drug product specifications are set based on clinical and nonclinical data; therefore, when there is an out-of-specification (OOS) result, the safety and efficacy of the drug product may be impacted. Title 21 CFR 211.192 requires that the failure of a batch or any of its components that meets any of its specifications shall be thoroughly investigated, even if the batch has already been distributed. The investigation must be extended to other batches of the same drug product and other drug product that may have been associated with the specific failure or discrepancy. Prompt initiation of the investigation is essential, and retest is not allowed until an investigation is initiated.

The FDA guidance document issued in October 2006 describes the investigation procedure with the lab phase and full phase applicable to APIs, excipients and other components, in-process materials, and finished drug products. Figure 8.5 shows the two phases of the investigation of suspect analytical data.

All companies must have an SOP for OOS investigation. The SOP must consider the history of the product and determine the root cause. The retest policy and protocol define the point where the testing stops, and evaluation occurs.

8.10.1 Lab-Phase Investigation

The immediate course of action when an OOS result is discovered is to confirm that the OOS is not an analytical error. Therefore, the first phase of the investigation occurs in the lab and is focused on the identification of possible assignable

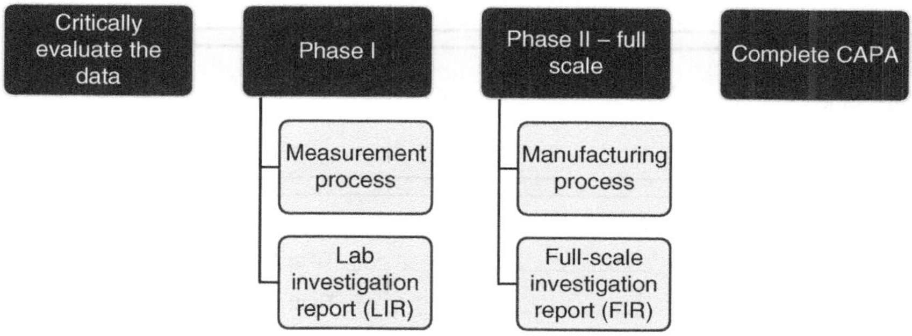

Figure 8.5 Quality investigation of suspect analytical data.

lab errors. The responsibilities of the analyst and supervisor are listed in the draft guidance. Figure 8.6 shows an example of a laboratory investigation report (LIR).

All test solutions, reagents, and standard solutions should be kept for the duration of the investigation, if possible. The quality of these solutions should be supported with solution stability data. Equipment used in the testing that generates the OOS should be reserved until the root cause is identified. Table 8.19 gives an overview of the analyst's and the supervisor's role during an investigation.

The critical part of the investigation is a review of related documents to identify the cause of the OOS. Checklists are used to aid in reviewing all relevant facts and serves to speed up the investigation process [1, 4, 8, 11].

8.10.1.1 Identifying an OOS Result

If the analytical result of the sample does not meet the specification of the underlying test procedure, then a potential OOS is identified for that sample and an investigation is initiated to determine the cause of the OOS. If a batch is rejected based on an OOS result, the investigation still needs to be done to determine if the result is associated with other batches of the same products. If the sample is past its expiry, the investigation still needs to be done. It is important to recognize that there are three different OOS data:

- *Validation data*
  When a result of a validation study does not meet the established acceptance criteria listed in a pre-approved protocol, the result needs to be investigated. Procedures may need to be changed after the investigation. Negating this type of OOS may disqualify the complete procedure.
- *Release data*
  When data generated for batch release does not meet specification, this lot must be quarantined until the investigation is completed with an assignable cause.

Laboratory Investigation Report (LIR)

Product:
Strength:
Batch number:
Manufacturing Date:

Name of the analyst reporting the occurrence: ______________________

Name of the supervisor: ______________________ Title: ______________

Name of the investigator: ______________________ Title: ______________

Name of the investigator supervisor: ______________ Title: ______________

Notebooks/worksheets inspected (Y/N): ________________

Discussion of test procedure (Y/N): ________________

Verification of calculations (Y/N): ________________

Verification of instruments (Y/N): ________________

Findings:

Recommendations:

Signature and date: ______________________ Date: __________

______________________ Date: __________

______________________ Date: __________

Figure 8.6 Example of a Laboratory Investigation Report (LIR).

Table 8.19 Responsibilities of analyst and supervisor.

Position	Responsibilities
Responsibilities of analyst	• Ensure that the equipment used is calibrated and meets the required acceptance criteria, • Report data only if the required system suitability tests pass acceptance criteria, • Check the data for compliance to specifications before discarding any test solutions, • Inform the supervisor if any unexpected results are obtained • Stop testing if an obvious error occurs; they should not knowingly continue testing when they expect to invalidate the data at a later time for an assignable cause, except when the sole purpose is to see what results are obtained when obvious errors are known.
Responsibilities of supervisor	• Perform an objective and timely assessment • Confirm that the analyst's knowledge and performance of correct procedures • Examine the raw data and identify anomalous or suspect information • Confirm the performance of the instruments • Examine the solutions, reagents and standards to confirm that they were appropriate for use during testing • Evaluate the performance of the test method • Document and preserve evidence of the assessment

Data of previous released batches or batches manufactured at the same period may be reviewed to determine the root cause.

- *Stability data*
  When stability data do not meet specifications, an investigation must be initiated immediately. If data for a sample stored at accelerated conditions do not meet the specification for the intended storage condition, then samples of intermediate conditions may need to be pulled and tested. If samples of intended storage conditions do not meet specification, a Field Alert Report may need to be issued and lots may need to be recalled.
  The PhRMA CMC Statistics and Stability Expert Teams published a 2-part article "Identification of Out-of-Trend Stability Results, A Review of Potential Regulatory Issue and Various Approaches" to provide a simple way to monitor

out-of-trend (OOT) results for stability data. Out-of-trend results should be investigated to monitor the stability profile of the drug product. If OOT data is confirmed, additional stability points should be considered to closely monitor the stability profile. And if necessary, shorter expiration dating should be considered until the profile can be determined [17, 18].

8.10.1.2 Laboratory Error

If the error is obvious, the analyst should document what happened immediately, and then data can be invalidated. The test can now be repeated due to analytical errors. Lab or analyst errors should be relatively rare. Frequent occurrence of laboratory errors can be an indication of an inadequate training program, deficient metrology program, poorly written analytical procedure, or a poorly written SOP. The cause of a system malfunction should be identified and, if possible, corrected before a decision is made whether to use any data before the suspect period.

8.10.1.3 Determining Root Cause

If the error is not obvious and testing results appear to be accurate, the main objective is to identify the root cause of the OOS. It may be necessary to test some prepared solutions to identify the root cause. These *diagnostic tests* must be written in the procedure and supervised. A common approach is to reanalyze the working sample solution and work backwards to test products of each step systematically to identify the source of error. If the source of lab error is identified, the OOS result can be invalidated or rejected. The retest results will substitute for the original OOS test result.

If an assignable cause is not identified, a determination of the root cause must be initiated. The root cause of analytical results typically comes from one of the five areas: methods, materials, equipment, people, and environment. When a source of lab error is identified, the organization must determine the corrective action to prevent a recurrence. Documentation of corrective action and preventive action is also important for GMP compliance.

If the root cause is not determined, and there is no scientific basis to invalidate the initial OOS results, all test results should be reported and considered in the batch disposition.

8.10.1.4 Retesting

If justified, retesting can be done, and new results will be used for the disposition of the sample. A second analyst, who is as experienced and qualified as the original analyst, should do the retesting. The sample used for the retesting should be taken from the same homogenous materials collected from the lot that yielded the OOS results. The number of retest samplings should be specified in the SOP. The retesting can be determined based on product variability and method variability. This should be done before retesting and cannot be adjusted based on the results obtained.

Retesting contains several phases:

1) Reanalyzing the working sample solution.
2) Retesting a different portion of the original sample.
3) Resampling of the product by obtaining a different sample (sub-lot) of the testing lot.

For a stability sample, the resampling option may not be feasible as the stability sample set represents the entire batch.

8.10.1.5 Outlier Test

An outlier test is a statistical test intended to identify the unexpected result in the tested sample. There should be an SOP to define how the outlier test is done, what product or test procedure that will apply to, and conducted. The outlier test cannot be used alone to invalidate data [11].

8.10.1.6 Averaging Results

According to the FDA OOS guidance, OOS and in-specification results cannot be averaged to obtain in-spec results for the chemical testing. However, this procedure is acceptable for microbiological testing [11].

8.10.1.7 Field Alert Report

If the OOS result is confirmed, then a field alert report (FAR) must be issued within three working days. Action depends on the outcome of the investigation.

8.10.2 Full Scale Investigation

If the OOS result is not caused by a measurement error, then Phase II (see Figure 5.8) must be started to show if a production error can be identified. Therefore, when a laboratory error cannot be determined, and the OOS is confirmed, the investigation must proceed to a full-scale investigation, including a review of production and sampling procedures and the impact of the OOS result on pre-distributed batches. The investigation is typically conducted by the QA unit, including multiple departments such as manufacturing, production, process development, maintenance, and engineering. Also, the full-scale investigation must be done timely, thoroughly, and well documented.

An OOS is an indication of a flaw in the product or process design. Therefore, changes may be necessary because of the investigation to assure product quality [11].

8.11 Compendial Testing

The United States Pharmacopeia (USP) is an independent standard-setting organization. It was established in 1820 and gained legal recognition in 1906 with the passage of the Pure Food and Drug Act.

USP monographs contain specifications and test procedures to which products must comply throughout the product's shelf life to be marketed in the United States. They also include storage conditions, nomenclature, and formulae.

USP general chapters provide instructions for performing test methods referenced in monographs such as Loss on Drying, Residue on Ignition. These general chapters carry numbers from <1> to <999>. General chapters numbered <1000> and higher, are informational. These chapters discuss methodology and concepts not referenced by an individual monograph. They provide guidelines or recommendations in different analytical areas.

The General Notices and Requirements section in USP address the basic concept that monograph requirements are applicable through the expiration period of an item in commerce. Generally, the information in the monographs takes precedence over the information in the general test chapters, which takes precedence over the information in the General Notices and Requirements.

8.11.1 Validation and Verification of Compendial Procedures

The USP analytical procedures contained in the monographs are considered validated. USP <1225> addresses the validation requirements for USP monographs. This chapter is consistent with international guideline ICH Q2. It guides validating chemical and physical procedures intended for submission as an official standard. Although analytical procedures listed in USP monographs are considered validated, verification must be performed on the official articles to make sure it is suitable for its intended use. Verification of the compendial procedure is addressed in USP <1226> *Verification of Compendial Procedures.*

It should be noted that analytical procedures listed in the monographs, though considered validated, are not stability indicating. If they are used for stability samples, then stability indicating characteristics must be studied [19–21].

USP is an invaluable source of test methodology and information applicable for monitoring and confirming product conformance to standards. These validated procedures can be easily verified and determined to be suitable for measuring the overall quality of pharmaceutical products through their approved expiration period. Chapters 5 and 6 will further discuss validation, verification, and transfer of analytical procedures [19–21].

Table 8.20 is an example of a checklist for an HPLC analytical procedure used to evaluate the validation characteristics to ensure that the properties are studied, and data are available. If validation data are not available, validation experiment

Table 8.20 Example of a checklist for HPLC method validation of a solid formulation.

Validation parameter	Checked	Validation activity
Specificity – API – solid		Heat
		Humidity
		Light
		No interference to API and degradation products
Specificity – API – solution		Acid
		Base
		Oxidation
		Light
		No interference to API and degradation products
Specificity – drug product		Heat
		Humidity
		Light
		Placebo (Heat, humidity, and light)
		Lack of interference from placebo and placebo stress
Linearity and range		Determine for standard
		Determine for the analyte in a formulation
		Regression R-square = 1.000 (or close)
		Y-intercept = 0 (or close)
		Typically, 5 concentrations around 100%
Precision		System – injection of standard solution (n = 6)
		Method – preparation of samples (n = 6)
		RSD of system precision <0.73%
		RSD of method precision
		Intermediate – different day, analyst, equipment
Accuracy/recovery		Spiked samples with active
		Typically, 3 concentrations in triplicate
Detection limit (DL)		Measure s/n ratio of blank
		Multiply background response by 3
		Validated by analysis of an analyte

(Continued)

Table 8.20 (Continued)

Validation parameter	Checked	Validation activity
Quantitation limit (QL)		Accuracy of QL
		Precision (%RSD) of QL
		Multiply background response by 10
		Validate by analysis of an analyte
Robustness		Column temperature
		Different batches of column
		Flow rate
		Mobile-phase composition
		Mobile phase pH
		UV wavelength
		Gradient parameters

API, active pharmaceutical ingredient; RSD, relative standard deviation.

should be conducted to ensure the procedure is valid for the intended use of the method [19–24].

8.11.2 USP Reference Standards

Many USP methods require the use of USP Reference Standards (RS) to determine the identity, strength, purity, and potency of official articles. The reference standards are authentic, highly purified, and characterized substances. These standards are typically employed in the monograph tests for identification, potency, and impurities regardless of whether the item being tested is the active ingredient or the final product [1, 4, 8].

8.12 Standard Operating Procedures

8.12.1 Control of SOPs

Procedures are established for the consistency of operations. An SOP system requires procedures for numbering, preparing, processing, reviewing, approving, distributing, and archiving SOPs. There should be a master file containing the master copy of a document, change request, and supporting data. Superseded document editions, to provide the history of a document, are maintained by a history file.

All SOPs should be reviewed and signed biennially. The effective date should be used as the tracking date for the biennial review. SOPs are created, revised, and archived whenever necessary and reasons should be documented [1, 4, 12].

8.12.2 Format of SOPs

The pharmaceutical industry relies on SOPs to manage their operations consistently in compliance with GMPs. Therefore, all critical GMP operations that impact the quality and safety of pharmaceutical products are required to have written SOPs. The SOP is a set of written instructions for a routine activity. The development and implementation of SOPs are an integral part of a successful quality system.

All SOPs should be written clearly and concisely. Many SOPs can include too much information, making them difficult to follow or too few details, causing them to be ambiguous. These SOPs can lead to audit observations or serious GMP violations. Therefore, each company should have comprehensive training on how to develop an SOP, defining critical elements of a structured SOP to make it easier to understand and comprehend by another analyst. The SOP maintenance system should include writing, formatting, tracking, execution, and management of SOPs. It also discusses the consequences of SOP misuse or inadequate SOPs.

All SOPs must contain a table of contents. The title of SOP should be brief, direct, and identify the purpose of the procedure. Keywords should be used to aid in locating the procedure in a list of SOPs. Each page of the SOP must include a header with the title, SOP tracking number, version, effective date, pagination, and authorized signature.

Related documents for an SOP could be other SOPs (referenced by identification number) or referenced validation documents. As for equipment SOPs, related documents could be the manufacturer's manual (referenced by identification number), literature information, or corresponding qualification documents.

A history section is also important to track the development process of the SOP. It should include the reason that the SOP was developed, any changes made to the SOP (with reasons), and the date and reason for its archival [4, 7, 12].

Attachments can be a form, sample template, figure, or even scanned images of equipment. The SOP tracking number should be included on these documents. Table 8.21 lists sections that should be included in a procedural SOP and a metrology SOP.

There are different types of SOPs:

- an SOP for a routine GMP activity (e.g. investigation SOP)
- an SOP for specific procedures (e.g. HPLC testing of ABC in XYZ tablets)
- an SOP for equipment (e.g. Calibration of pH meters)

Table 8.21 Comparison between a procedural SOP and metrology SOP.

Procedural SOP	Metrology SOP
1) Purpose	1) Purpose
2) Scope (include safety)	2) Scope
3) Responsibility	3) Safety (include dress code)
4) Related documents	4) Responsibility
5) Definitions	5) Related documents
6) Procedure	6) Definitions
7) History	7) Installation
8) Attachments	8) Maintenance/cleaning/repair
	9) Procedure
	10) History
	11) Attachments

8.12.3 Flow of Documents

Documents are created based on business needs and could serve only for a limited time. The flow of the document is defined when a need for a document is identified, and it will trace the document from the draft stage to the approval stage. It should also include minor and major revisions to an approved document for traceability. The author should identify the business need for the document; create a draft, and route the document for approval. The reason for creating the document should be documented. Document reviewers will approve the document, and the distribution list should be identified. The author is also responsible for facilitating the review and resolving all comments before the issuance of the document. The reviewer of the document is responsible for reviewing the document for scientific content and accuracy, grammar, clarity, and for resolving all comments with the author. The QA representative verifies the authenticity and completeness of the data and compliance with regulations and company SOPs.

Official copies to document the record distribution are kept. In cases where there is no distribution records, reference copies can be retained and marked as such. Official copies of Test Procedures, Specification Documents, Specification Monographs, and Equipment Procedures are maintained on company logo paper. Controlling these documents entails distributing the current version of the document and retrieving the superseded version for destruction.

8.13 Analytical Documents

8.13.1 Hierarchy of Documentation Systems

Figure 8.7 illustrates the hierarchy of the documentation system. At the top level, conformance standards are found; these are written from regulations that govern the directives of the sublevels. Next are Level 1 documents such as the Quality Manual. This document breaks the regulations into parts specific to the requirements the company must follow. These documents establish overall principles and guidelines on how the company plans on developing, documenting, and implementing a compliance quality system.

Level 1 – These documents typically do not contain any specific information.

Level 2 – These documents further breakdown the regulation parts into specific topics or subjects and become company policies, which provide direction to all activities to ensure consistency across departments. Level 2 documents typically do not provide specific instructions or forms for documenting data but rather provide the overall intention and guidelines governing the development of critical programs or systems. These documents will apply to all departments within a GMP compliant company.

Level 3 – These documents are typically SOPs developed specifically for a department or a function. These documents provide specific step-by-step instructions for performing operational activities. In the QC lab, analytical procedures can be distributed as an SOP.

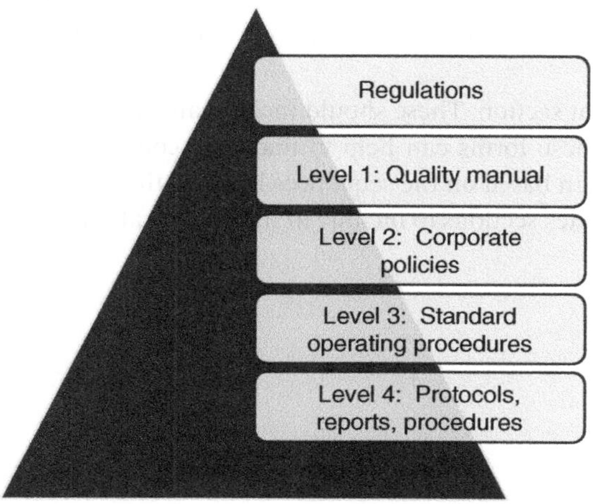

Figure 8.7 Hierarchy of analytical documentation systems.

Level 4 – This group of documents are the most specific. These are documents such as test procedures, validation protocols, or batch records that apply to a specific department, product, equipment, or process. They provide step-by-step instructions for laboratory or process-related activities and provide templates or forms for documenting data. The details may override directions given in other higher-level documents, and the levels can be adjusted to meet the company's specific needs for documentation.

8.13.2 Analytical Protocols

An analytical protocol is a document that has instructions and acceptance criteria needed to complete a study. The scope of the protocol must be clear and concise. It is helpful to include a background or development section to include information on how the directions or acceptance criteria are derived. The protocol includes the responsibilities of all those involved and a list of materials and equipment.

The procedure section includes detailed step-by-step instructions, and they must be clear and precise. Concentrations or calculations should be spelled out, including units. All numbers must be accompanied by appropriate units and decimal places. Pre-determined acceptance criteria should be included in all protocols to determine whether the step passes or fails. Signatures are important in the protocol, and at least two signatures are required: one technical signature and one quality signature.

It is recommended that exceptions and references are included. It is also important to include information on how to handle failures or related SOPs for investigation.

Attachments are an important section. These should include any form or template used to report results. These forms can help to maintain consistency for reporting data and can be written based on the sequences listed in the procedure section. The following list includes sections to include in the analytical protocol:

- Scope/objectives
- Purposes
- Background/development
- Responsibilities
- List of materials
- List of equipment
- Procedures
- Acceptance criteria
- Exceptions (if any)
- Responsible signatures

- References
- Attachment

The protocol can be amended with an appropriate explanation and approval levels. It is important to include justification for the amendment. If a deviation occurs, it must be documented with an appropriate investigation to ensure that the deviation does not affect the study purpose. Signatures must be included with all amendments or deviations.

8.13.3 Analytical Reports

Analytical reports are used to document the results and supporting data of a study. All documents should be prepared, reviewed, and approved according to internal procedures by appropriate departments. These documents should be written in an orderly fashion.

Once approved, these documents should not be changed without appropriate authorization and approval. A history of issuance, revision, or withdrawal of all documents should be maintained with reasons for each. All superseded documents should be retained for a specific period according to internal procedures or applicable regulations.

Analytical reports should include a conclusion drawn from the data presented and any references as available. Copies of these documents should be tracked to ensure that any superseded documents are not reintroduced.

These reports should not be handwritten aside from data entries on the templates, in which case sufficient space should be available for such entries. Any alteration should be accompanied with a signature and date [7, 12].

Reports can be generated and stored electronically. Systems must be developed to ensure the integrity and accuracy of these reports. Limited access is recommended and a record of changes or deletions should be available. If electronic signatures are used on documents, they should be authenticated and secured. The following sections should be included in an analytical report:

- Purpose
- Background/development
- Materials and equipment
- Experiment/procedure
- Results/data
- Evaluation of data/statistical analysis
- Conclusion(s)
- References
- Responsible signatures
- Attachment(s)

8.13.4 Annual Product Review

A written procedure must be available for an annual product review including trending for every drug product. Data from manufacturing and testing records should be evaluated to assess the processes for continuous quality improvements. This review also lists changes to manufacturing processes, testing procedures, and commercial specifications. The following list contains typical data included in this review. If this set of data were generated from different sites, then a cross-site report would be necessary to provide an overall review and trend evaluation across sites. The following data should be evaluated in the review of the annual program for drug products:

- Batches manufactured
- Comparison to previous annual product review
- Deviations and investigations
- Rejection rates
- Calculated percent yield
- Rework or reprocessing
- Validation
- Complaints or adverse events
- Change authorization approved or implemented
- Product specific observations from regulatory agencies
- Field alerts or recalls
- Returned or salvaged goods.

8.13.5 Changes to Documentation

Any changes made to documentation must be recorded and justified. There are different levels of changes and different types of documentation. Companies must have a change control program to manage documentation changes throughout the lifecycle of the product. Revisions must be tracked with appropriate approval.

Minor corrections, such as grammar or typographical errors in a final approved document, that do not affect the quality and integrity of the study may not need to be reissued or go through an approval process again. A note can be included with the revised copy, documenting the change and reason for the change.

Any additions, deletions, or corrections that affect the integrity of the final document are considered major revisions. These major revisions require a new revision number and undergo the full review and approval process. The history of the document must be retained, including reasons for changes, change of dates, and change of author [1, 4, 7, 11].

8.14 Quality Assurance

8.14.1 Six Quality Systems and Their Supporting Programs

Pharmaceutical products that are developed and manufactured must be done in compliance with CGMPs regulations. In the early 2000s, the FDA adopted six systems as the GMP inspection model. This approach is based on the merger of science-based knowledge management and risk-based management to focus on the adequacy of six critical areas that assure product quality and safety. These inspections apply to both domestic and foreign inspections and can be either full or abbreviated under this model. The goals of the program are to focus regulatory resources on the aspects of manufacturing, which pose the greatest potential risks to consumers, and to enable the agency to apply a consistent and predictable approach to inspections. Because companies cannot choose when and where to apply GMPs, a quality culture is critical to maintaining compliance and to ensuring the safety and efficacy of the drug product manufactured. Figure 8.8 shows all the different quality systems that are the foundation directly impacting the pharmaceutical products. Critical process parameters, critical material attributes, and critical quality attributes must be well evaluated and understood using pharmaceutical risk management to establish control strategies [25–27]. Other chapters in this book discuss in detail many of these critical areas.

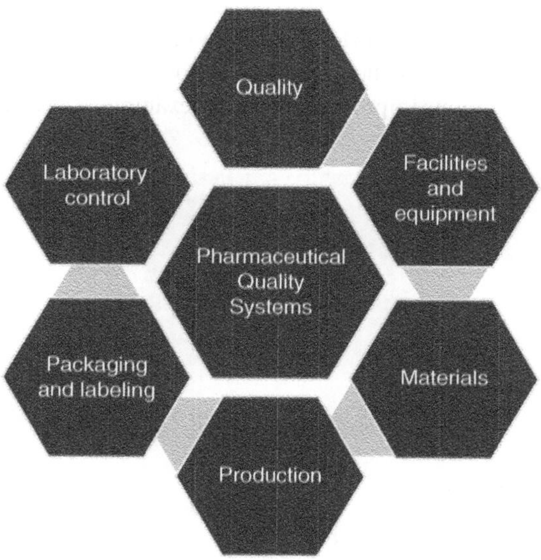

Figure 8.8 Six quality systems.

The main idea of these systems is to look for a pattern of failure. During the inspection, the inspector would review documentation of activities against the SOPs related to such activities, and then identify how the company handles unapproved changes, omissions, or improper corrections.

ICH Q10 describes this comprehensive model based on International Organization for Standardization (ISO) quality concepts. This approach can be used throughout different stages of the product life cycle. It also emphasizes the accountability of management and includes control elements for process performance and product quality monitoring such as corrective and preventive action, change control management, and product quality review [27].

These systems interact dynamically and influence the quality of the drug products to ensure the drug products achieve the established specifications for their intended use. Written procedures are necessary for the manufacture, supply, use, and testing of starting materials, packaging systems, different equipment, drug products, and APIs. Analytical procedures must be validated for their intended use. Analysts and operators must be trained. Equipment must be qualified, and the use of starting materials, intermediate products, and in-process materials must be established [23–27].

To better manage these quality systems as a quality foundation, there are supporting programs to document the cross-functional impact. Figure 8.9 illustrates these programs as a part of GMP requirements.

Corrective and Preventive Actions (*CAPA*) – This program provides a process to correct issues that leads to nonconformity and the level of action needed to address the non-conformance. It also establishes the corrective actions for the non-conformance or encourages continuous improvement. As an example, a CAPA

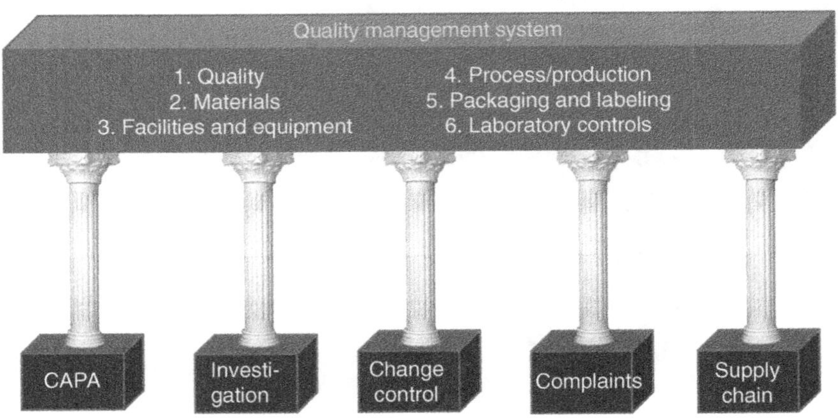

Figure 8.9 Supporting quality systems to support overall QMS.

system can identify improper failure to follow a procedure, inadequate SOP, or personnel training. It can also identify areas for improvement of processes.

Investigation – This program provides a systematic process to collect and analyze data to determine the root cause of the non-conformance or failure. Unexpected events often happened and great efforts are made to minimize the frequency and severity of these unexpected events. Therefore, when it occurs, an investigation is necessary to determine the root cause of these deviations. These deviations can be planned or unplanned. For example, an investigation is needed to identify the root cause of an OOS release result or a deviation of a procedure.

Change Control – This program provides a structured system to manage changes to prevent unintended consequences that may occur. Change is an intrinsic part of the drug product lifecycle; thus, implementing an effective change control program is mandatory for the GMP environment to ensure the product, facilities, equipment, procedures, processes, and systems continue to remain in a validated and compliant state. A change can be an addition, a deletion, an improvement, a modification, or an upgrade that could impact the product quality or regulatory obligations.

Complaints – This program manages complaints received after the products have been distributed in compliance with the GMP requirements. Reported complaints are typically due to unexpected issues with the product such as defective or not meeting expected quality. All complaints must be tracked and thoroughly investigated because a confirmed issue may put the patient at risk.

Supply Chain – This program assures that the product is delivered properly from the manufacturers to the patients. This program helps the company manage its inventory by tracking the operations, forecasting future needs, and optimizing their policies or procedure in response to the needs. It will provide feedback to other programs to manage the market effectively. As examples, it will identify temporarily outsourced laboratory, alternate suppliers, and different routes of distribution.

Overall, these supporting systems must be managed effectively to uphold the quality management system and sustain GMP in manufacturing pharmaceutical products.

8.14.2 Quality System Performance

It is important to develop an effective quality system to monitor and control process performance and product quality. Areas such as validation, training, change management, investigations, and documentation should be carefully monitored. Corrective actions should be developed based on audit findings and implementation should be tracked for timely completion. A system should be in place to verify the effectiveness of implemented corrective actions to prevent reoccurrence.

ICH guidance Q10 defines four critical elements needed to maintain a pharmaceutical quality system:

Process performance and product quality monitoring system
Corrective action and preventive action system
Change management system
Management review of process performance and product quality

These elements must be tracked and trended to provide guidance to upper management. Performance metrics should be established to continuously identify opportunities for quality improvement. Key performance indicators should be identified for all critical activities to monitor effectiveness as well as to alert management. Table 8.22 lists areas of quality for which selected metrics can be established. Please note that this is not a comprehensive list. If there is any deficiency in any of these areas, corrective actions should be taken to ensure the laboratory complies [25–27].

Table 8.22 Quality assurance checklist.

Quality area	Questions to evaluate
Analytical procedure	• Are SOPs available for critical test procedures? • Are current analytical procedures validated according to current standards? • Are all method transfers completed? • Is method performance verified with system suitability?
Laboratory equipment	• Are SOPs available for all equipment used? • Are all pieces of equipment qualified, calibrated, and well maintained?
Analyst training	• Does analyst have training record? • Has necessary training been conducted for specific instrument? • Has analyst trained on procedures performed?
Documentation system	• Are all documents comply with current SOPs? • Are all spreadsheets qualified? • Is audit trail being reviewed with the data? • Have lab data been reviewed and evaluated according to SOPs? • Have audit activity been conducted? • Are all CAPA completed and verified?
Sample management	• Are current reference standards available? • Are control samples available? • Are reagents and solutions maintained according to SOPs? • Are retention samples taken and monitored?

8.14.3 Lifecycle Management Approach for Analytical Procedures

Many quality systems impact the management of the pharmaceutical lab; among them, analytical procedures are important to provide a total control strategy of product performance because analytical data are used to monitor the quality of the product. These procedures include critical parameters such as operating conditions, method parameters, sample characteristics, equipment settings, and quality attributes related to the analysis of drug substance and drug product materials. Table 8.23 lists different studies to be examined in method development to establish the analytical control strategies of the procedures. These controls aim to verify that the method developed can maintain the designed quality performance throughout the routine use of the application [25–27].

A robust method must consist of an understanding of the sources of variability to manage procedure changes through the product lifecycle. It is recommended that a lifecycle approach is taken to understand the change of method variabilities through the lifecycle of the procedure and establish analytical control strategies. Figure 8.10 shows how the method is developed and qualified. Method development starts with defining a target of measurement and selection of appropriate technologies. A systematic approach with an analytical target for the method development is preferred over a trial-and-error process. Chapter 5 provides additional detail on method development and validation as part of method qualification for the intended use of the analytical procedure. The information provides analytical control strategies to manage method changes and method improvement until the discontinuation of the procedure [23, 24, 28–30].

Table 8.23 Planning for a method validation.

Validation parameters	Procedure
Principle/scope	What is the scope? Principle of the procedure?
Apparatus/equipment	Which apparatus is used?
Operating parameters	Operating parameters to be studied?
Reagents/standards	List of reagents, sources, availability?
Sample preparation	How is a sample prepared?
Standards control	What reference standard is used?
Solution preparation	How are solutions prepared? Which volume?
Procedure	Step-by-step procedure
System suitability	Acceptance criteria for system suitability test?
Calculations	All different calculations including the units of measurement
Data reporting	How are data to be reported officially

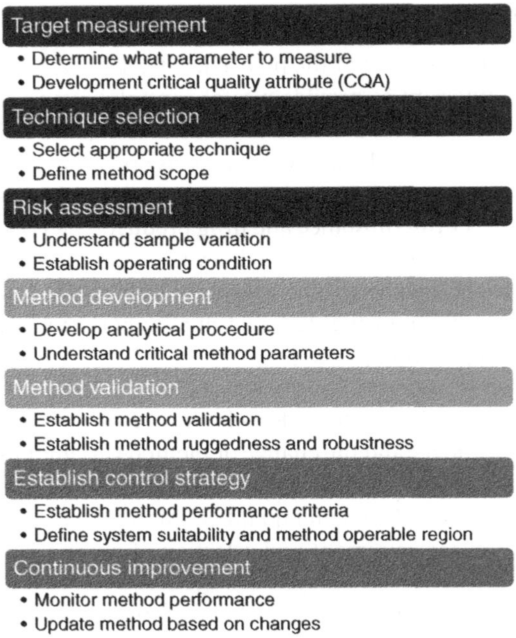

Figure 8.10 Managing the lifecycle of analytical procedure. *Source:* Food and Drug Administration [28].

Changes to the analytical method are inevitable; this activity must be managed in the change control program. Most of these changes have resulted from a CAPA or continuous improvement. It may require prior approval from the regulatory authority; therefore, it must be considered carefully. Typical reasons for changes are listed as follows:

- Synthetic process changes
- Formulation changes
- Out-of-specification investigations
- Specificity problems (e.g. new impurity)
- Ruggedness issues
- Post-approval changes (e.g. site change)
- Improving method performance
- Modification of method parameters to improve robustness

Method development and validation of a procedure are documented and approved prior to the product approval for its intended use. Therefore, any change

Table 8.24 Checklist to evaluate and prepare for chromatographic method validation.

Validation parameters	Applicable Y/N	Study range	Acceptance criteria
Specificity – identification			
Specificity – no interference			
Specificity – stress studies			
Linearity/range			
Accuracy			
Precision-repeatability			
Precision–Inter. Precision			
Precision – reproducible			
Detection limit			
Quantitation limit			
Solution stability			
System suitability			
Ruggedness			
Robustness			

Source: Huynh-Ba [30].

to the analytical procedure would impact the validation of the method and must go through a change control program to evaluate and justify. Neglecting the evaluation of current validation can disqualify the data generated and exert the risk to the products distributed. This is especially critical when most of the QC testing is supporting the stability program; therefore, change of methods must be closely monitored. Table 8.24 is an example of a checklist to evaluate different validation parameters of a chromatographic procedure to ensure that data are available to support the change. If not, additional validation work would be necessary. It is imperative to understand that method validation is an ongoing process, thus the validation report is a living document. Analysts should be aware of method validation each time the procedure is performed. Any change should be noted and evaluated. Analysis should not knowingly be continued if there is a discrepancy [10, 11, 23, 24, 28, 30].

8.14.4 Audit and Inspection program of the Laboratory

Audit and inspection programs are developed to assure quality performance and to drive continuous improvement. They are performed periodically to determine the weak elements of the quality system for compliance with Good

Manufacturing Practices (GMPs) and corporate standards. They are designed to detect any shortcomings in the implementation of GMP and to recommend necessary corrective or improvement actions. The internal audit and inspection of the laboratories are typically handled by the QA group or a third-party service provider. The laboratory area is part of the quality system that is often subjected to inspection and audit; therefore, technical personnel must understand well GMP regulations, internal practices, and procedures to address questions during an audit or inspection [31–33].

Table 8.25 lists different areas that may be reviewed during a laboratory audit. There should be written programs with appropriate training procedures. Analysts should also be trained to handle an inspection or audit.

The quality audit should always be conducted by a team of independent, experienced, and qualified persons from within or outside the company. Upper management must be aware of the level of compliance each area maintained and provide adequate support to maintain or improve them. Violation of maintaining an effective quality system is typically the result of a lack of resources, lack of knowledge, lack of support, and will impact the product quality leading to the regulatory warning letters and loss of customer trust.

Procedures for inspection should be in place and analyst training should be provided, for all or parts of a system, with the specific purpose of improving, documenting results, evaluations, conclusions, and recommended corrective actions. Table 8.26 presents an example of a checklist to audit a pharmaceutical laboratory. This list shows many, but not all, critical areas to be evaluated in an audit [32, 33]. Companies are recommended to develop their own checklist to evaluate their laboratory according to the intended use of the laboratory [34].

A detailed report should be issued with all deficiencies noted including a recommendation for correction of those deficiencies. A corrective action plan to rectify the deficiencies should be developed with the responsible individuals and to include timelines. These corrective actions should be verified in a subsequent audit to prevent reoccurrences. These deficiencies will be part of the investigation, CAPA, and change control programs as appropriate.

8.15 Summary

Good documentation practices allow one to track all GMP activities performed during batch manufacturing, from the receipt of raw materials to the final product release; they provide a history of the batch and its distribution. Keeping accurate records is an essential part of GMP, and it helps to convey the message that procedures are being followed during an audit or inspection disregarding whether it is in paper format or electronic system. It also demonstrates that the processes are

Table 8.25 Quality program to be reviewed in a laboratory audit.

Program	Areas to be audited
Procedure-related	• Analytical procedures are adequate to support the work submitted to the laboratory. • Methods are validated based on current regulations. For stability testing, method must be stability indicating. • Compendial testing using the most current monograph with supporting verification to the test article.
Sample-related	• Appropriate chain of custody of samples. • Adequate sampling plan to obtain representative samples for testing • Adequate program to establish and maintain reference standards • Storage and labeling of samples and reagents.
Training-related	• Adequate number of analysts and supervisors allocated for the work. • Adequate supervisory for the laboratory • Adequate SOPs and analytical procedure to guide the work • Training documentation of analysts.
Instrument-related	• Adequate number of instruments is available to be used. • Appropriate system for instrument qualification and calibration. • Appropriate qualification for computer systems • Is there unique identification and password to use computer systems
Data documentation	• Adequate system to generate and maintain analytical data • Procedures are in place for data recording and evaluation. • Adequate system to maintain data integrity for paper and electronic documentation. • Program to trend and analyze analytical data • Limited access to quality system maintenance

known and controlled and can serve as a tool for training. The fundamental keys for good documentation are:

- Do not trust your memory.
- Do not document someone else's work.
- Do not sign for anyone else.
- Do not assume without documentation.

Table 8.26 Example of a checklist to audit a pharmaceutical laboratory.

A. Laboratory control system			Always	Often	Sometimes	Never
1	211.42	Is the laboratory suitable in size, construction and location to facility proper operations, cleaning, and maintenance?				
2	211.42	Does the laboratory have adequate space for orderly placement of equipment and materials to prevent mix-ups and cross-contamination?				
3	1301.73	Is the security of the laboratory adequate (e.g. building perimeter, internal areas within the building, limits on employee access, etc.)?				
4	211.56	Are sanitation and maintenance procedures and ventilation equipment being followed?				
5	211.16	Are there adequate written calibration procedures? Is there a record of the calibrations being performed?				
6	211.16	Is all equipment identified? Remedial action?				
7	211.42, 211.56, 211.67	Is there a formal maintenance schedule for manufacturing equipment? Is it kept visible near each piece of equipment?				
8	211.56	Does a written program for insect and rodent control exist?				
9	211.56	Hazardous materials and waste procedures in place and followed?				
10	211.42	Temperature and humidity controls procedure(s) in place and followed?				
11	211.5	Facility cleaning and maintenance procedures in place and followed?				
12	211.22	Are the lab facilities adequate?				
13	211.182	Does all equipment have a logbook? Are logbooks maintained?				
14	211.22 211.182	Do the facilities have adequate equipment for the facilitation of function?				
15	211.11	In cases of computer systems for control of processes, data collection, etc. have they been validated?				

Table 8.26 (Continued)

A. Laboratory control system			Always	Often	Sometimes	Never
16	211.165	Have the accuracy, sensitivity, specificity, and reproducibility of test methods been demonstrated via validation?				
17	211.16	Are test procedures documented?				
18	211.192	Is there an investigation procedure? Are the lab tracking OOS/OOT infractions?				
19	211.17	Are raw material file samples maintained for future reference?				
20	211.165	Are the laboratory reagents and other chemical supplies identified and given expiration dates?				
21	211.16	Are the laboratory instruments requiring calibration maintained according to the calibration procedures?				
22	211.190	Is there as adequate change control system? Are the change completed?				
23	211.180	Do protocols include acceptance criteria? Are protocols signed by the QA?				
24	211.180	Are records and reports maintained with Good Documentation Practices				
25	211.192	Are all records reviewed in a timely manner? Are all discrepancies documented and evaluated?				
26	211.194	Are all calculations include graphs, chart, spectra from instruments as appropriate?				
27	211.194	Are all records being reviewed by second analyst for accuracy, completeness, and compliance?				

List of Abbreviations

API	Active Pharmaceutical Ingredient
CAPA	Corrective and Preventive Actions
CoA	Certificate of Analysis
FDA	Food and Drug Administration

GDocP	Good Documentation Practices
GMPs	Good Manufacturing Practices
HPLC	High-performance Liquid Chromatography
LIMS	Laboratory Information Management System
LIR	Laboratory Investigation Report
LOQ	Limit of Quantitation
OOS	Out-of-Specification
OOT	Out-of-Trend
QA	Quality Assurance
QL	Quantitation Limit
QT	Qualification threshold
RL	Reporting Limit
RS	Reference Standards
SOP	Standard Operating Procedures
UHPLC	Ultra-High-Performance Liquid Chromatography
USP	United States Pharmacopeia

References

1 Nally, J.D. (2007). *Good Manufacturing Practices for Pharmaceuticals*, 6e. Informa Healthcare: New York, NY.

2 Food and Drug Administration (1978). 21 CFR Part 211 – current good manufacturing practice for finished pharmaceuticals. Rockville, MD: Food and Drug Administration. https://www.accessdata.fda.gov/scripts/cdrh/cfdocs/cfcfr/CFRSearch.cfm?CFRPart=211(accessed 28 December 2020).

3 Patel, K.T. and Chotai, N.P. (2011). Documentation and records: harmonized GMP requirements. *Journal of Young Pharmacists* 3 (2): 138–150.

4 Huynh-Ba, K. (ed.) (2009). *Handbook of Stability Testing for Pharmaceutical Products: Regulations, Methodologies, and Best Practices*. New York, NY: Springer Science & Business Media.

5 Food and Drug Administration (2003). Guidance for industry, Title 21 Part 11- Electronic records; electronic signatures – scope and application. Rockville, MD: Food and Drug Administration. https://www.fda.gov/regulatory-information/search-fda-guidance-documents/part-11-electronic-records-electronic-signatures-scope-and-application (accessed 28 December 2020).

6 Food and Drug Administration (2016). Guidance for industry: data integrity and compliance with drug CGMP questions and answers. Rockville, MD: Food and Drug Administration. https://www.fda.gov/regulatory-information/search-fda-guidance-documents/data-integrity-and-compliance-drug-cgmp-questions-and-answers-guidance-industry (accessed 28 December 2020).

7 USP <1029> Good documentation guideline. *USP 43-NF 38*. Rockville, MD: United States Pharmacopeial Convention.

8 USP General Notices Section 7.20 *USP 43-NF 38*. Rockville, MD: United States Pharmacopeial Convention.

9 International Council for Harmonisation of Technical Requirements for Pharmaceuticals for Human Use (2006). ICH Harmonised Tripartite guideline: impurities in new drug substances Q3A(R2). https://database.ich.org/sites/default/files/Q3A%28R2%29%20Guideline.pdf (accessed 3 December 2020).

10 International Council for Harmonisation of Technical Requirements for Pharmaceuticals for Human Use (2006). ICH Harmonised Tripartite guideline: impurities in new drug products Q3B(R2). https://database.ich.org/sites/default/files/Q3B%28R2%29%20Guideline.pdf (accessed 3 December 2020).

11 Food and Drug Administration (2006). Guidance for iindustry: investigating out-of-specification test results for pharmaceutical production. Rockville, MD: Food and Drug Administration. http://www.fda.gov/downloads/Drugs/Guidances/ucm070287.pdf (accessed 28 December 2020).

12 Huynh-Ba, K. (ed.) (2010). *Pharmaceutical Stability Testing to Support Global Markets*. New York, NY: Springer Science & Business Media.

13 International Council for Harmonisation of Technical Requirements for Pharmaceuticals for Human Use (2000). ICH guideline Q6A on specifications: test procedures and acceptance criteria for new drug substances and new drug products: chemical substances. https://database.ich.org/sites/default/files/Q6A%20Guideline.pdf (accessed 28 December 2020).

14 International Council for Harmonisation of Technical Requirements for Pharmaceuticals for Human Use (1999). ICH guideline Q6B on specifications: test procedures and acceptance criteria for biotechnological/biological products. https://database.ich.org/sites/default/files/Q6B%20Guideline.pdf (accessed 28 December 2020).

15 USP <711> Dissolution testing. *USP 43-NF 38*. Rockville, MD: United States Pharmacopeial Convention.

16 USP <724> Drug release. *USP 43-NF 38*. Rockville, MD: United States Pharmacopeial Convention.

17 PhRMA CMC (2003). Statistics and stability expert teams, identification of out-of-trend stability results, Part I. *Pharmaceutical Technology* 27 (4): 38–52.

18 PhRMA CMC (2005). Statistics and stability expert teams, identification of out of trend stability results, Part II. *Pharmaceutical Technology* 29 (10): 66–79.

19 USP <1225> Validation of compendial methods. *USP 43-NF 38*. Rockville, MD: United States Pharmacopeial Convention.

20 USP <1226> Verification of compendial procedures. *USP 43-NF 38*. Rockville, MD: United States Pharmacopeial Convention.

21 USP <1224> Transfer of analytical methods. *USP 43-NF 38*. Rockville, MD: United States Pharmacopeial Convention.

22 Food and Drug Administration (2015). Guidance for Industry: Q2 (R1) analytical procedures and methods validation for drugs and biologics. Rockville, MD: Food and Drug Administration. https://www.fda.gov/files/drugs/published/Analytical-Procedures-and-Methods-Validation-for-Drugs-and-Biologics.pdf (accessed 28 December 2020).

23 Huynh-Ba, K., Wong, A.W., and Dong, M.D. (2020). Validation of stability-indicating HPLC methods for pharmaceuticals: overview, methodologies, and case studies. *LCGC North America* 38 (11): 606–618.

24 Huynh-Ba, K. and Dong, M.D. (2020). Stability studies and testing of pharmaceuticals: an overview. *LCGC Europe* 33 (6): 289–302.

25 Food and Drug Administration (2009). Guidance for industry: Q8 (R2) pharmaceutical development. Rockville, MD: Food and Drug Administration. http://www.fda.gov/downloads/Drugs/Guidances/ucm073507.pdf (accessed 28 December 2020).

26 Food and Drug Administration (2006). Guidance for industry: Q9 pharmaceutical risk management. Rockville, MD: Food and Drug Administration. https://www.fda.gov/media/71543/download (accessed 28 December 2020).

27 Food and Drug Administration (2009). Guidance for industry: Q10 pharmaceutical quality system. Rockville, MD: Food and Drug Administration. http://www.fda.gov/downloads/Drugs/Guidances/ucm073517.pdf (accessed 28 December 2020).

28 Food and Drug Administration (2018). Guidance for industry: Q12 technical and regulatory considerations for pharmaceutical product lifecycle management core guideline. Rockville, MD: Food and Drug Administration. https://www.fda.gov/regulatory-information/search-fda-guidance-documents/q12-technical-and-regulatory-considerations-pharmaceutical-product-lifecycle-management-core (accessed 28 December 2020).

29 Huynh-Ba, K., Dong, M.D., and Ayers, J.T. (2020). Development of stability indicating analytical procedures by HPLC: an overview with best practices. *LCGC North America* 38 (8): 440–456.

30 Huynh-Ba, K. (2020). Analytical method validation for quality assurance and process validation professionals. *American Pharmaceutical Review* 23 (6): https://www.americanpharmaceuticalreview.com/Featured-Articles/569734-Analytical-Method-Validation-for-Quality-Assurance-and-Process-Validation-Professionals/.

31 Martin, G.P., Barnett, K.L., Curry, P.D. et al. (2013). Lifecycle management of analytical procedures: method development, procedure performance qualification, and procedure performance verification. *USP Stimuli to the Revision Process*. https://www.uspnf.com/notices/stimuli-article-lifecycle-management-analytical-procedures-posted-comment (accessed 28 December 2020).

32 McDowall, R.D. (2020). Understanding the lifecycle approach for analytical procedures. *LCGC North America* 38 (4): 233–240.

33 Russell, J.P. (ed.) (2013). *The ASQ Auditing Handbook*, 4ee. Milwaukee, WI: ASQ Quality Press.

34 Food and Drug Administration (2014). Guide to inspections of pharmaceutical quality control laboratories. Rockville, MD: Food and Drug Administration. http://www.fda.gov/ICECI/Inspections/InspectionGuides/ucm074918.htm (accessed 28 December 2020).

9

Stability Program Supporting Pharmaceutical Products

Kim Huynh-Ba

Pharmalytik LLC, Newark, DE, USA

TABLE OF CONTENTS

Analytical Testing for the Pharmaceutical GMP Laboratory, First Edition. Edited by Kim Huynh-Ba.

9.1 Introduction

The stability program is a critical part of the successful development of a drug product or active pharmaceutical ingredient (API). The data from stability studies are used for justifying the product specifications and establishing the product shelf life. The quality attributes, such as identity and strength, are defined and monitored throughout the product lifecycle using appropriate analytical testing procedures. This chapter will cover the crucial aspects of a stability program, which requires complying with the Code of Federal Regulations (CFR), specifically regulations from 21 CFR Part 211, ICH, and World Health Organization (WHO) guidelines. The chapter will also discuss the operational aspects of the stability program to comply with Good Manufacturing Practices (GMPs).

9.2 Regulatory Requirements for Stability Testing

The development of a new medicine relies heavily on compliance with the Code of Federal Regulations (21 CFR Part 211). The Key Information section below lists the GMPs requirements for stability testing stated in 21 CFR Part 211.166 of the Federation Regulations [1].

Section 21 CFR 211.166 stability testing

a) There shall be a written testing program designed to assess the stability characteristics of drug products. The results of such stability testing shall be used in determining appropriate storage conditions and expiration dates. The written program shall be followed and shall include:
   1) Sample size and test intervals based on statistical criteria for each attribute examined to assure valid estimates of stability;
   2) Storage conditions for samples retained for testing;
   3) Reliable, meaningful, and specific test methods;
   4) Testing of the drug product in the same container closure system as that in which the drug product is marketed;
   5) Testing of drug products for reconstitution at the time of dispensing (as directed in the labeling) as well as after they are reconstituted.

b) An adequate number of batches of each drug product shall be tested to determine an appropriate expiration date, and a record of such data shall be maintained. Accelerated studies, combined with basic stability information on the components, drug products, and container-closure system, may be used to support tentative expiration dates provided full shelf-life studies are not available and are being conducted. Where data from accelerated studies are used to project a tentative expiration date that is beyond a date supported by actual shelf-life studies, there must be stability studies conducted, including drug product testing at appropriate intervals, until the tentative expiration date is verified, or the appropriate expiration date is determined.

c) For homeopathic drug products, the requirements of this section are as follows:
   1) There shall be a written assessment of stability based at least on testing or examination of the drug product for compatibility of the ingredients and based on marketing experience with the drug product to indicate that there is no degradation of the product for the normal or expected period of use.
   2) Evaluation of stability shall be based on the same container-closure system in which the drug product is being marketed.

d) Allergenic extracts that are labeled "No U.S. Standard of Potency" are exempt from the requirements of this section.

Source: Food and Drug Administration. Code of Federal Regulations. Title 21, Part 211, Sec 166 [1].

A good stability program must be managed with written Standard Operating Procedures (SOP) that define requirements for conducting the studies, handling, and storing samples, reporting results, analyzing data, and investigating deviations. Once documented, the studies are initiated to assess the product behavior on storage and support the drug product expiry. Section 211.166 contains specific information to be included in this program. Typically, at a minimum, companies establish stability protocols with details pertaining to sample sizes, testing intervals, analytical procedures, acceptance criteria, and storage conditions for each study placed in a stability program.

Analytical methods must be developed to monitor critical characteristics of a drug product that is assumed to change over time; thus, the procedures must be stability-indicating and validated to ensure the reliability of data. Mass balance is critical while developing a stability-indicating method. Impurities and degradation products are monitored throughout the shelf-life durations to assure the quality and safety of the drug products.

The International Council for Harmonisation of Technical Requirements for Pharmaceuticals for Human Use (ICH) proposes the frequency of stability testing and the conditions under which samples are stored and evaluated. Table 9.1 lists

Table 9.1 ICH time points for stability protocols.

	Time on stability (in months)								
Conditions	**TZ**	**1 mo**	**3 mo**	**6 mo**	**9 mo**	**12 mo**	**18 mo**	**24 mo**	**Annually thereafter**
Intended storage condition (long-term)	x	x	x	x	x	x	x	x	x
Intermediate			(x)	(x)	(x)	(x)			
Accelerated		x	x	x					
Photo-stability		Q1B							

TZ: Time Zero.
LTS: Long Term Stability - the intended long-term storage condition.
Intermediate Condition: Test only if signification change is found for accelerated condition (through 12 m).
Q1B: ICH condition required for photostability.

Table 9.2 ICH storage conditions for stability studies.

Label condition	**Stability studies**	**Storage condition**	**Submission requirements (months)**
Room temperature	Long term	25 °C/60%RH	12
	Intermediate	30 °C/65%RH	6
	Accelerated	40 °C/75%RH	6
Refrigerator	Long term	5 °C/Ambient	12
	Accelerated	25 °C/60%RH	6
Freezer	Long term	−20 °C/Ambient	12

the time points, and Table 9.2 lists the storage conditions recommended as per the ICH stability guidelines. Temperature and humidity are controlled to within ±2 °C and ± 5%RH, respectively. For products that are to be stored at room temperature, if the samples stored in accelerated condition (see Table 9.3) do not meet the acceptance criteria, samples stored in the intermediate condition will be evaluated to support a tentative expiration dating period, storage excursions, and to allow limited data extrapolation [2].

Current Good Manufacturing Practice (CGMP) requires that the drug product be tested in the same container closure system as proposed in the product registration (i.e. final commercialized packaged container as appearing at the pharmacy's counter). Therefore, stability studies must be done on stability stations, setting the

Table 9.3 Definitions of significant changes in data stored at accelerated conditions.

API (drug substance)	A significant change is a failure to meet the predefined specification.
Drug product	1. A 5% potency change from the initial assay value; 2. Any specified degradation product exceeding its acceptance criterion 3. Failure to meet acceptance criteria for appearance and physical properties (e.g. color, phase separation, resuspendability, delivery per actuation, caking, hardness); and as appropriate to the product type; 4. pH exceeding its acceptance criteria; and 5. Dissolution exceeding the acceptance criteria for 12 dosage units.

Table 9.4 Typical conditions for semipermeable container.

Stability conditions	Temperature and humidity
Long term	25 °C/40% RH
Intermediate	30 °C/35%RH
Accelerated	40 °C/NMT 25%RH
	40 °C/75%RH

time point in each specific storage condition in their actual marketed storage containers. It may be an issue if there is not enough material available to be placed on stability. For drug substances, a functionally similar container is used to mimic the cardboard or plastic drum storage of raw material.

21 CFR 211 suggests that data from accelerated studies are used to support tentative expiration dates; however, the real-time studies are to be ongoing and continued until the actual projected expiration date to verify the expiry. It is a common practice to continue the stability studies a year beyond the set expiry. The regulatory agency separately addresses the samples claimed to be sterile or pyrogen-free.

For products stored in a semipermeable container, stability studies are conducted at a lower humidity level to assure that no weight loss would affect the potency and quality established (see Table 9.4).

9.3 Types of Stability Studies

Stability characteristics of a drug product are established through physical, chemical, biological, and microbiological testing, that are performed during the development process. Stability testing allows a company to set the retest periods for

APIs, evaluate product shelf life and storage conditions of the formulation developed and the marketing package. It must be determined whether the drug is sensitive to environmental factors such as humidity, light, temperature, and oxygen exposure during product development. Data are collected through clinical formulation phases, process development, packaging selection, or post-marketing activities to establish the product expiry, storage conditions, and commercial specifications. There are different types of stability studies.

9.3.1 Stability Studies of Materials Used in Clinical Development

Product specifications are essential as safety and efficacy data are established during clinical phases. Therefore, if the product does not meet specifications during the commercial phase, safety and efficacy may no longer represent the marketed product. Consequently, patient treatment outcomes may be affected. The storage conditions used in these studies depend on where the clinical studies are conducted. Typically, only data from samples stored in the intended storage conditions are needed to support clinical development; however, accelerated and stress studies are also conducted on clinical materials to understand the drug product developed. Limited samples may be available in the early phases. Formulation and analytical procedures may change rapidly during development. Table 9.5 shows an example of a typical study duration, which shows that the studies typically end as the clinical studies end. However, the company may decide to continue the stability studies to build a stability database to support registration, especially when the formulation is close to the final formulation that will be marketed. Data from these studies are used to support the expiry of clinical materials and included in the registration package as supporting data. Because the formulation of clinical materials frequently changes during development, stability studies can be terminated if there is no need to support the associate clinical study or if supporting data is unnecessary. This data set is used to support the expiry of clinical materials and is included in the registration package as supporting data.

Table 9.5 Example of stability protocol to monitor clinical supplies.

	Time on stability (in months)								
Conditions	**TZ**	**1 mo**	**3 mo**	**6 mo**	**9 mo**	**12 mo**	**18 mo**	**24 mo**	**36 mo or EoS**
25 °C/60%RH or intended LTS	x	x	x	x	x	x	x	x	x

TZ: Time zero.
EoS: End of studies (to support clinical trial).
LTS: Long-Term Storage.

9.3.2 Stability Studies of an Active Pharmaceutical Ingredient (API) or Drug Substance (DS)

Specifications are established for purity, physical and chemical properties of the materials; therefore, understanding the API stability profile is essential to developing any formulation and supporting the packaging selection. It is also important to understand the impurities profiles and their levels existing in the drug substance because changes to the impurity profiles can potentially affect the product's safety and efficacy. The degradation profile is established by subjecting the drug substance to various accelerated and stress storage conditions to establish storage conditions that minimize the formation of degradation products and, if necessary, to characterize degradation products exceeding the allowable levels as indicated in ICH Q3. Such information is also crucial to develop suitable processing conditions and design packaging components needed to protect the drug product during shipping and distribution. Stability data are used to establish the retest period for the raw material used in the process.

Analytical procedures for stability samples must be stability-indicating to monitor the purity of the API and quantitate the impurities if they appear. These impurities must be monitored, reported, and identified as per ICH Q3A applicable for API. It is not necessary to monitor process impurities that do not increase over time on long-term stability. Impurities shown to increase during storage times are referred to as "degradation products"; these degradation products can be attributed to heat, humidity, or light. Specifications of individual and total impurities must be set based on available stability data. One must also be aware that if any of these impurities are classified as genotoxic, a more vigorous surveillance program must be implemented.

Table 9.6 lists a typical stability protocol of API. Accelerated and stress conditions are used to establish the API stability profile. Critical testing is performed for chemical, physical, biological, and microbiological properties, as appropriate, to set the API stability specifications.

9.3.3 Stability Studies to Support Formulation Development

The drug product is developed by adding excipients or non-active constituents to a drug substance, also known as an API. The excipients or non-active constituents help the dosage deliverability and maintain product performance criteria or marketability, such as adding color, improving taste, maximizing solubility, or controlling pH, moisture, or oxygen content. Stability studies are necessary to monitor the quality of the clinical materials. Different formulations are placed on stability studies to better understand the interactions of the excipients with the drug substance, between excipients, or with the packaging materials. Changes to excipients can inherently change the established quality of the drug product;

Table 9.6 Example of stability protocol for API/drug substance.

	Time on stability (in months)								
Conditions	**TZ**	**1 mo**	**3 mo**	**6 mo**	**9 mo**	**12 mo**	**18 mo**	**24 mo**	**Annually thereafter**
25 °C/60%RH or intended LTS	x	x	x	x	x	x	x	x	x
30 °C/65%RH or intermediate		(x)	(x)	(x)	(x)	(x)			
40 °C/75%RH or accelerated		x	x	x					
50 °C		x							
60 °C		x							
70 °C		x							
Photostability		Q1B							

TZ – Time Zero.
LTS – Long-Term Storage.
30 °C/65%RH or intermediate – Test if signification change is found for accelerated condition.
Q1B – ICH condition required for photostability [3].

therefore, the stability studies are performed to ensure no change negatively impacts the final drug product.

Many companies will manufacture smaller batches of the drug product at the extreme end of the range that their manufacturing process is designed for this product. These batches are then placed on stability studies to determine the stability profiles of the drug product. These studies typically include varied accelerated stress conditions to obtain a better understanding of process capabilities, the degradation mechanism of the drug substance, and the drug product quality in the presence of excipients.

Forced degradation studies are also completed to have a better understanding of the degradation mechanism. The Arrhenius equation may be limited as temperature and humidity may cause different impacts on the drug product. Another stability evaluation program can be used, such as the Accelerated Stability Assessment Program (ASAP) [2, 4–6]. More detailed information on forced degradation is available from many references, and it is outside of the scope of this chapter [4, 6].

9.3.4 Stability Studies to Support Drug Product (DP) Registration

Stability data are required for the registration of a drug product. Stability studies of the final formulation in the marketing packages must be conducted to determine the shelf life and the intended storage condition of the commercial product.

Table 9.7 Example of stability protocol for drug product.

	Time on stability (in months)								
Conditions	**TZ**	**1 mo**	**3 mo**	**6 mo**	**9 mo**	**12 mo**	**18 mo**	**24 mo**	**Annually thereafter**
25 °C/60%RH or intended LTS	x	x	x	x	x	x	x	x	x
30 °C/65%RH or intermediate			(x)	(x)	(x)	(x)			
30 °C/75%RH for Zone IVb		x	x	x	x	x	x	x	x
40 °C/75%RH or accelerated		x	x	x					
50 °C		x							
Photostability		Q1B							

TZ – Time Zero.
LTS – Long-Term Storage.
30 °C/65%RH or Intermediate – Test only if signification change is found for accelerated condition (through 12 m).
Q1B – ICH condition required for photostability.

Similar to the API, Table 9.7 lists a typical stability protocol for the global registration of drug products. Critical testing should be done for chemical, physical, biological, and microbiological properties. Stability data are used to set the specifications of the drug product.

At this time, the stability protocols are constructed to follow the requirements of the Q1A(R2) ICH Guideline. This ICH guideline requires at least a 12-month long-term stability data for three batches of drug products needed for drug product registration. Stability testing will then be continued until the end of the expiry period that the product is approved, if not yet available. Also, accelerated, stress, and photostability studies are conducted to obtain a tentative expiration date [2, 3].

9.3.5 Stability Studies to Support Marketed Products

After receiving approval from the regulatory agency, stability studies are continued to support the commercial product through their annual marketed product stability program. A sampling of newly manufactured production lots must be monitored continually by placing representative lots on stability programs annually. Degradation products must be monitored through expiry against approved product specifications. Validated stability-indicating analytical procedures are used to monitor the product's chemical, physical, biological, and microbiological characteristics [2].

Stability studies need to be conducted when changes are made to the manufacturing process, formulation, or packaging process after approval. These changes include modifications to processes, raw material suppliers, packages, excipient sources. The size of the stability program depends on the change made to the product. Because stability studies are to support post-approval changes, they usually include accelerated conditions. Analytical procedures should be verified or revalidated concerning the new material or condition to support the product lifecycle.

9.3.6 Bulk Stability

Bulk stability is typically done before commercialization, with samples stored in packages mimicking those used in the production area. For drug substances, materials are stored in a simulated package, such as a double polyethylene bag in a small fiber drum. For a drug product, it depends on how the drug product is stored before packaging. These studies can run from six months to a year in a controlled environment, depending on the labeled storage condition of the drug substance. Critical attributes are tested and monitored every three months. Data from these studies are also helpful to support bulk package storage.

9.3.7 In-Process Testing

Stability studies are conducted to support the storage of in-process or intermediate materials if they are held for more than 30 days. Critical attributes are tested if they are held for a longer period to avoid unintended changes that impact the quality of the materials before the next step of the process.

9.3.8 In-Use Testing

In-use stability studies are necessary to provide support to products after the container is opened, such as multi-dose types of products or products that need to be reconstituted before use. This type of study should simulate the typical use of the product as it pertains to both usage and storage conditions. The length of the study depends on how long the product is used. Testing includes physical, chemical, and microbiological tests to focus on changes after the container is opened. Different regulatory agencies may have a different emphasis on in-use testing requirements, primarily depending on the distribution practices in their country.

9.3.9 Stability Studies to Support Excursions

Drug products go through different transportation paths and storage during distribution; hence, the product may periodically encounter extreme environmental stress conditions in short periods. Brief stability studies of unpackaged and packaged products are necessary to provide data to support these excursions and determine a proven temperature range for the product. Typically, the following studies are conducted:

Thermal Cycling – These studies are initiated for all drug products that are exposed to many temperature cycles. For example, drug products are exposed to the temperature that is cycled between 40 °C for three days and 25 °C for three days. Typically, three cycles of temperature are performed. This condition provides data to support distribution and storage excursions.

Freeze–Thaw Cycling – These studies are done for a product where stability characteristics may change due to low temperature, especially for liquid drug products. In this study, samples are stored where the temperature is cycled between room temperature and freezing conditions. For example, drug products that are stored, where the temperature is cycled between −10 to −20 °C for three days and 25 °C/ambient RH for three days. Typically, three cycles are conducted.

Extreme Temperature Studies – These studies are conducted at 50, 60, 70 °C in a brief time, such as seven days to two weeks.

Physical and chemical testing is done to evaluate any physical or chemical changes that impact drug product quality. Data can be used to support excursions during the storage or distribution of the drug product. After commercialization, these data also address questions received from the sales force, distribution centers, or product complaint office. Additional guidance is found in the Parental Drug Association (PDA) Technical Report No. 53: Stability Testing to Support Distribution of new Drug Products [4, 6].

Figure 9.1 illustrates how cycling studies can be set up. Most companies have the excursion study and the long-term study done separately. If the drug product is sensitive to temperature and humidity, the samples are placed on long-term stability storage for the expiration period after exposure to the duration of temperature/humidity excursions. The data will show the effect of environmental factors after excursions. The company may choose to expose the drug product to the full duration of temperature/humidity excursions at the end of its expiry to ensure the stability of the aged product does not exhibit any unexpected change. These studies should be repeated if there is any critical change to the product.

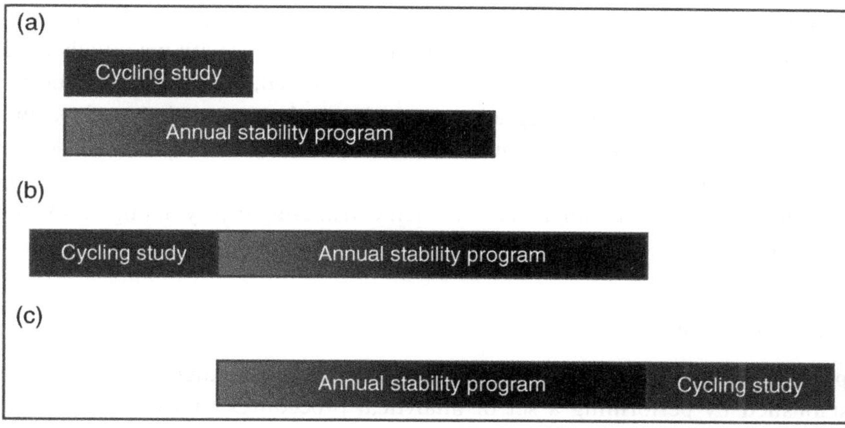

Figure 9.1 Setup excursion study and long-term stability study. (a) Cycling study is done separately from the long-term stability study. (b) Cycling study *before* the long-term stability study on the same set of samples. (c) Cycling study *after* the long-term stability study on the same set of samples.

9.4 Stability Program

9.4.1 Fundamental Principle of Stability

ICH Guideline Q1A(R2) was issued in 2003 to provide the minimum stability data package for the registration of a new drug substance or new drug product in Zone I and Zone II climatic conditions. This guideline does not cover abbreviated or abridged applications, variations, or clinical trial applications. For a listing of the ICH regulatory members, refer to http://ICH.org (members and observers). Few other non-ICH countries adopted this guideline with some modifications more specific for their climate region. The guideline indicates "the purpose of stability testing is to provide evidence on how the quality of a drug substance or drug product varies with time under the influence of a variety of environmental factors such as temperature, humidity", and "light, and to establish a retest period for the drug substance or a shelf life for the drug product and recommended storage conditions [2, 4]."

Stability studies are set up for the intended storage condition, accelerated, and stress conditions to provide data to support the shelf-life of the drug product. For drug substances, such studies establish the retest date in addition to the storage condition of the raw material. The stability profile must be ascertained as any inherent instability could lead to chemical degradation and, thus, a loss of drug potency and the possible formation of new chemical species with potentially toxic side effects. Various degradation studies (e.g. stress, forced, accelerated, or intermediate) predict the potential degradation products, determine their safety, and develop analytical procedures to quantitate these new chemical species. Different temperature and humidity storage conditions are required to gain market access to countries with different climate zones when considering expiry dating and long-term stability.

9.4.2 Specifications

Specifications are used to evaluate the quality of a drug substance or drug product established by performing a set of analytical procedures that determine and measure corresponding *physical, chemical, biological, and microbiological* attributes. The acceptance criteria are set based on the data obtained from validated testing procedures for the material used in pre-clinical and clinical studies. Different acceptance criteria can be set for release and stability purposes. Levels of impurities and degradation products must be quantitated and reported for stability samples [7, 8].

9.5 Stability Chambers

An important aspect of the stability program is the stability chamber. These chambers are commercially available and come in a variety of sizes, from small bench-top reach-in chambers to custom-designed walk-in rooms. Reach-in chambers are used when the number of samples is small or when the conditions are likely to be changed more often. Walk-in chambers can be built to any size depending upon the amount of space needed or when a larger space is required for sample storage. It is also convenient to use walk-in chamber size if the storage conditions are likely to remain constant for an extended time.

Regardless of the size, chambers require qualification, calibration, preventative maintenance, and monitoring. For typical chambers, the qualification process traditionally includes three stages identified as the Installation Qualification (IQ), Operation Qualification (OQ), and Performance Qualification (PQ). Chambers must be qualified for their specific uses and configuration prior to being placed in operation.

The ICH Q1A(R2) guideline requires the stability chambers controlled within ±2 °C and ± 5% relative humidity (RH) from the setpoints. Refrigerated chambers with a set point of 5 °C have a tolerance of ±3 °C (2–8 °C). Freezers have a tolerance of ±5 °C or, in some cases, ±10 °C. If chamber conditions exceed these ranges, an investigation must be initiated to determine any negative impact on the stability samples stored in that chamber. An accurate inventory is necessary for all chambers; therefore, all chambers should have limited and controlled access and restricted to authorized personnel only. Chambers are monitored continuously using a qualified monitoring system that can be done with electronic equipment or chart recorders. The monitored data should be reviewed frequently for any chamber trends to evaluate equipment performance. Chamber logbooks are available to document the daily usage, repairs, calibrations of the chambers.

The SOP should be available for stability chambers detailing requirements for qualification, monitoring, maintenance. Training is conducted for any person working with these chambers. The SOP should address circumstances when samples need to be moved to an alternate chamber and how stability chambers are handled under such excursions. These activities must be documented and justified [4, 6].

9.6 Stability Sample Management

Sample management for stability studies is a critical function of GMP operations; any deviation that is not handled appropriately can negate the validity of the study and affect the product registration. Figure 9.2 shows the flow of stability samples from the time the study is initiated until the data is evaluated. Studies are set up based on approved stability protocols. Samples are received, staged, labeled, and placed in appropriate storage conditions. The number of units needed to be placed on stability depends on the study length, conditions, and the testing that needs to be performed. It is a frequent practice that at least 50% extra samples should be available for each study at each condition in case additional testing or investigation is needed. The purpose of the study must be documented and understood by the stability studies team to determine the impact of the study data when change occurs. Most companies have an electronic sample management system, such as a Laboratory Information Management System (LIMS). This system will monitor the testing schedule approved for the study. The Key Information section below lists the information needed to create a stability study. A checklist can be created to ensure all necessary information is collected. Each study is given a unique number with a defined purpose when the study is initiated [4].

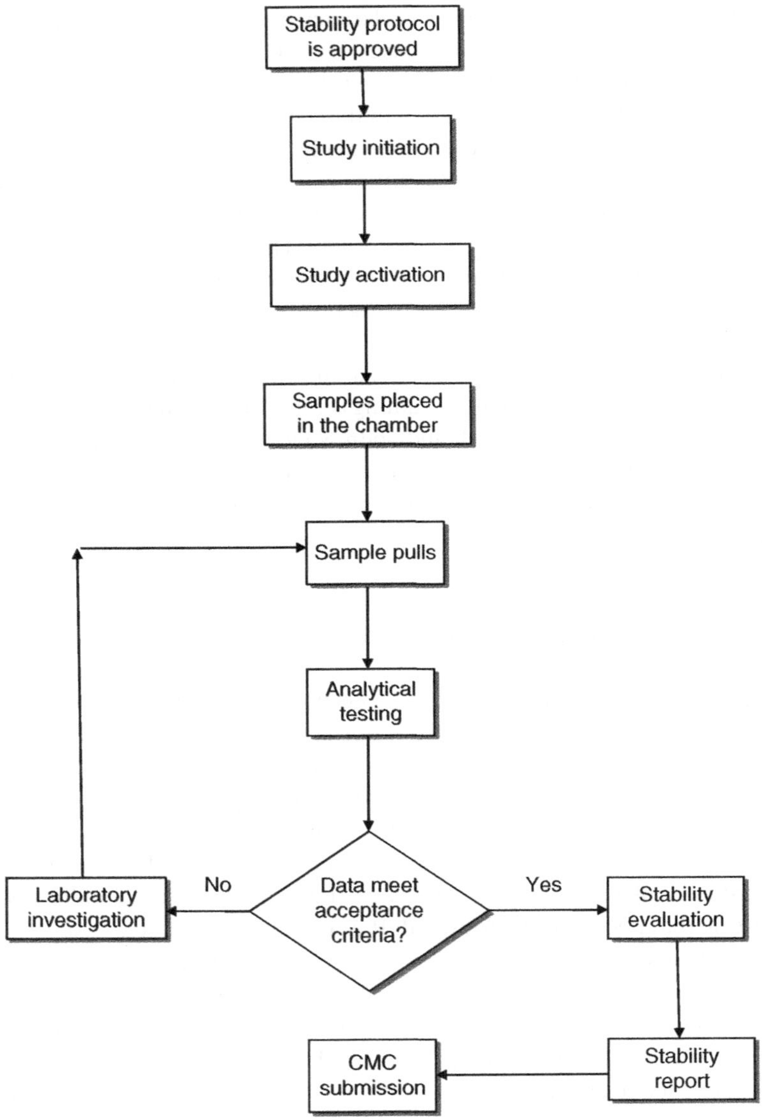

Figure 9.2 Flow of a stability study.

The sequential order of events is important for every stability study. The SOP should describe in detail how these dates are determined. Figure 9.3 lists the chronological order of stability dates [4].

Information needed to initiate a stability study
1) Study number 2) Protocol to be followed 3) Formulation description 4) Lot number 5) Dosage form 6) Strength 7) Packaging description 8) # of pulls (samples per test period) 9) Units to be pulled 10) Other information, if applicable

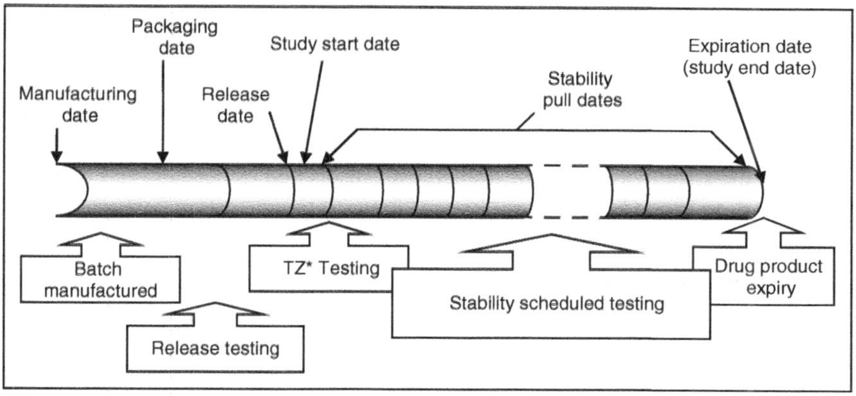

*TZ: Time zero

Figure 9.3 Chronological order of stability activities to support a drug product study. *TZ: Time Zero. *Source:* Huynh-Ba [4].

9.6.1 Study Start Date

Typically, the study starts when the samples are placed in the temperature and humidity-controlled chambers. An alternate study start date is also seen in Industry for marketed product stability as the "date of manufacture." Samples are inspected, counted, and labeled properly according to the SOP in place. Time zero testing is typically performed at this time. Depending on the product stability profile and the amount of time elapsed since manufacture, the initial release data could be used for the Time Zero timepoint, if the release testing is not more than 30 days before samples are placed in the chamber.

9.6.2 Sample Pull Dates

Samples are removed from the storage chambers based on the established time points of the stability schedule. Once the samples are pulled from the storage chamber, the samples should be maintained in conditions that are not more stressful than the storage condition from where they were removed. It is also helpful to track the location of the samples in the chamber. Typically, samples are pulled from the stability chambers within certain calendar days from the pull date. For example, stability samples in 25 °C or 60%RH chamber can be removed within 3 days after the scheduled pull date for time points less than or equal to 3 months, 7 days after the scheduled pull date for those less than a year, and 14 days after the scheduled pull date for those equal or greater than one year. The stability program SOP should describe these pull windows and the management of pulling outside the windows.

If the pull needs to occur outside of the allowable window, then a deviation of the stability sample pull date should be initiated. Figure 9.4 lists an example of this type of deviation.

Practically, sample testing should be done within 30 days of the pull date; however, it is advisable to complete testing as soon as possible for those samples with time points less than six months. If testing needs to be canceled, the reason should be documented in the stability report to be available for inspection as needed [4]. Figure 9.5 shows an example of a form that can be used for a study cancelation purpose.

9.6.3 Study End Date

The study end date is when the last sample is pulled, tested, and data are reported. The study is not considered complete if there is an open investigation on any of the results. Studies that support clinical trials can end after the last patient uses the clinical materials or at the expiry of the clinical materials. If the length of a clinical study is extended, the stability study needs to be amended accordingly to cover the extended times of the clinical study.

Once the study ends and all data are reported, excess stability samples should be removed from the chambers, and the stability inventory is updated. This activity is part of the routine maintenance of stability chambers.

9.6.4 Sample Inventory

All samples must be accounted for; therefore, sample inventory is required for the stability program. Figure 9.6 depicts a typical inventory form for stability chambers. The number of samples should be reconciled annually with the LIMS system. Discrepancies must be promptly investigated and documented. Figure 9.7 is an example of a form that can be used to reconcile the stability sample inventory.

STABILITY SAMPLE PULL DATE DEVIATION FORM

Part 1: To be Completed by Stability Personnel

Product:	Batch Number(s): Study Number:	Time Point:
Storage Condition(s):	Required Pull Date:	Actual Pull Date:
Reason for Deviation:		
Supporting Document:		

Stability Personnel:	Title:
Signature:	Date:

Part 2: To be Completed by Analytical Department

Impact on the Current Study:	
Analytical Department:	Title:
Signature:	Date:

Part 3: To be Completed by Stability Management

Analytical Department:	Title:
Signature:	Date:

Figure 9.4 Example of a sample pull date deviation form.

9.7 Stability Protocol

Each stability study should be structured from an approved stability protocol. The protocol should include purpose, sample description, the origin of drug products, and package configurations. It also lists sample size, test intervals, test methods and acceptance criteria, storage conditions, and storage orientation of the samples, as applicable. The stability protocol must be approved by the Quality Assurance (QA) department. An example of a stability protocol to set up a study is listed in Figure 9.8. The QA department must approve the stability protocol.

A protocol template can be established for multiple studies. Storage conditions depend on the product label and the regions where the product is registered for marketing; thus, standard protocols can be set up to provide consistency

STABILITY TEST CANCELLATION FORM

Part 1: To be Completed by Stability Personnel

Product:	Batch Number(s):	Study Number:
Storage Condition(s):	Time Point:	Tests to be canceled:
Reason for Test Cancellation:		
Impact on the Stability Program:		
Supporting Document:		

Stability Personnel:	Title:
Signature:	Date:

Part 2: To be Completed by Analytical Department

Analytical Department:	Title:
Signature:	Date:

Part 3: To be Completed by Stability Management

Analytical Department:	Title:
Signature:	Date:

Figure 9.5 Example of a stability test cancelation. *Source:* Based on Huynh-Ba [4].

throughout the stability program. Standard protocols are a set of stability protocols developed and pre-approved by Stability Management and QA to provide optimal stability information for specific and designated purposes. Standard protocols can be developed for a drug substance or dosage form at different phases of clinical development, registration, or post-approval changes. In addition to the storage conditions and time points, tests and specifications are also assigned to the study. Typically, additional studies are initiated when there is a change in formulation, manufacturing, packaging components, or dosage forms, depending upon the significance of the change [4, 9, 10].

Study information is necessary to identify the samples. It contains specific information such as study purpose, lot information, manufacturing date, expiry date, packaging information, manufacturing site, and packaging site.

SAMPLE INVENTORY FORM

Study Number: ____________ Product Name: ____________

Strength: ____________ Lot Number: ____________

Activation Date: ____________ Manufacturing Date: ____________

Last Pull Date: ____________ **Storage Condition:** ____________

Package Description: Amount/Package: ____________

Excess Amount/Package: ____________

Unit (Circle): Bottle Blister Vials Open dish Poly Bag in Fiber Drum

Date	Timepoint	Amount Removed	Amount Available	Comments	Analyst Initial

Sample Pending Disposition

Transfer Information

Amount Moved	Current Status*	Current Condition	Initials Date	Comments Sent to/Destroyed	Initials Date

Figure 9.6 Example of an inventory form for a stability chamber.

Testing information includes all tests to be performed on the pulled stability samples. Different tests may be required at different time points and storage conditions. If more than one lab is designated for the testing, then the testing labs should be listed in the study information and the report. It is important to emphasize that the analytical procedures used for stability testing must be validated and stability-indicating [2].

9.7.1 Deviation of Stability Protocol

Changes made to the protocol after the start of the study are called *protocol deviations*. There are planned and unplanned deviations. If planned, justification must be recorded with appropriate approval. Sufficient samples to accommodate the planned change should be evaluated and documented. Unplanned deviations can also occur throughout the study. When they occur, investigations must be conducted, and corrective actions and preventive actions (CAPA) must be determined to avoid recurrence and maintain the integrity of the stability study. The impact of all deviations on the study must be assessed and justified [2, 4].

INVENTORY ADJUSTMENT FORM

Study Number: ____________________ Product Name: ____________________

Strength: ____________________ Lot Number: ____________________

Last Pull Date: ____________________ Storage Condition: ____________________

Condition	Amount needed to complete study	LIMS inventory	Chamber inventory	Difference	Impact of adjustment

Reason for Difference:

Impact on the current study:

Reconciliation Activities:

Name:	Title:
Signature:	Date:
Approved by:	Title:
Signature	Date:
Stability Approval:	Title:
Signature:	Date:

Figure 9.7 Example of an inventory adjustment form.

9.7.2 Study Cancelation

If stability data are no longer needed, the stability study can be terminated. Appropriate approval must be secured, and the remaining samples should be removed from the stability chambers, once approval is received. As well, a study can be put on hold if needed. Again, appropriate documentation is needed, and the impact of the stability program must be addressed before samples are removed from their storage conditions. These activities should be included in the stability program SOP.

9.7.3 Reduce Testing with Bracketing and Matrixing

Conducting stability programs is expensive. To reduce the cost of the stability program while maintaining the regulation requirement, many companies employ *bracketing* or *matrixing* options to reduce the amount of testing and resources. The

STABILITY PROTOCOL FORM	
Description:	Date:
Type of Study:	Lot Number:

Storage Time	25°C/60%RH	30°C/65%RH	40°C/75%RH	Photo	

Tests to be done	Specifications:	Comments:

Stability Department:	Title:
Signature:	Date:
Quality Assurance:	Title:
Signature:	Date:

Figure 9.8 Example of a stability protocol.

ICH Q1D guideline has provided specific principles for bracketing and matrixing with minimal regulatory consultation. Regulatory agencies also encourage the use of these matrixes to reduce testing and minimize redundant testing [11].

Bracketing refers to the design of a stability schedule, which at any time point includes the samples on the extremes of certain design factors, for example, strength, container size, or filled volume. The bracket design assumes that the extreme levels to be tested represent the stability of the sample at the intermediate levels; therefore, it is not necessary to generate another similar data set. Thus, the use of a bracketing design would not be appropriate if the samples selected for testing are not of the extreme configuration. Table 9.8 is an example of a bracketing configuration.

Matrixing is a statistical design of a stability schedule. Table 9.9 is an example of a matrixing configuration. At a specified timepoint, a selected subset of the total number of possible samples is tested for all factor combinations. At a subsequent timepoint, a different subset of samples for all factor combinations is tested. The design assumes that the stability of each subset of samples tested represents the stability of all samples at a given time point. Long-term trends are approximately linear across the studied presentations; thus, the comparative stability of each presentation can be evaluated [5, 11].

Table 9.8 Example of a simple bracketing design.

Strength		50 mg			100 mg			250 mg			500 mg		
Batch		**A**	**B**	**C**	**A**	**B**	**C**	**A**	**B**	**C**	**A**	**B**	**C**
Container size	50 mL	T	T	T	—	—	—	—	—	—	T	T	T
	100 mL	—	—	—	—	—	—	—	—	—	—	—	—
	250 mL	T	T	T	—	—	—	—	—	—	T	T	T

Table 9.9 Example of a two-thirds factorial matrixing design.

Time point		**Months on stability**							
Batch		**0**	**3**	**6**	**9**	**12**	**18**	**24**	**36**
Strength 1	A	T	T	—	T	T	—	T	T
	B	T	T	T	—	T	T	—	T
	C	T	—	T	T	T	T	T	T
Strength 2	A	T	—	T	T	T	T	T	T
	B	T	T	—	T	T	—	T	T
	C	T	T	T	—	T	T	—	T

Unlike bracketing, where the extremes are evaluated, matrixing is applicable where differences are identified. The differences in the samples for the same drug product must be identified as, for example, different batches, different strengths, different sizes of the same container and closure, and different closure systems. The ICH Q1D guideline establishes a series of possible scenarios where matrixing could be applied without prior approval. Matrixing can be applied across packaging systems when a secondary packaging system is used to add to the drug product stability. The advantage of matrixing is that it can revert to full testing, if necessary, because all samples are placed on stability.

Different packages, different dosage forms, and different tests can also use different matrices depending on the risk involved with the configurations. Reduced testing presents a significant saving of resources. However, it also poses serious limitations. For example, data evaluation can be more complex, confidence intervals may be wider, and the design may not be as sensitive to differences as to when full testing is done. It is strongly recommended to consult a stability statistician for these applications. Also, all samples must be placed in the stability chambers and scheduling the study for testing is time-consuming and labor-intensive.

There are numerous references to discuss these two practices. The Food and Drug Administration (FDA) encourages matrixing; however, the experience with this type of reduced testing is limited globally. Therefore, discussion with regional regulatory agencies is advisable if a matrixing application is to be submitted globally [4, 6, 11].

9.8 Stability Report

Stability results from all time points are collated into tables called stability datasheets. Figure 9.9 presents a typical example of a stability data record, including the batch information, study information, and testing information. These data sheets are compiled into a stability report. Statistical analysis is performed to determine the retest date of API and the expiry dating period of the drug product [4, 6, 12]. Typically, a tentative expiration date of twice the data available can be requested.

The stability report is a vital component in any regulatory submission. The following Key Information section lists critical sections of the stability report [4, 6].

Sections of stability report

- Executive summary
- Stability protocols
- Primary stability data
- Secondary stability data
- Stability commitment
- Stability evaluation
- Report summary

9.9 Annual Product Review (APR)

Section 21 CFR 211.180 requires an annual product review (APR) to be done as part of the CGMP requirements. This process includes a compilation and review of records for quality characteristics of API and drug products. The compiled data are trended and summarized to verify the manufacturing and packaging processes and corresponding quality systems. The review schedule is a fixed annual schedule that is approved by the QA department. This schedule is typically aligned with the anniversary of the regulatory filing. Product families such as ones with different dose strengths or packages, or different levels of active ingredients, can be included in the same annual product review. Products with different formulations or manufacturing processes are reviewed separately. The Key Information sections below list

STABILITY ANALYTICAL RECORD

Sample Name: Lot#: Study #: Protocol #: Study Start Date: Study Purpose:	Manufacturing Date: Manufacturing Site: Expiration Date: Testing Site: Packaging Site:	Storage condition: Sample Orientation (if applicable): Packaging Information: Packaging Date:

Test Name	Method	Acceptance Criteria	Time Zero	1 Mo	2 Mo	3 Mo	6 Mo	9 Mo	12 Mo
Pull Date									
Test Date									
LIMS ID									
Appearance									
Assay									
Impurities Individual Total									
Dissolution Average % RSD Range									
Moisture									

Completed By: ____________________ Date: __________

Reviewed By: ____________________ Date: __________

Approved By: ____________________ Date: __________

Figure 9.9 Example of a stability data record.

the minimum key elements found in an APR, and the data that should be included in an APR. These are data generated during this period of time [13].

All data generated for the drug product should be reviewed and included in the APR. Data of the rejected batches are also included with the additional investigation even if they are not distributed commercially. Batches with a final disposition

Key elements of an APR

- Table of contents
- General product information including a brief process review
- Product name/strength(s)
- Regulatory filing number (if available)
- Review period
- Product code
- Primary/secondary manufacturing site
- Primary/secondary packaging site
- Packaging configurations

Data included in the APR

- Batch numbers
- In-process control data
- Analytical release data
- Yields of all validation steps
- Rework or reprocessing batches
- Quarantine materials
- Rejected batches during this period
- Stability data
- Visual inspection of retained materials
- List of deviations
- Laboratory investigations
- Customer complaints including all adverse events
- Change control impacting product quality, validation, and regulatory filing
- Validation information
- Product-specific audit observations and responses
- Field alerts and recalls
- Salvaged APIs and drug products
- Returned APIs and drug products

(either approved or rejected) should undergo trend analysis. Current data are reviewed against previously obtained data, and results should be presented in tabular and graphical formats [13].

Additional resources provide a thorough discussion of this process, including a review of stability data and an assessment of the product stability profile [4, 6, 9]. Performing this type of an assessment will help the company determine if changes are needed for product specifications, formulation, process, or analytical procedures. An updated report of annual stability data is submitted to the regulatory agency, and it should include any stability trends, deviations, or changes observed from the previous review. A discussion of any investigation of out-of-specification (OOS) or out-of-trend (OOT) data of the stability program must also be included. It is helpful to present this data and have it updated graphically. Statistical analysis is also helpful to demonstrate if the stability program continues to support the approved product expiry. Review of stability data for the annual product review is also an excellent quality tool to monitor the product quality and verify its safety and efficacy through its shelf life.

List of useful questions when preparing the APR
• Are data consistent throughout the review period? • What is the range of the data? • How are data compared to the specifications? • What is the trend in potency? In impurities? Is there an increase or decrease? • How is the rate of increase or decrease as compared with historical data? • Is there any abrupt change in values? • Does the trend continue, or is it an isolated incident? • What is the change over time? • Is there an assignable cause? • How is this data set compared with the previous year? • Does any corrective action still require following up?

The APR conclusion must include a summary of all assessments, highlighting any product trend compared with the previous evaluation. It is also necessary to include a statement of recommendation for further assessment, as appropriate. The QA unit coordinates this activity and approves the APR report.

9.10 Summary

Stability operations are critical to any GMP organization. The stability program determines characteristics of APIs and drug products in development, registration, or commercial distribution on long-term storage. Stability data are crucial to determine the appropriate storage condition, shipping requirements, release and stability specifications, and product expiry. Forced degradation studies can now identify stability problems for drug substances and drug products [10, 14]. The development of the stability program depends on available resources and quality systems. Other chapters within this book discuss quality systems such as training programs, metrology, materials, testing program, all of which are vital to support stability testing on an ongoing basis.

List of Abbreviations

API	Active Pharmaceutical Ingredient
APR	Annual Product Review
ASAP	Accelerated Stability Assessment Program

CAPA	Corrective Actions and Preventive Actions
CFR	Code of Federal Regulations
DS	Drug Substance
GMP	Good Manufacturing Practices
ICH	International Council for Harmonisation of Technical Requirements for Pharmaceuticals for Human Use
IQ	Installation Qualification
LIMS	Laboratory Information Management System
OQ	Operation Qualification
PDA	Parental Drug Association
PQ	Performance Qualification
QA	Quality Assurance
SOP	Standard Operating Procedures
WHO	World Health Organization

References

1 Food and Drug Administration (2019). Stability testing. Code of Federal Regulations. Title 21, Part 211, Sec 166. Rockville, MD: Food and Drug Administration. https://www.accessdata.fda.gov/scripts/cdrh/cfdocs/cfcfr/CFRSearch.cfm?CFRPart=211 (accessed 2 December 2020).

2 International Conference on Harmonization (2003). Stability testing of new drug substances and products (second revision) Q1A (R2). https://database.ich.org/sites/default/files/Q1A%28R2%29%20Guideline.pdf (accessed 2 December 2020).

3 International Conference on Harmonization (1998). Photostability testing of new drug substances and products Q1B. https://database.ich.org/sites/default/files/Q1B%20Guideline.pdf (accessed 3 December 2020).

4 Huynh-Ba, K. (ed.) (2008). *Handbook of Stability Testing for Pharmaceutical Products: Regulations, Methodologies, and Best Practices*. New York: Springer Science and Business Media.

5 Rack, C. (2017). An introduction to the accelerated stability assessment program (ASAP). *Pharmaceutical Technology* 20 (5): 86–90.

6 Huynh-Ba, K. (ed.) (2010). *Pharmaceutical Stability Testing to Support Global Markets*. New York: Springer Science and Business Media.

7 International Conference on Harmonization (2000). Specifications: test procedures and acceptance criteria for new drug substances and new drug products: chemical substances Q6A. https://database.ich.org/sites/default/files/Q6A%20Guideline.pdf (accessed 3 December 2020).

8 International Conference on Harmonization (1999). Specifications: test procedures and acceptance criteria for biotechnological/biological products Q6B. https://database.ich.org/sites/default/files/Q6B%20Guideline.pdf (accessed 3 December 2020).

9 Barron, M. (2003). Implementing successful stability testing operations. *Pharmaceutical Technology* 1: 14–21.

10 Bruemmer, H. and Belikova, N. (2017). Stepping up the pace of drug stability studies. *Pharmaceutical Technology* 41 (9): 46–51.

11 International Conference on Harmonization (2002). Bracketing and matrixing designs for stability testing of drug substances and drug products Q1D. https://database.ich.org/sites/default/files/Q1D%20Guideline.pdf (accessed 3 December 2020).

12 International Conference on Harmonization (2003). Evaluation for stability data Q1E. https://database.ich.org/sites/default/files/Q1E_Guideline.pdf (accessed 3 December 2020).

13 Henson, E. (2005). Conducting an effective annual product review. *Journal of GXP Compliance* 6 (2): 15–23.

14 International Conference on Harmonization (1996). Stability testing for new dosage forms Q1C. https://database.ich.org/sites/default/files/Q1C%20Guideline.pdf (accessed 3 December 2020).

10

Laboratory Information Management System (LIMS) and Electronic Data

Parsa Famili and Susan Cleary

Novatek International, Montreal, Canada

TABLE OF CONTENTS

Analytical Testing for the Pharmaceutical GMP Laboratory, First Edition. Edited by Kim Huynh-Ba.

10.1 Introduction

Everywhere you look, every process you are involved with is now somehow partially or fully computerized or automated to facilitate the process. Changing the processes from manual to automated is inevitable, and Good Manufacturing Practice (GMP). The impact of the COVID-19 pandemic validates the need for computerized system in pharmaceutical manufacturing and healthcare system. The automated systems are reproducible thus can be more easily validated. Manual systems are more error-prone, and with the new data integrity regulations, harder to justify and defend.

The manual system, even though a good process, is only as good as the individuals who are following the process versus the automated process where every person must follow the same process. Additionally, paper-based systems cannot support remote working requirements and the new ways of working going forward.

Automation can substantially improve the laboratory process, especially those laboratories working in a Current Good Manufacturing Practices (CGMPs) environment. A Laboratory Information Management System (LIMS) is a software-based information management system with features that facilitate efficient laboratory operations. Key features include process-specific workflows and data management to ensure the security and integrity of the records housed in the database. A contemporary LIMS system will have a flexible architecture and data exchange interfaces with laboratory equipment that fully support use in regulated industries with features which include user access management and audit trails. The modern LIMS is dynamic because the laboratory's requirements are rapidly evolving, and different labs often have different data management needs based on the type of samples being tested. Laboratories can be dedicated to specific sample types or can test a variety of different samples. These samples can be raw material samples, which include active pharmaceutical ingredients (APIs), excipients or fillers, components, solutions, standards, in-process or bulk products and finished products, and stability samples.

The features and uses of a LIMS have evolved over the years from simple sample tracking to an enterprise resource planning tool that manages multiple aspects of laboratory informatics and operations. LIMS functionality has grown to include assay data management, data mining, data analysis, equipment integration, sample inventory control, as well as integration to an electronic laboratory notebook (ELN), enabling the realization of virtually complete laboratory quality control within a single software solution.

A robust LIMS system is necessary for all modern laboratories as part of quality management (QM). Quality is an integral part of the pharmaceutical processes ensuring that the outcome meets the intended requirements rather than testing the product after manufacture. Quality management can be divided into four core primary categories: design, control, assurance, and continuous improvement, and focuses on product quality and on the processes by which it is achieved. Quality management uses control of processes to realize consistent quality; it is the company's procedures, processes, infrastructure, and resources required to implement quality management.

Quality management practices are put in place to ensure every aspect of the system functions harmoniously. The data generated by these practices are best managed and maintained in a quality management system (QMS). Similar to LIMS, a database driven QMS has features that facilitate efficient quality operations. Robust QMS software will include features to manage Standard Operating Processes (SOPs) and other documents, simplify the creation of batch records and annual product reviews, and control quality procedures such as change control, investigations, and corrective and preventive actions (CAPA). Contemporary QMS will enable quality management personnel to produce quality work in a controlled software system in the same way a LIMS empowers laboratory analysts to produce quality work.

10.2 Analytical and Quality Data Management

Because LIMS and QMS software systems house the data on which quality decisions for production, quality control, product release, and operations are based, this system is critical. Four types of data analytics are included in big data analysis [1]:

1) Descriptive, which reports on the performances or events of the past;
2) Diagnostic which uses data mining and correlation techniques to understand why an event occurred;
3) Predictive, which uses various algorithms and scenarios based on previous events and available data to predict future outcomes; and
4) Prescriptive, which uses various scenarios to specify the best path forward and optimal outcomes.

Figure 10.1 illustrates the four types of data leading to decisions and corresponding actions. These decisions may relate to the quality of a product, the release/reject of a batch based on minor or major issues at various levels of manufacturing or packaging, CAPA processes, environmental monitoring, controls, calibration and preventive maintenance, and production batch review among several others.

Increased machine learning and artificial intelligence are incorporated into automated manufacturing to streamline the processes, detect flaws, and correct them before they impact the product or become a risk for patients.

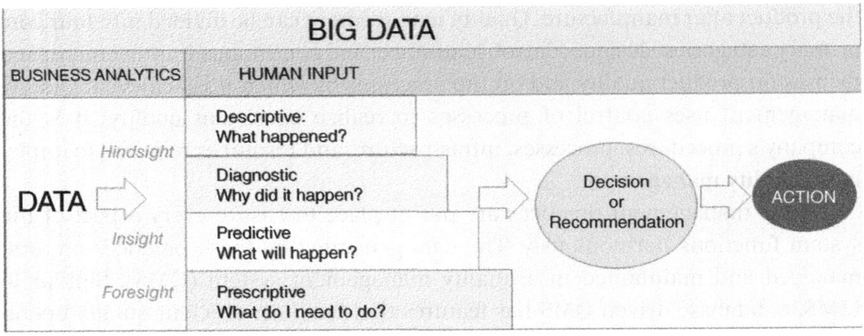

Figure 10.1 Big data – types of analytics. *Source:* Barbero et al. [1].

Each part of the quality process deals with several factors relating to big data. For example, environmental monitoring trending is an ongoing process generating a massive amount of data. Based on the data, one needs to assess many types of trends.

Another example, where big data analytics can be very useful in decision-making processes, is the routing testing of the incoming materials or finished products and stability studies of each product with several (Stock Keeping Units) SKUs. Many drug products are available in several dosage forms using different excipients to change the color, taste, compressibility, microbial control (preservatives), packaged with different packaging components. Each of these excipients and packaging components may interact with the APIs; therefore, big data is necessary to have a full view of all products and processes used in the manufacturing of the drug product.

Big data follows a 6V-Model, which consists of velocity, volume, variety, veracity, viability, and value of the data [1] (see Figure 10.2). Big data analysis allows the quality unit to view the entire quality operation process allowing the strategic decision-making process.

10.3 Quality System

The Food and Drug Administration (FDA) has concluded that modern quality systems, when coupled with the manufacturing process and product knowledge, and the use of effective risk management practices, can handle many types of changes to facilities, equipment, and processes. The quality system provides the foundation for the manufacturing systems that are linked and function within it. The quality system model described in this guidance does not consider the five manufacturing systems (production system, facility and equipment system,

Where

Velocity is the frequency at which data generated,

Volume is the growth of data in time,

Variety is the types of data (either structured or unstructured),

Veracity is the trust in the accuracy of data,

Viability is the relevance of data to a specific process and,

Value is the return on investment (ROI)

Figure 10.2 Big data 6V-model. http://www.ngi-summit.org/wp-content/materials/EU_papers/DG_digit_study_big_data_analytics_for_policy_making.pdf. *Source:* Barbero et al. [1].

packaging and labeling system, materials system, and laboratory control system) as discrete entities, but instead integrates them into appropriate sections of the model.

Effective decision-making in a quality system environment is based on an informed understanding of risks. Elements of risk should be considered relative to the intended use of a product and, in the case of pharmaceuticals, patient safety and availability of medically necessary drug products. Management should assign priorities to activities or actions based on an assessment of the risk, including both the probability of occurrence, harm, and the severity of the harm. It is important to engage appropriate parties in assessing the risk. Such parties include customers, appropriate manufacturing personnel, and other stakeholders. Implementation of quality risk management includes assessing the risks, selecting, and implementing risk management controls commensurate with the level of risk, and evaluating the results of the risk management efforts. Because risk management is an iterative process, it should be repeated if additional information is developed that changes the risk management.

In a manufacturing quality systems environment, risk management is a tool in the development of product specifications and critical process parameters.

10.4 Process Control in Quality Management

Metrics must pass the "So what?" test [2]. To accomplish this, consider creating a metric profile for every metric. The profile approach enforces the requirement of a clear definition that can be used to help train people on the metric itself.

Furthermore, the metric profile should be clear enough to allow users to deduce conclusions. Metrics are a work-in-progress; thus, it should be updated, improved, and clarified to increase its usage.

What should be included in a metric profile?

1) *Process metric name*: The metric name should be obvious enough that the person who is viewing it understands what he/she is looking for.
2) *A representation of the process metric*: The type of graph used will also assist in clarifying the meaning of the metric (e.g. a line graph, a stacked bar chart, Weibull graph, etc.).
3) *The idea behind the metric*: It is crucial to understand the reason for the metric. For example, a cumulative-production run chart provides insight into the level of production for load balancing.
4) *The implications of the metric*: To continue the example from above, if the cumulative-production run chart reflects less than the estimated upper and lower control limits, then that should lead to a root-cause analysis. Potential root causes can include demand variation, overproduction, quality problems, production problems, capacity constraints, etc.
5) *The process metric objective*: It is important to understand the expectations and the magnitude of performance gaps to improve any process.
6) *The metric's data source*: The source and the consistency of data, where the data reflected within the metric, ensure the accuracy of the metric.
7) *Calculation*: Often, data can be taken directly from the source; however, occasionally, the metric requires a calculation using the source data. For example, the weekly productivity chart (i.e. number of units/person/week) may require someone to take the prior week output, divided by the number of staff. There should be no guesswork on how to perform the calculation.
8) *Frequency*: Metric "actionability" identifies metric frequency. The more dynamic the environment, the more metrics are required (e.g. continuous environmental monitoring of a sterile facility or yearly employee performance review).
9) *Roles and responsibilities*: It is imperative to have an owner for the metric. The owner is responsible for updating the metric, publishing, and distributing the metric, and initiating the root-cause analysis.
10) *Company-specific requirement*: Any other metrics necessary that should be included in the metrics to provide a clear picture of the operations or quality processes.

"To err is human" but to what degree can one deviate, and what is the impact of error? To ensure adequate data is available to make quality decisions, the quality department looks at the following metrics among others listed in Table 10.1.

Table 10.1 List of quality metrics to be evaluated for LIMS and QMS.

LIMS	QMS
• Materials • Product release • Stability studies • Cleanrooms, controlled areas • Cross contamination and cleaning validations • Equipment CPM (metrology)	• Risks • SOPs and document • Compliance and audit • CAPA/Remediation • Training • Change management • Validation

10.5 Material Management

Testing raw materials is important to ensure APIs, excipients, or components used in the production and manufacture of pharmaceutical products are suitable for their intended use. Conducting raw material analysis using appropriate tests and compendial methods can prevent costly production problems and delays.

Managing to test and releasing APIs, excipients, or components for use in production is a complex process. Factors such as vendor approval status, received lot size, number of containers, number of units, and the nature of the product, play a role in determining the specific regulatory requirements for sample collection, sample composition (pooled/non-pooled), testing requirements, and ultimately the release or reject decision.

The general material is registered, and then each lot is logged in under its material code as it is received. Samples are then collected and distributed to the Chemistry and Microbiology departments for testing. Qualitative and quantitative results from testing are entered in the system for review, approval, and then used for release decisions, as well as quality reporting, trending, and analysis for annual vendor reviews and process control verifications. The product release/rejection process is where the product is either deemed acceptable for production or investigated and quarantined for further processing or returned to the supplier.

10.6 Product Release

Good Manufacturing Practice (GMP) batch/lot release testing is a requirement to ensure high-quality pharmaceuticals and biopharmaceuticals before releasing for sale, supply, or export. The question of how to release a batch is very often a discussion point, especially when you must fulfill European and U.S. regulatory requirements.

Batch (Lot) release and comparability testing are necessary for evaluating the purity, potency, and identity products; analytical assays are available for quality control testing that is regulatory compliant.

Product release testing is a crucial part of ensuring drug product quality and safety. Regulatory bodies such as the (FDA) and the European Medicines Agency (EMA) require lot release tests for each biopharmaceutical batch. This testing is performed at all stages of drug development:

- Raw materials
- Unprocessed bulk materials
- Purified bulk materials (drug substance)
- Final finished product (drug product)

Lot release testing relies on previously established specifications to demonstrate product identity, purity, and potency. Similarly, comparability testing is aimed at identifying whether changes in the manufacturing process led to downstream changes in product quality.

The product code is registered in the system, and then each batch is logged under its product code as it is assigned when manufactured. Samples are then collected during production for the in-process and bulk samples, and from the warehouse when production is complete for the finished products, these are distributed to the Chemistry and Microbiology departments for testing. Qualitative and quantitative results from testing are entered in the system to be reviewed, approved, and then used for batch release decisions as well as for quality reporting, trending, and analysis for annual product reviews and process control verifications.

10.7 Stability Studies

Some companies manufacture significant numbers of various products with different SKUs, and many of those are common products in different dosage forms and package sizes. These need to be placed in the stability program for both initial drug approval and as part of an ongoing quality program. It is impractical, if not impossible, to manage large numbers of stability studies without using a computerized system to manage the pulling, testing, and disposing of product samples at different times, as well as the entry of test results and the statistical evaluation of the data to derive the API or product shelf life.

Missed samples are one of the biggest issues in the stability program. Having a large number of sample pulls at different dates and times substantially increases the risk of missed samples. A computerized stability system will alert the users, thus reducing or eliminating the risk of missed samples.

10.8 Cleaning Validation – Contamination Control

According to the World Health Organization (WHO) [3], cross-contamination is "Contamination of a starting material, intermediate product, or finished product with another starting material or product during production."

The cleaning validation process is created to ensure equipment is cleaned to the level to prevent material and product contamination. This process has significantly been improved since the days of the Barr Laboratories Court Case [4], and since the first FDA guidelines referencing the subject of cleaning validation were published in 1991.

At that time, requirements for cleaning validation were scarcely a page of the Pharmaceutical, Chemical, and Biopharmaceutical guidance documents. Now, with the abundance of guidelines, regulations, and technical reports, cleaning validation has evolved into a mature and complex process. This increased complexity made the next step of automating the process a clear necessity. Even with the technology, which is readily available today, many companies still rely on paper-based systems, or a hybrid of spreadsheets and paper forms to manage their cleaning validation programs.

The cleaning validation is the act of proving the process of cleaning equipment will work in the worst-case scenario, where the worst-case scenario is defined as the most toxic API, and the most difficult to clean product and equipment, among other criteria. The amount of data generated from cleaning validation is substantial and, therefore, can be difficult to manage manually without proper tools. The process of managing cleaning validation is illustrated in Figure 10.3 [2].

Automation of this process is beneficial as the key components of the process are based on executing calculations. Data are entered into a series of master lists for analytical methods, APIs, cleaning agents, equipment and its subgroups, products, and the production lines used. Once the master list data is entered and grouped, the system will enforce the verification of any change to the data, which may affect the validated state of the cleaning process. This verification includes an evaluation to see if the worst-case scenario has changed and the recalculation of the maximum allowable carryover of Product A into Product B.

Qualitative and quantitative results from testing are entered in the system to be reviewed, approved, and then used for batch release decisions as well as for quality reporting, trending, and analysis for annual facility reviews and process control verifications.

10.9 Equipment Management (Metrology)

Pharmaceutical companies invest substantial amounts of capital in procuring their laboratory assets, which includes capital equipment such as high-performance liquid chromatographs (HPLC), gas chromatographs (GC), mass spectrometers

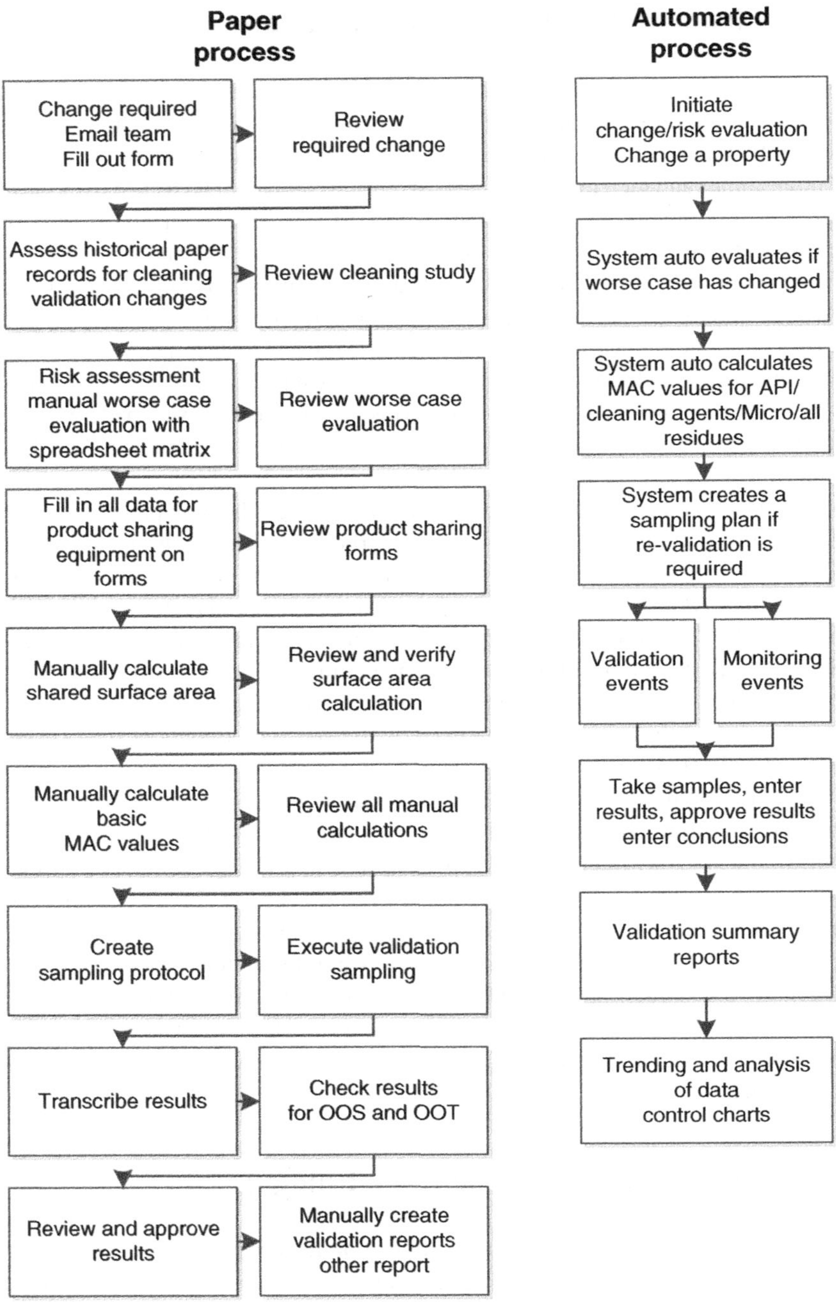

Figure 10.3 Paper vs computerized cross contamination. *Source:* Cleary and Famili [2].

(MS), ultra-performance liquid chromatographs (UPLC), and hybrid systems (LC/MS, GC/MS, and MS/MS), as well as other medium and small end equipment such as balances, centrifuges, etc.

Proper management of the equipment in the laboratory is necessary to ensure accurate, reliable raw data, and timely testing of samples, so equipment management is one of the essential elements of a quality management system. Records should include instrument related information such as asset number, manufacturer, model, serial number, descriptions of operation, instrument, and equipment performance measures, and both preventive maintenance and calibrations requirements and schedule.

The calibration process for precision and accuracy commonly involves using the equipment to verify one or more samples of known values; these are called calibrators or standards. The results confirm the equipment is producing accurate measurements of the prescribed precision to the intended purpose. Adjustments are made to the equipment if it is found to be out of the acceptable range based on the results produced not being within a certain margin of error from the known values.

Preventative maintenance is an activity that is regularly performed on equipment to reduce the probability of it failing. Preventative maintenance is performed while the equipment is functioning so that it does not break down unexpectedly.

Calibration and preventive maintenance procedures for the equipment define the frequency at which these activities take place with a scheduling pattern for executing the procedures on the equipment. Both calibration and preventive maintenance work orders are created using the defined scheduling pattern to determine the work that should be executed on each date.

In pharmaceutical operations, manufacturing, and laboratory are critical in the production process. Tens of products are manufactured and hundreds of tests are performed daily. These products need to be of a specific quality, and the tests are used to certify and accept the quality of the product. Therefore, the instruments and equipment used in the manufacturing and testing process must be calibrated to perform at their optimal setting. If there is an error in the output of the equipment, then the final product may be substandard and rejected, wasting hundreds of working hours and substantial operational loss.

10.10 Laboratory Operations

Laboratory quality could be defined as the accuracy, reliability, and timeliness of the reported test results. All laboratory testing results must be as accurate as possible, all aspects of the laboratory operations must be reliable, and reporting must be timely. Furthermore, in a regulatory compliant laboratory, individuals who perform analytical methods must be different than those reviewing and approving the raw and processed data to ensure the data is accurate and reliable.

Automating the quality process in the laboratory impacts all the subprocesses, which result in samples being sent to the lab. In the case of cleanroom and utility monitoring, or starting materials and components, or in-process and finished products, and stability studies produce samples that will be tested in the same microbiology or chemistry laboratories, only the method of collection is different.

Currently available off-the-shelf (OTS) or configurable software systems can enforce the requirements of both peer-reviewed test results and other higher-level approvals right up to final batch approvals and ultimately release decisions and release execution. Furthermore, the harmonization of the testing, result management, and approval of executed testing, and the data itself being centralized lend to efficient laboratory operations.

Having robust LIMS systems in modern laboratories is now a basic requirement to ensure efficiency, control, and compliance.

10.11 Automation of Risk Management

Risk management principles are successfully utilized in many areas of business and government, including public health, pharmaceutical, medical device, biotechnology, and by agencies regulating these industries. The importance of quality systems has been recognized in the pharmaceutical, medical device, biotechnology industries, and it has become evident that quality risk management is a valuable component of an effective quality system. In the pharmaceutical industry, there are numerous stakeholders, including the company, the patients, the medical practitioners, as well as government and the industry itself, with the protection of the patient by managing the risk to product quality being considered of principal importance. The risk is frequently defined as the combination of the probability of occurrence of harm, the severity of that harm, and the ease with which that harm can be detected.

There are many risk management models and tools; one of the most well-known models is Failure Mode and Effects Analysis (FMEA) [5] and Failure Modes, Effects and Criticality Analysis (FMECA). These methodologies are designed to pinpoint possible failure modes for either a product or a process, to evaluate the risk related to those failure modes, to rank the issues regarding importance or severity, and to identify and carry out corrective actions to address the most major issues.

The FMEA/FMECA analysis process is a tool that has been adjusted in various ways for many unique purposes, and it can contribute to better quality designs for both products and processes, resulting in higher reliability, increased quality, enhanced safety, and reduced costs. The tool can also be used to create and fine-tune maintenance plans and can contribute to controlling strategies and other quality assurance procedures, as can be found in cleanroom contamination control and other pharmaceutical manufacturing, testing, and laboratory processes.

$$\text{Risk priority number}\left(\text{RPN}\right)=\text{Severity}\left(\text{S}\right)\times\text{Probability}\left(\text{P}\right)\times\text{Likelihood of detection}\left(\text{D}\right)$$

A tabular example of this risk tool would include the following information for each different risk area (see Figure 10.4).

Automation of risk management is still in the early stages, and most companies continue to manage risk evaluation in a manual or paper-based process. Companies which have implemented electronic risk management systems have a system where the risk scores for severity, probability, detectability are manually entered, as are all the other text fields, illustrated in Figure 10.4. The evolution of true automated risk-based systems is in the infancy stage; these systems are the predictive systems, expert systems, and artificial intelligence systems which will be the next generation of technology that will learn from the data, predict the risks, and provide mitigation ideas and strategies.

10.12 Automated Training Management

The FDA regulatory requirements (i.e. 21 CFR 211.25 and 820.25) and the quality management standards from the International Organization for Standardization mandate companies to perform and document employee training. These requirements ensure that employees fully understand how to perform their tasks and responsibilities factoring in both company work and quality procedures, as well as the industry guidelines and regulations. Effectively managed and well-documented training programs lessen the risk of noncompliance, thus improving the overall quality of the products. For most of regulatory agencies, a properly documented training record for each employee, whose job affects product quality, should cover such things as tasks that the employee performs, proof that the employee has the proper skills and training for his or her job, and evidence that the employee can perform his or her tasks competently.

Potential failure mode	Potential effects	Potential cause	S	P	D	RPN	Actions: risk reduction strategy	After action taken	S	P	D	RPN

Figure 10.4 An example of the FMEA table. *Source:* Based on Food and Drug Administration [5].

Automating training record management involves electronically storing and accessing the training records of all employees. This training can be from both internal company training and external courses. Automation delivers increased efficiency with a centralized repository for all employee training records and associated documentation, thus making tracking and retrieval of records faster and easier and allowing for linking with other quality processes to prevent untrained people from executing tasks they should not. A good example is where the LIMS system is connected to the training system, and it checks if the analyst is trained on specific test protocol and allows or disallows the analyst to perform the test, ensuring only those with proper training are performing the test.

10.13 SOPs and Document Management

Pharmaceutical, biotech, medical device, and many other industries are extremely sensitive to how documents and information in document form are managed. Whether these documents are Standard Operating Procedures (SOPs), work instructions (WIs), template forms for investigations, sterility failures procedures, change management, risk assessments models, or any other process control or support document, document creation, control, and efficient document reviews are critical from both a regulatory standpoint and for organizational efficiency as well.

Stringent regulations, and complex procedures, have forced quality professionals and compliance officers to evaluate and update their document management practices. Most companies that are under regulatory scrutiny need to ensure the correct and approved versions of documents are being used, thus avoiding erroneous or outdated versions of documents still being available to the analyst and potentially being used.

Computerized document management systems control all aspects of the document lifecycle. Starting with writing the first draft, circulating it for editing, and then review and approval is the basic process flow to produce version one (1) of a document. Once approved, the document cannot be used until all personnel affected are trained on the new document. Over time, the document will be edited, and the updated version will be approved, trained on, and released. The automated system should control this lifecycle and allow access to the document and interacting with the document only to eligible users including limiting access to edit, review, approve, and retire document. These control mechanisms include restricted access to printing and viewing the documents, or only access to the currently approved versions. Furthermore, the system should enforce the sequencing of steps to ensure the lifecycle process is followed. Finally, integration into the training management system is critical to ensure that all users are automatically notified of training requirements and managers can follow up to confirm that training is successfully executed.

10.14 Audit Management and Compliance

Audits are an indispensable quality management tool for verifying the objective evidence of processes to ensure compliance. An audit is a scheduled or surprise inspection for the verification of activities, records, documents, procedures, and other elements in a quality system to evaluate their conformance with the requirements of a given quality standard such as 21 CFR 211. The FDA defines a quality audit in section 820.3 (t)

> (t) Quality audit means a systematic, independent examination of a manufacturer's quality system that is performed at defined intervals and at sufficient frequency to determine whether both quality system activities and the results of such activities comply with quality system procedures, and that these procedures are implemented effectively, and that these procedures are suitable to achieve quality system objectives.

Companies which fall under the scrutiny of a regulatory agency and are subject to external audits will typically have an internal audit or self-audit system to ensure a state of audit readiness that will ensure quality and minimize risks; companies strive to reach a state of continuous audit readiness. Without an effective audit program, an organization is at higher risk for non-conformances, which can result in regulatory actions, warning letters, poor product quality, loss of certification and registration, as well as increased product liability.

Most organizations have implemented audit management software systems to facilitate and document both their internal and external audits. Internal audits can assess how effective processes have been executed; the audit will provide evidence verifying the reduction or elimination of problems in the quality management system. Internal audits assist the organization in promoting quality by reporting non-conformances, verifying the efficacy of corrective or preventive actions, and by citing areas of good and best practices. In this way, different departments can share data, modify their operational practices, and contribute to continuous improvement.

10.15 Corrective and Preventive Actions (CAPAs)

Corrective and preventive actions (CAPAs) are changes to an organization's processes to eliminate the actual or potential cause(s) of non-conformance, which compromise product and overall quality, operational productivity, and efficiency. Typically, CAPA is a set of activities that are implemented in an organization and relate to manufacturing, documentation, procedures, or other systems to correct the problem(s) and prevent the recurrence of the problem(s).

A CAPA can be initiated from an audit observation or based on a deviation or non-conformance in any of the quality system or laboratory processes. An out of specification result, contamination in a cleanroom, or any failure may result in a CAPA to address the issue.

Software systems supporting the CAPA process usually facilitate the initiation of a CAPA based on a deviation or incident. Once initiated, the user can perform and document an investigation leading to root-cause analysis. A robust system will support root-cause analysis tools and will integrate directly with a change management system.

10.16 Change Management

Change management is an integral part of any quality management program. Before any change is made, one must evaluate the reason for the change, the impact of change, the associated risk with the change, and finally, the sustainability of the change. Because introducing a change to a process or a system can also introduce problems, the change control process measures the impact and risks of all changes. The primary purpose of change control is to evaluate all changes regarding the established requirements and determine the actions needed for implementing the change, or deciding that the change should not be implemented. Change control, therefore, ensures that a system remains in the validated state before, during, and after the change.

Changes made in an organization can have widespread effects. Any given change can require modification to training procedures and require personnel retraining to include the new change, or the change can require an internal audit to be performed to ensure affected departments are still in compliance with the new change. Any changes to procedures, analytical methods, or raw material and component changes can necessitate an update to documents, such as SOPs and WIs, and require securing approval of new versions.

The source of these changes can be from either corrective or preventive actions. Change control is not specific to a department or a process, it is a fundamental part of the operations of the whole company due to the extensive area of application of properly controlling change.

When considering the automation of change control in conjunction with other quality management processes, including CAPA, document management, training management, audit management, and an LIMS, we achieve a better state of control through these integrations. The flow of data and actions through a looped system, where each process phase can be the input for a subsequent phase, and the output of information from a preceding process phases are combined with the evolving changes, and controlling the changes is at the center of the loop. The open looped system can be the basis for continuous improvement when properly implemented.

10.17 Computer Systems Validation

According to many regulatory agencies, any software which will be used to house GMP data, or that will be used to house any data on which quality and release decisions are made, or has any impact on product quality, should be validated. In the case of the software, validation is the process of establishing documentary evidence proving the software system will perform as per its intended functionality and can be operated as per the user requirements specification. Validation is much more than software system testing; the scope is larger than just verifying system features and emphasizes achieving a validated state and then maintaining that validated state.

GAMP 5: A risk-based approach to Compliant GxP Computerized Systems is a guidance document released by the International Society for Pharmaceutical Engineering (ISPE), it provides a process to achieve compliance for computerized systems, and it points to the future of computer systems compliance by focusing on principles in major industry documents including ICH Q8, Q9, Q10; and ASTM E2500 among others [6] (see Figure 10.5).

GAMP 5 has classified software under various categories based on its complexity and recommends evaluating the reliability based on extended use in the industry, and the likelihood of failure. GAMP recommends the identification and evaluation of risks throughout the life cycle of the software. Critical risk areas must be identified, and efforts concentrated on the point of risk.

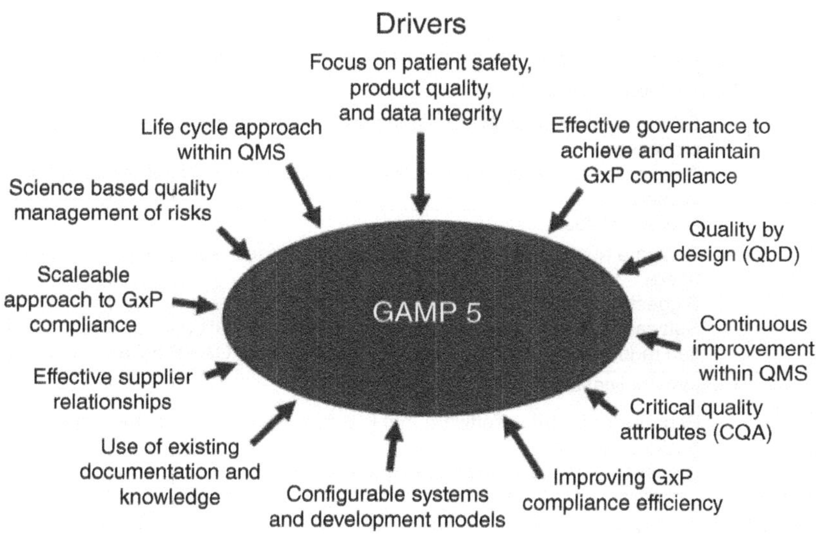

Figure 10.5 Validation drivers. *Source:* International Society for Pharmaceutical Engineering [6].

Validation efforts can be optimized by focusing on the key and critical features and the functionality that is the highest risk, which is determined using both Risk Analysis and GAMP 5 categorization. If the core of the software system is OTS, and 95% of the user requirements are met and fall into category 3, the remaining 5%, which can be considered a requirement gap could be customized (new development) by the software vendor to meet the needs of a specific process, this part would be considered Category 5 (see Figure 10.6).

The main validation effort can be focused on the following areas: the bespoke 5% based on category 5, and the electronic signatures, 21 CFR part 11 requirements, and the statistical analysis portion of the software based on a risk analysis citing these areas as high risk.

Regardless of the feature focus of the validation activities, a complete validation document package including design specification, user specification, functional specification, installation guide, user guide, installation qualification, operation qualification, and performance qualification make for the key documents to assemble for a complete set. Finally, a traceability matrix linking all requirements and testing being executed should be included.

Category	GAMP 5
1	Infrastructure software – Established or commercially available layered software, including operating systems, databases, office software, and the like. – Infrastructure software, including virus protection, network management tools, and so forth.
2	Discontinued – Firmware is treated as software in one of categories 3, 4, or 5.
3	Non-configured products – Off the shelf products which cannot be changed to match the business processes. – Also, includes configurable products where only the default configuration is used.
4	Configured products – Software applications that are configured to meet user-specific business needs. – Biggest and most complex category. – Software can be configured to return different outputs and, as a result of can require a much higher level of validation than GAMP 5 Category 3.
5	Custom applications – Bespoke software is software that is written from scratch to fulfill the business needs. – Highest risk of the software categories.

Figure 10.6 GAMP categories. *Source:* Based on International Society for Pharmaceutical Engineering [6].

The validation approach illustrates the standard validation v-model (Figure 10.7). The left side of the V represents the specifications; the right side represents the system testing against the specifications; the bottom represents the code modules.

- User Requirement Specifications (URS) – is the documented specification used in software engineering that specifies what the user expects the software to be able to do.
- Functional Requirement Specifications (FRS) – is the documented specification used in software engineering that specifies what the user expects the software to be able to do.
- Design specifications (DS) – is the documented specification used in software engineering that specifies what the user expects the software to be able to do.
- Installation qualification (IQ) – confirms complete documentation, which includes checking purchase orders, proper hardware installation, and software verification according to the manufacturer's specifications.
- Operational qualification (OQ) – confirms the system operations by testing the design requirements that are traced back to the function specifications, including software and hardware functions under normal load, and under realistic stress conditions to assess whether equipment and systems are working correctly.

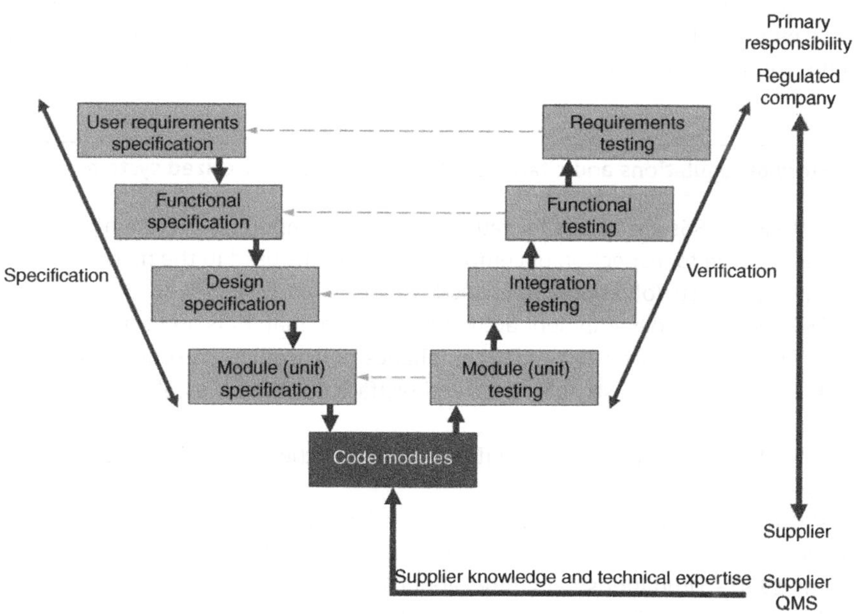

Figure 10.7 GAMP validation V-model. *Source:* International Society for Pharmaceutical Engineering [6].

- Performance qualification (PQ) – confirms that a system is capable of performing or controlling the activities of the process while operating in a specific environment – namely, a series of checks by the user against the original requirement specifications of the system.
- Risk Assessment (RA) – verifies the impact of changes being made to the system to ensure quality efforts are focused.

Some small organizations may not have adequate means to purchase compliant quality software such as LIMS or QMS. Those companies may use Microsoft Excel, other types of worksheets, or homegrown systems to automate certain processes and calculations. The worksheets must be validated and locked to comply with data integrity regulations until a validated compliant quality computerized system can be qualified.

10.18 Evolution in the Data Integrity Regulation

The usage of the computerized system in the regulated environment is not new. Since 1979 when the first computerized systems were used in various aspects of compliant manufacturing, the industry has been putting in place guidelines and regulations to govern the use of these systems while promoting their use for the betterment of the industry. The history of the regulations and citations related to the computerized system are outlined in the Key Information section below.

History of regulations and citations related to the computerized system

1979 –FDA 483s are issued for faulty control of computerized systems
1983 – Guide to inspection of computerized systems used in the manufacture of drug products (bluebook) published
1987 – Software development activities reference materials and training aids for FDA investigators 1987 – FDA compliance policy guide on source code
1988 – Japanese ministry of health and welfare (MHW) issues GLP inspection of computer systems
1989 – UK department of health (DoH) publishes the application of GLP principles to computer systems
1990 – The Australian therapeutic goods administration (TGA) publishes the code of GMP for therapeutic goods, includes expectations for computer systems 1990 – Drug industry inquires about "going paperless."
1991 – European Union (EU) publishes good clinical practices (GLP) for trials on medicinal products, includes expectations for computer systems

1992 – Japanese MHW issues computer control guidelines for drug manufacturing and electronic records/electronic signatures (ER/ES) proposed rule published
1993 – EU publishes GMP for medicinal products, includes computer system expectations
1995 – Good automated manufacturing practice (GAMP) first published; Annex 11 (computerized systems) added to EU GMP and FDA publishes glossary of computer terminology
1996 – Software development activities reference materials and training aids for investigators updated
1997 – Draft general principles of software validation; Japanese MWH issues expectations for retention of electronic records; MCA issues guidance on electronic signatures, and 21 CFR Part 11 becomes effective
1998 – Guide to inspections of computerized systems in the food industry; GAMP update
1999 – Guidance for industry "Computerized Systems Used in Clinical Trials."
2000 – ICH publishes GMP for API with a section on computer systems
2001–2002 – GAMP4 and general principles of software validation; final guidance for industry and FDA staff
2003 – Guidance for industry part 11, ER/ES scope and application, and PIC/S issues good practice for computerized systems in regulated "GxP" environments
2004 – Pharmaceutical CGMPs for the twenty-first century – a risk-based approach, final report
2008 – GAMP5 A risk-based approach to compliant GxP computer systems
2011 – EU GMP Annex 11 update
2013 – EU data integrity guidance
2016 – FDA data integrity guidance
2015 – EU data integrity guidance update

10.19 Data Integrity

Compliance, security, and data integrity go together and must be built into the system, and these important regulations equate to numerous features. Each user has his/her unique system login/password to represent their electronic signature. No user has access or can interact with the system with any login except his/her own. Each user is a member of a permission grouping that allows or disallows access to the features, windows, and reports throughout the system. These permissions groups include every feature in the system without exception.

Audit trails record all changes to data within the system. They include the timestamp, the unique name of the user executing the operation affecting the data, the nature of the operation, and the modified data (both the old and new values). Finally, both the reason for the change and, where applicable, a change control number is captured. The Key Information section below lists specific software requirements that are specified in 21 CFR 11 (FDA 2016a).

List of specific software requirements specified in 21 CFR 11

1) Validation of systems to ensure accuracy, reliability, consistent intended performance, and the ability to discern invalid or altered records
2) The ability to generate accurate and complete copies of records in both human-readable and electronic form
3) Protection of records to enable their accurate and ready retrieval throughout the record retention period
4) Limiting system access to authorized individuals
5) Use of secure, computer-generated, time-stamped audit trails
6) Use of operational system checks to enforce permitted sequencing of steps and events
7) Use of authority checks to ensure that only authorized personnel can use the system, electronically sign a record, access the operation or computer system input or output device(s), alter a record, or perform the operation at hand
8) Use of device/system checks to determine the validity of the source of data input or operational instruction
9) Confirmation that personnel who create, maintain, or use electronic record/electronic signature systems have the education, training, and experience to execute their assigned task
10) The establishment of, and adherence to, written policies and procedures that hold individuals responsible and accountable for actions executed under their electronic signatures
11) Use of appropriate controls over system documentation

Source: Modified from Food and Drug Administration [7].

In recent years, the number of FDA citations, both 483s and warning letters, has increased substantially (see Figure 10.8). Data management has been a focus of many of the citations because the quality and quantity of data is challenging for most organizations.

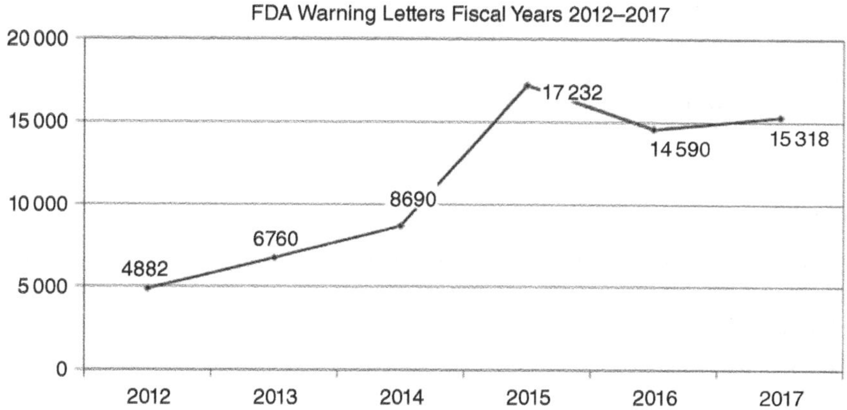

Figure 10.8 Number of FDA warning letters from 2011 to 2017. *Source:* Based on Food and Drug Administration [8].

FDA, Health Canada, and MHRA have all introduced new guidance for data integrity [9]. Data integrity refers to the completeness, consistency, and accuracy of data. The main points of the data integrity guidance are regarding:

- Audit trail/e-signatures/data validation
- All data must be included in CGMP decision-making (including raw data, in-process data), or a proper justification provided
- Data changes to be performed by authorized personnel, with justification
- Audit trail to be reviewed (not just data) before final approval
- Control of blank forms or cells, printing, missed samples
- "Testing into compliance" is prohibited
- Personnel should be trained in detecting data integrity issues – if an issue is found, ensure steps are taken to remedy the situation

Any data created as part of a CGMP record must be evaluated by the quality unit as part of release criteria (§Section 211.22 and 212.70) and maintained for CGMP purposes (Section 211.180). Electronic data generated to fulfill CGMP requirements should include relevant metadata. Excluding data from the decision-making activity requires a valid, documented, scientific justification for its exclusion (see the guidance for industry Investigating Out-of-Specification [OOS] Test Results for Pharmaceutical Production, and 211.188, 211.192, and 212.71[b]). The requirements for record retention and review do not differ depending on the data format; paper-based and electronic data record-keeping systems are subject to the same requirements [2].

When the FDA talks about the "State of Control," they expect the industry to use the generated data to evaluate their processes and make quality decisions to improve them. To demonstrate its emphasis on data integrity risks, the FDA has cited various pharmaceuticals and biotech companies. The Key Information section that follows lists a few common risks associated with data integrity and related citations from FDA (FDA 2017).

List of common risks associated with data integrity and related citations from FDA

Citation example 1: "Failure to prevent unauthorized access or changes to data, and to provide adequate controls to prevent manipulation and omission of data."

Your quality control unit did not have basic controls to prevent changes to your electronically stored laboratory data. During our inspection, we requested that you display original electronic data for the analysis of heparin and heparin-related drug samples. Your analyst was unable to retrieve requested data and explained that he deletes older data to make space for newly acquired data [10]. https://www.fda.gov/ICECI/EnforcementActions/WarningLetters/2017/ucm565030.htm

Citation example 2: "Failure to have laboratory control records that include complete data derived from all laboratory tests conducted to ensure compliance with established specifications and standards."

For example, our investigator reviewed the audit trail from your assay testing for (b)(4) lot (b)(4) and found that you tested the same sample set three times over several days without documentation or investigation. You reported only the result of the third and final test for purposes of completing your certificate of analysis and releasing this batch of API.

https://www.fda.gov/ICECI/EnforcementActions/WarningLetters/2017/ucm558917.htm

Citation example 3: "Your firm does not have laboratory records that include complete data derived from all tests necessary to assure compliance with established specifications and standards, including examinations and assays."

Specifically, our investigator observed your microbiologist using the antibiotic zone reader. Our investigator verified your microbiologist recording the correct value as read from the plate reader. Then our investigator copied the handwritten zone diameter test results taken at the time of testing for zones of the standard series from the microbiologist's issued worksheet. On the completed form, the test results for the zones were different from observed by our Investigator. Your response does not provide documentation, which you allege differs from our investigators' direct observation. You have not subsequently verified complete raw data was maintained.

https://www.fda.gov/ICECI/EnforcementActions/WarningLetters/2017/ucm560653.htm

Citation example 4: "Failure to prevent unauthorized access or changes to data, and failure to provide adequate controls to prevent omission of data."

a) Our inspection found your laboratory systems lacked controls to prevent the deletion of and alterations to raw electronic data.
b) Our review of audit trail data revealed that your analysts manipulated the date/time settings on your high-performance liquid chromatography (HPLC) systems. During the inspection, your analysts admitted to setting the clock back and repeating analyses for undocumented reasons. Initial sample results were overwritten or deleted, and unavailable for our investigators' review. Your firm reported only the passing results from repeat analyses. When test results are overwritten, the quality unit is presented with incomplete and inaccurate information about the quality of the drugs produced by your firm.
c) Your quality control analysts used a shared login account to access HPLC systems. This shared account allowed analysts, without traceability, to change the date/time settings of the computer, to modify file names, and to delete original HPLC data.
d) Seven of your firm's HPLC systems used for API testing had the audit trail feature disabled, although all had audit trail functionality.

FDA introduced the concept of ALCOA in April 2016 draft guidance, "Data Integrity and Compliance with CGMP Guidance for Industry" [9] for different documentation factors (Figure 10.9).

Data integrity is critical to all systems, from a simple system like a pH meter to a more complex system such as a LIMSs (see Figure 10.10) [9].

Attributable (data can be assigned to the individual performing the task)
Legible (data can be read by eye or electronically and retained in a permanent format)
Contemporaneous (data is created at the time the activity is performed)
Original (data is in the same format as it was initially generated, or as a "verified copy," which retains content and meaning)
Accurate (data is true/reflective of the activity or measurement performed)

Figure 10.9 Data integrity ALCOA. *Source:* Based on [9].

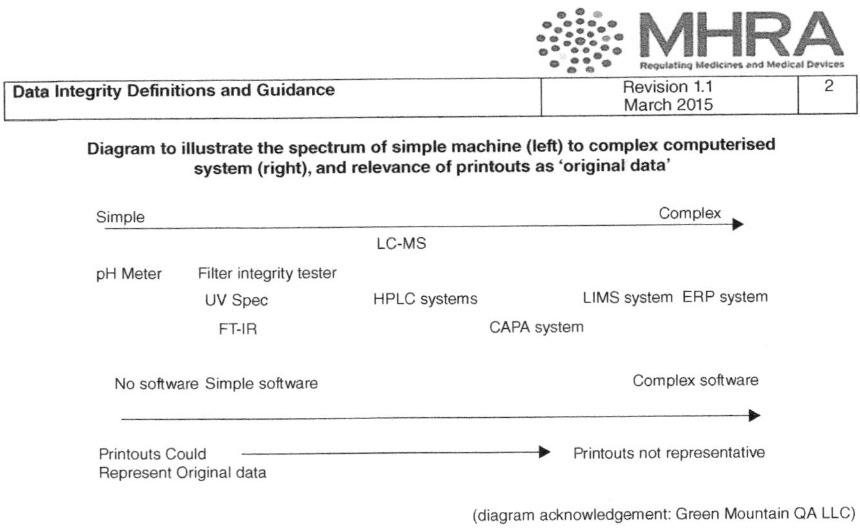

Figure 10.10 MHRA data integrity guideline. http://academy.gmp-compliance.org/guidemgr/files/MHRA_GxP_data_integrity_consultation.pdf. *Source:* Modified from [9].

10.20 Quality Data

Your data are much more than just a number! More than ever, the ability to manage vast amounts of quality data is critical to a company's survival and ultimately business success. But even with the emergence of data-management functions and data integrity regulations, most companies remain behind the curve.

Only small amounts of an organization's structured data are used in making quality decisions, and most of the analysts' time is spent simply generating and preparing data. Data defense and offense are terms used to differentiate different business objectives and the events designed to address them [10].

Data defense is about alleviating, if not diminishing risks. Activities include ensuring compliance with guidelines (such as regulations governing data integrity, patient privacy, confidential data, etc.), using analytics and audit trails to detect and limit unauthorized access, and building processes to prevent data manipulation.

Data offense focuses on supporting business objectives such as increasing productivity, workload, and internal and external customer satisfaction. It typically includes activities that generate productivity insights (i.e. data analysis and modeling).

Every corporation must use both data offense and data defense to succeed, but getting the right balance is important. As budgets are reduced, and departments must do more with fewer resources, they must ensure the integrity of data in each of the processes and avoid shortcuts that reduce the quality.

10.21 Benefits of Computerized Systems

Nowadays, we rely on the computerized system in most aspects of our lives, and it is evident that this has become an important part of the work environment. Analytical data are collected in a LIMS, often referred to as LIMS, and it becomes crucial in day-to-day laboratory functions listed in the Key Information section below.

Laboratory functions that can be supported by LIMS

- Trending of analytical data
- Investigation of OOS/OOT analytical results
- Querying and mining data to support quality systems or risk management
- Reviewing and approving data remotely
- Storing and managing all data in the same database
- All data can be easily analyzed expediting problem detection
- Assuring data integrity and security
- Addressing 483s warning letters and/or citations
- Managing quality big data

10.22 Big Data

The term "Big Data" is usually associated with consumer data trends and behavior to predict consumer buying patterns, market share, and larger corporate profits. In the laboratory setting, "Big Data" is also being applied to quality operations such as compliance, quality assurance, root-cause analysis, investigation, risk mitigation, and prevention. An example of "Big Data" would be the review of all finished product lots manufactured using a specific lot of API that was found to be contaminated with microbial organisms. Big Data can be helpful in establishing control strategies in the development and production of API and drug products. Many companies depend on the analysis of these databases to support their quality systems.

List of Abbreviations

API	Active Pharmaceutical Ingredient
CAPA	Corrective and Preventive Action
CGMPs	Current Good Manufacturing Practices
ELN	Electronic Laboratory Notebook

ER/ES	Regulations on Electronic Records and Electronic Signatures (ERES)
EU	European Union
FDA	Food and Drug Administration
GAMPs	Good Automated Manufacturing Practices
GCP	Good Clinical Practice
GLP	Good Laboratory Practice
GxP	Good (Anything) Practice
IQ	Installation Qualification
LIMS	Laboratory Information Management System
OQ	Operational Qualification
PIC/S	Pharmaceutical Inspection Co.-operation Scheme
PQ	Process Qualification
QM	Quality Module
QMS	Quality Management System
SKUs	Stock Keeping Units
SOPs	Standard Operating Procedures
WIs	Work Instructions

References

1 Barbero, M., Coutuer, J., Jackers, R. et al. (2016). Big data analytics for policy making. A study prepared for the European Commission DG Informatics (DG DIGIT). http://www.ngi-summit.org/wp-content/materials/EU_papers/DG_digit_study_big_data_analytics_for_policy_making.pdf (accessed 16 November 2020).

2 Cleary, S. and Famili, P. (2017). Chapter 6: automation of environmental monitoring: achieving efficiency and control environmental monitoring. In: *Environmental Monitoring: A Comprehensive Handbook*, vol. 8 (ed. J. Moldenhauer). PDA/DHI Publishing.

3 World Health Organization (2011). Supplementary guidelines on good manufacturing practices for heating, ventilation, and air conditioning systems for non-sterile pharmaceutical dosage forms. https://www.who.int/medicines/areas/quality_safety/quality_assurance/SupplementaryGMPHeatingVentilationAirconditioningSystemsNonSterilePharmaceuticalDosageFormsTRS961Annex5.pdf (accessed 16 November 2020).

4 Food and Drug Administration (1993). *Guidance Document: Dosage Form Drug Manufacturers cGMPs*. Rockville, MD: Food and Drug Administration https://www.fda.gov/inspections-compliance-enforcement-and-criminal-investigations/inspection-guides/dosage-form-drug-manufacturers-cgmps-1093 (accessed 16 November 2020).

5 Food and Drug Administration (1995). *Document: Failure Mode and Effects Analysis*. Rockville, MD: Food and Drug Administration https://www.fda.gov/inspections-compliance-enforcement-and-criminal-investigations/inspection-guides/glossary-computer-system-software-development-terminology-895 (accessed 9 December 2020).

6 International Society for Pharmaceutical Engineering (2008). *GAMP 5: A Risk-Based Approach to Compliant GxP Computerized Systems*. ISPE Publishing https://ispe.org/publications/guidance-documents/gamp-5 (accessed 10 December 2020).

7 Food and Drug Administration (2016). *Electronic Records; Electronic Signatures*. Rockville, MD: Food and Drug Administration.

8 Food and Drug Administration (2016). *Guidance for Industry: Data Integrity and Compliance with cGMP*. Rockville, MD: Food and Drug Administration https://www.fda.gov/downloads/drugs/guidances/ucm495891.pdf (accessed 16 November 2020).

9 Data integrity and compliance with CGMP guidance for industry Draft Guidance. U.S. Department of Health and Human Services FDA, April 2016/MHRA GxP Data Integrity Definitions and Guidance for Industry "Draft version for consultation" (April 2016). https://www.fda.gov/media/97005/download (accessed 9 December 2020).

10 DalleMule, L. and Davenport, T.H. (2017). What's your data strategy? *Harvard Business Review* 05: 12–121. https://hbr.org/2017/05/whats-your-data-strategy (accessed 14 November 2020).

Index

d

e

m